HIGH-POWER LASER RADIATION IN ATMOSPHERIC AEROSOLS

# ATMOSPHERIC SCIENCES LIBRARY

# High-Power Laser Radiation in Atmospheric Aerosols

## Nonlinear Optics of Aerodispersed Media

by

V. E. ZUEV, A. A. ZEMLYANOV,
Yu. D. KOPYTIN, and A. V. KUZIKOVSKII

*Institute of Atmospheric Optics, U.S.S.R. Academy of Sciences,
Siberian Branch, Tomsk, U.S.S.R.*

D. Reidel Publishing Company

A MEMBER OF THE KLUWER ACADEMIC PUBLISHERS GROUP

Dordrecht / Boston / Lancaster

**Library of Congress Cataloging in Publication Data**

CIP

Main entry under title:

High-power laser radiation in atmospheric aerosols.

(Atmospheric sciences library)
Bibliography: p.
Includes index.
1. Aerosols–Effect of radiation on. 2. Laser beams–Atmospheric effects. I. Zuev, V. E. (Vladimir Evseevich) II. Series.
QC882.H54 1984 551.5 84–29828
ISBN-13: 978-94-010-8809-1 e-ISBN-13: 978-94-009-5219-5
DOI: 10.1007/978-94-009-5219-5

---

Published by D. Reidel Publishing Company
P.O. Box 17, 3300 AA Dordrecht, Holland

Sold and distributed in the U.S.A. and Canada
by Kluwer Academic Publishers,
190 Old Derby Street, Hingham, MA 02043, U.S.A.

In all other countries, sold and distributed
by Kluwer Academic Publishers Group,
P.O. Box 322, 3300 AH Dordrecht, Holland

## TABLE OF CONTENTS

# INTRODUCTION

Unique properties of laser radiation including its monochromatic properties, polarization, high spectral intensity, coherence, narrow beam divergence, the possibility of controlling the pulse duration and radiation spectrum and, finally, the fact that extremely high power and energy create very favorable conditions for the extensive application of lasers to communication systems, systems for the lidar sensing and ultra-high-precision ranging, navigation, remote monitoring of the environment, and many other systems operating in the atmosphere.

The operative efficiency of the above systems depends significantly on the state of the atmosphere and the corresponding behavior of laser radiation propagating through it. This circumstance has stimulated the studies of the above regularities during the passt 10-15 years. For the investigations to be carried out the scientists were forced to develop new theories and methods for stŭdying the problem experimentally. Moreover, during such investigations some previously unknown phenomena were observed, among them the nonlinear effects accompanying high-power laser radiation propagating through the atmosphere are of paramount importance.

Among the nonlinear effects caused by high-power laser radiation interaction with the atmosphere, the effects accompanying the propagation of high-power radiation through the atmospheric aerosols are of particular interest. Aerosols always occur in the atmosphere. It should be noted that the microphysical and optical characteristics of atmospheric aerosols vary widely, this fact causes a great variety in the features of their interaction with radiation.

Many works devoted to the problems of investigating the propagation of high-power laser radiation through the atmosphere have already been published, e.g., monographs by V. E. Zuev Laser radiation propagation in the atmosphere (Radio i svyaz, Moscow, 1981), Laser Beams in the Atmosphere (New York, Plenum Publishing, 1982); V. E. Zuev, et al., Nonlinear Optical Effects in Aerosols (Nauka, Siberian Branch, U.S.S.R. Acad. Sci., Novosibirsk, 1980); O. A. Volkovitskii et al., Propagation of High-Power Laser Radiation in Clouds (Gidrometeoizdat, Moscow, 1982).

The present monograph generalizes the most important results of both the theoretical and the experimental investigations of the effects of high-power laser radiation propagation in atmospheric aerosols not described in the above monographs. The bulk of this book contains the results of investigations carried out at the Institute of Atmospheric Optics, Siberian Branch, U.S.S.R. Acad. Sci. under the scientific guidance of, and with the participation of, the authors.

This book consists of an introduction and eight chapters. The first chapter briefly describes the results of experimental and theoretical investigations of microphysical and optical characteristics of atmospheric aerosols, such as clouds, mists, hazes, and precipitation. The models of atmospheric aerosols of practical importance are also presented here.

The second chapter describes the investigations of the kinetics of evaporation and combustion of a single aerosol particle under the effect of moderate-intensity radiation, when there is no heat explosion of a particle. This chapter also contains the results of the study of nonthermal mechanisms of interaction.

The third chapter studies the energetics of the propagation of high-power infrared radiation through clouds and mists. Here, an analysis is carried out of the formation and movement of dissipation waves, as well as of the influence of wind and turbulence. Experimental data concerning the dissipation of natural mists are presented.

The fourth chapter describes the self-action of laser beams in droplet media when their parameters undergo refraction and fluctuation distortions, also taking into account the fluctuations of the meteorological parameters of the medium. A quantitative solution is derived for the problems associated with the beam's selfaction in the vaporized aerosol when the refraction, diffraction, and fluctuation distortions of the beam jointly affect the process of dissipation. The influence of wind velocity fluctuations on the dissipation of the aerosol is considered. This chapter also presents the results of investigating the sounding beams propagating along the clear channels created by the radiation.

In the fifth chapter the propagation of high-power laser radiation through an aerosol is considered under the conditions of droplet explosion. The authors concentrate on the explosion of a single liquid particle. The classification of droplet explosions used here is based on the optothermodynamic approach and the analysis of experimental data. The basic models of droplet explosion evaporation are discussed. These models adequately describe real physical situations. The calculations and estimates of the optical characteristics of the exploding droplets are presented. Some of the most important experimental results are considered concerning the effect of a pulsed $CO_2$-laser on artificial mists.

In the sixtth chapter the results of investigations into the effects of the self-action of high-power laser radiation are considered with respect to atmospheric hazes of different types. The classification of thermal nonlinear interactions in solid-phase hazes is suggested, based on the analysis of characteristic times of thermo-acoustic relaxation processes in the medium with discrete sources of heat release. The effects of nonlinear light scattering on the thermal aureoles around the radiation-heated particles are considered. The results of the theoretical analysis and the data from laboratory experiments concerning investigations into pulsed-radiation beam self-broadening due to a joint effect of the processes of nonlinear scattering and regular refraction are discussed. The basic features of beam self-action for continuous and quasi-continuous radiation are described for the models of conservative and nonconservative atmospheric admixtures.

In the seventh chapter an analysis of the processes of the optical breakdown of aerodispersed media is made, including the problems associated with bare cascade ionization in the vapor aureole of an aerosol particle and an estimate of the effect of statistical spikes of laser radiation intensity in a turbulent atmosphere on the probability of the appearance of breakdown sources. The mechanisms of water-aerosol optical breakdown are considered. The most important results of both laboratory and field measurements of optical breakdown and the related effect of blocking the high-power radiation transmission by a gas-dispersion medium are presented.

The eighth chapter discusses some applications of the nonlinear effects described to the problems of laser ranging and navigation, as well as remote sensing of the atmosphere based on the use of the phenomenon of the laser spark and luminescence.

## NOMENCLATURE

In the manuscript there are two types of nomenclature: basic - which is used throughout the book, and specific nomenclature - different for each chapter. The basic nomenclature is given below.

Constants: $c$ - speed of light; $k_B$ - Boltzmann constant; $R_\mu$ - universal gas constant.

Thermodynamic characteristics of the medium: $T$ - temperature; $\rho$ - density; $V = \rho^{-1}$ - specific volume; $p$ - pressure; $U$ - specific internal energy; $Q$ - specific heat; $H$ - specific enthalpy; $Q_e$, $Q_m$ - specific heats of vaporization and fusion, respectively; $v$ - velocity of movement; $C_p$, $C_V$ - specific heats at constant pressure and at constant volume, respectively; $\mu$ - molecular weight; $\lambda_T$, $\chi$, $\chi = \lambda_T/c_p\rho$, $D$ - coefficients of thermal conductivity, temperature conductivity, and diffusion, respectively; $C_s$ - velocity of sound; $\gamma$ - adiabatic exponent; $\eta$ - dynamic viscosity; $\nu$ - kinematic viscosity.

Optical and microphysical characteristics of the aerosol medium: $\varepsilon$ - dielectric constant; $\varepsilon_0$ - unperturbed dielectric constant of air; $m_a = n_a - i\kappa_a$ - complex index of refraction of particle matter; $k_{ab}$ - volume coefficient of particle matter absorption; $\sigma$, $\sigma_s$, $\sigma_{ab}$ - cross-sections of extinction, scattering, and absorption of the particle, respectively; $K$, $K_s$, $K_{ab}$ - efficiency factors of extinction, scattering, absorption of particle; $\alpha$, $\alpha_s$, $\alpha_{ab}$ - volume coefficients of aerosol extinction, scattering, and absorption, respectively; $\alpha_g$ - volume coefficient of gas absorption; $a$ - radius of particle; $N_0$ - particle concentration; $f(a)$ - particle size distribution function; $\tau$ - optical depth.

Space-time coordinates: $t$ - time; $\vec{r}$ - space coordinate.

Characteristics of radiation: $\lambda$ - wavelength; $\underline{\vec{E}} = \vec{E}(\vec{r}, t)\exp(ikx + i\omega t)$ - electric field strength; $k = 2\pi/\lambda$ - wavenumber; $\omega$ - angular frequency; $\vec{E}(\vec{r}, t)$ - slowly-varying complex amplitude of field; $I$ - intensity; $W$ - energy; $P$ - power; $w$, $J$ - density of optical radiation energy; $R_0$ - initial effective radius of laser beam; $F$ - focal length of beam.

Indices: '0' - initial (equilibrium, boundary) values; '$\infty$' - value of physical magnitude at a distance from aerosol particle; 'cr' - critical point; 'b' - normal boiling point; 'S' - state of saturation; 'a' - values of magnitudes for an aerosol particle; 'g' - value of a magnitude for a gaseous medium; 'Π' - value of magnitude for the vapor phase; 'L' - value of magnitude for the liquid phase.

CHAPTER 1

# MICROPHYSICAL AND OPTICAL CHARACTERISTICS OF ATMOSPHERIC AEROSOLS

## 1.1. INTRODUCTION

The atmospheric aerosol is one of the main factors which causes the attenuation of optical waves in the atmosphere and it is also the most variable component of the atmosphere. This variability refers both to its microphysical parameters (such as its number density, size spectrum, complex refractive index, and shape of particles) and to its optical parameters (such as its coefficients of extinction, scattering, and absorption, its scattering phase function, and other components of the scattering phase matrix).

Many papers can be found elsewhere in the literature on the experimental and theoretical studies of the optical characteristics of individual aerosol particles, as well as of different ensembles of aerosol particles. These investigations have recently become more intensive; this is connected with an urgent need for quantitative data on laser light scattering by aerosols as a result of the extensive use of lasers in communication systems, in systems for navigation, and in other optical systems operating in the atmosphere.

Qualitative data on various aerosol parameters are, at the same time, of great significance in calculations of radiation fields of the atmosphere, since the aerosol component of the atmosphere plays an important role in the processes governing weather and climate.

The main results of investigations of aerosol microphysics and optics obtained during the last 10 to 15 years were analyzed in the monographs [1-6, 22, 24], where one can also find an extensive bibliography.

According to estimates made in [7], the total mass of the natural aerosol is about $2.3 \times 10^9$ tonnes per year that makes about 88.5% of the global mass of atmospheric aerosols. Although this quantity varies insignificantly from year to year, the entrainment of aerosol particles in moving bodies of air makes the aerosol characteristics inhomogeneous in space and time due to wind, circulation, and other mechanisms of air mass movement.

The mass of the aerosols of industrial origin is, according to the same reference [7], only 11.5% of the total mass of atmospheric aerosols, but it is generated from limited areas of the globe and thus plays a very important role in the formation of significant variability between aerosols. This is especially noticeable in highly industrialized regions.

The values of the microphysical and optical parameters of atmospheric aerosols can vary over a very wide range but, nevertheless, only three basic types of aerosols can be defined according to the typical scattering properties of each. These types are: (1) clouds and fogs; (2) hazes; (3) precipitation.

This chapter gives a brief description of the contemporary investigations into the microphysical and optical parameters of the aforementioned types of aerosols, but only in the context of the problems to be considered in the following chapters. Detailed information on the subject can be found in the monographs cited above.

## 1.2. PRELIMINARY DISCUSSION

### 1.2.1. Light Scattering by a Single Aerosol Particle

1.2.1.1. Scattering, Absorption, and Extinction Coefficients. The absorption coefficient $\sigma_{ab}$ is defined as the ratio of the Poynting vector flux of the total field through a sphere of radius R to the intensity of incident radiation, taken with a minus sign. It can easily be seen that the flux, thus defined, is equal in its absolute value to the amount of field energy absorbed by the volume of the medium under consideration [1].

The scattering coefficient is generally defined as the ratio of the flux of the energy scattered by a particle of radius a to the intensity of incident radiation.

The sum of the absorption and scattering coefficients determines, in accordance with the law of the conservation of energy, the extinction coefficient, i.e.,

$$\sigma = \sigma_{ab} + \sigma_s.$$

In the case of spherical particles the efficiency factors are normally introduced. Correspondingly the extinction, scattering and absorption efficiency factors are defined as follows

$$K = \sigma/\pi a^2, \qquad K_s = \sigma_s/\pi a^2, \qquad K_{ab} = \sigma_{ab}/\pi a^2. \tag{1.2.1}$$

As seen from (1.2.1), these factors are numerically equal to the amount of energy removed from the incident light flux due to extinction, scattering or absorption, divided by the amount of energy incident on the cross-sectional area of the particle $\pi a^2$.

General expressions for K, $K_s$ and $K_{ab}$ are given by the Mie theory. These expressions form infinite series over two arguments, one of which characterizes the relative size of particles $\rho_M$ and the other is the relative refractive index of the particulate matter, i.e.,

$$\rho_M = 2\pi a/\lambda, \qquad m_a = m_a'/m_\infty,$$

where $\lambda$ is the wavelength of the scattered radiation and $m_a'$ and $m_\infty$ are the complex indices of refraction of the particulate matter and the surrounding medium, respectively.

Under the conditions in the atmosphere, the relative refractive index $m_a$ can be considered to be equal to the complex refractive index of the particulate matter: $m_a = m_a' = n_a - i\kappa_a$, where $n_a$ is the refractive index and $\kappa_a$ is the absorption coefficient of the aerosol matter.

In certain asymptotic cases the factors K, $K_s$, and $K_{ab}$ can be expressed in an explicit analytical form.

The functions K, $K_s$, and $K_{ab}$, tabulated using Mie formulas, can also be found elsewhere in the literature. When tabulating these functions, one should truncate corresponding series at terms of serial numbers of the order of $\rho_M$ to provide sufficient accuracy.

Because of the great interest of specialists from different disciplines in the values of K, $K_s$, and $K_{ab}$, numerous calculations of the functions have been made for various values of $m_a$ and $\rho_M$.

The most complete information concerning the particles of atmospheric aerosols can be found in works cited in [1, 6]. In [6] one can find a vast amount of information concerning the above-mentioned calculations.

Some authors succeeded in constructing approximations for describing the behavior of K on $m_a$. Deirmendjian [8] constructed an approximating formula for the function $K(\rho_M, m_a)$ which is valid for any $\rho_M$ and $m_a$ if only the condition $|m_a| < 2$ is fulfilled:

$$K = (1 + D)K_1 \tag{1.2.2}$$

where $K_1$ is the function described by formulas valid for the asymptotic case of so called 'soft particles', and $(1 + D)$ is the approximating factor [6]. As shown in [8], the functions $K(\rho_M, m_a)$ calculated using (1.2.2) approximate to those obtained from exact Mie expressions within accuracy limits less than 4%.

1.2.1.2. Scattering Phase Function. Scattering phase functions of spherical particles were calculated in numerous papers, the list of which can be found, e.g., in [1-6]. The most comprehensive data concerning angular scattering functions have been obtained for scattering angles from 0 to 5° in 0.1° angular increments, from 5° to 90° in 1° increments, and in the angular region of the primary rainbow of water droplets 135-140° in 0.2° increments.

All components of the scattering phase matrix, as well as the components of the field of a scattered wave for water droplets, have been investigated most thoroughly for the spectral region from 0.4 to 12 µm.

It follows from these results that small particles (i.e., $\rho_M \to 0$) with $m_a \simeq 1$ have a symmetric Rayleigh scattering phase function, while for small particles with $m_a \to \infty$ the portion of backward-reflected radiation is larger than that of the forward scattered radiation. With the growth of spheres, i.e., for $\rho_M$ steadily increasing from 0 to $\infty$, the scattering phase function of aerosol particles continuously changes the shape, which becomes more and more asymmetric and elongated in the forward direction. This is known in the literature as the Mie effect.

The changes of the scattering phase function accompanying the growth of aerosol particles are controlled by the fact that their refractive index is a complex value, as well as by the oscillations in the intensity of the radiation scattered at different angles, which depends on $\rho_M$, $\theta$, and $m_a$.

1.2.1.3. Light scattering by nonspherical particles. Many investigators assume that the shape of atmospheric aerosol particles can be approximated by spheres, however, this question should be studied more thoroughly, since certain types of atmospheric aerosols, such as dust particles, ice crystals, and crystalline particles in clouds, can have arbitrary shapes. It is important in these circumstances to know how the shape of the particles can modify their optical properties. A considerable number of results were obtained and summarized in [9] for ellipsoids, solid cylinders, discs, and particles of other shapes with various $\rho_M$ and $m_a$ parameters and different orientations with respect to the direction of incident radiation.

As the analysis shows, the optical characteristics of particles of different shapes essentially depend on the aspect ratio of particles as well as on their orientations with respect to the incident radiation. The optical characteristics of such particles also depend on the degree of polarization of the incident radiation and on the complex index of refraction of particulate matter.

The optical characteristics of nonspherical particles can strongly differ from those of spherical particles having the same volume. Care is required, therefore, when applying the Mie theory to nonspherical particles. However, if the shape of the particles does not differ markedly from a sphere (as in the case of ellipsoids with an aspect ratio of about 1.5 to 2, cubic particles, or cylinders with height and diameter of equal length) then such characteristics as the scattering phase function and the extinction, scattering and absorption coefficients also do not differ strongly from those of spherical particles with the same volume.

### 1.2.2. Light Scattering by a System of Particles

The exact electrodynamic formulation of the problem of electromagnetic wave scattering by a system of particles, as well as the method for solving this problem, have been widely discussed in the literature. It is generally assumed that the solution of the problem of light scattering by a single particle is known and that all the physical parameters sought can be found by the corresponding statistical averaging over an ensemble of particles.

If we consider non-interacting particles, then for the intensity of radiation in the medium we can derive a conventional radiation transfer equation. In the case of interacting particles, there are some necessary corrections to this equation if the interaction is taken into account in an ordinary way.

The expression for the extinction coefficient $\alpha(\lambda)$, calculated for the unit path length along the beam direction, is written as follows

$$\alpha(\lambda) = N_0 \int_0^\infty \sigma(a, \lambda) f(a)\, da, \tag{1.2.3}$$

where $N_0$ is the number density of the particles; $\sigma(a, \lambda)$ is the extinction coefficient of a particle with radius a at wavelength $\lambda$ (here, and below, the particles are assumed to be spherical); f(a) is the particle size distribution function, the meaning of which is determined by the relationship $N_a\, da = N_0 f(a)\, da$, where $N_a$ is the number density of particles with radius from a to a + da.

Since $\sigma(\lambda, a) = \sigma_s(\lambda, a) + \sigma_{ab}(\lambda, a)$, then it follows from (1.2.3) that the polydispersed scattering and absorption coefficients are

$$\alpha_s(\lambda) = N_0 \int_0^\infty \sigma_s(\lambda, a) f(a)\, da, \tag{1.2.4}$$

$$\alpha_{ab}(\lambda) = N_0 \int_0^\infty \sigma_{ab}(\lambda, a) f(a)\, da. \tag{1.2.5}$$

Using the extinction coefficient $\alpha(\lambda)$, one can easily write the equation for changes of the intensity of the radiation propagating along some path in the form

$$dI(\lambda) = -I(\lambda)\alpha(\lambda)\, d\ell. \tag{1.2.6}$$

Integration of (1.2.6) gives a well known expression for the aerosol component of atmospheric transmission $T_{tr}(\lambda) = I/I_0$, where $I_0$ is the intensity of the incident radiation.

$$T_{tr}(\lambda) = \exp\,(- \int_{(\ell)} \alpha(\lambda, \ell)\, d\ell). \tag{1.2.7}$$

Integration in (1.2.7) is made over the propagation path. The variations of $\alpha(\lambda)$ along the path are assumed to be due to possible variations in the size spectrum and number density of the aerosol particles.

### 1.2.3. Scattering Phase Matrix

Scattering phase matrix M contains complete information on the light field scattered by the particles. The knowledge of the components of this matrix provides, in particular, the possibility of solving any problem associated with the scattering of waves by particles.

In the case of molecular light scattering, the matrix has the simplest form

$$M(\theta) = \frac{3}{4 + 3d_p} \begin{pmatrix} 1 + \cos^2\theta + d_p & -\sin\theta & 0 & 0 \\ -\sin^2\theta & 1 + \cos^2\theta & 0 & 0 \\ 0 & 0 & 2\cos\theta & 0 \\ 0 & 0 & 0 & 2\cos\theta \end{pmatrix}$$

where $d_p$ is the depolarization coefficient.

In the general case of atmospheric aerosols, the matrix can have 16 different components. However, the particles' symmetry and their orientation in space lead to the reduction of the number of independent components and to cancellation. Thus, for spherical particles, the scattering phase matrix has the following form

$$M(\theta) = \begin{pmatrix} M_{11} & M_{12} & 0 & 0 \\ M_{21} & M_{22} & 0 & 0 \\ 0 & 0 & M_{33} & M_{34} \\ 0 & 0 & M_{43} & M_{44} \end{pmatrix}$$

and, moreover, in this case $M_{11} = M_{22}$, $M_{12} = M_{21}$, $M_{33} = M_{44}$, $M_{34} = M_{43}$.

In [10] one can find the results of the calculations of all four components of $M(\theta)$ for water droplets. Detailed computations of $M_{43}(\theta)$ made in [11] revealed a high sensitivity of this parameter to variations of the size spectrum and to the complex refractive index of polydispersed ensembles of scattering particles.

The results of the calculations of scattering phase matrices made by Deirmendjian [12] for some typical models of polydispersed aerosols are now widely used in atmospheric optics research. In [13] the authors revealed changes of the matrix components when the aerosol model used is constructed on the basis of the experimental data in [14].

The most comprehensive experimental information on the elements of the scattering phase matrix was obtained at the Institute for Atmospheric Physics of the U.S.S.R. Academy of Sciences, as a result of many years investigations carried out at the Zvenigorod field base. These results are presented in [15-20]. It was shown during the analysis of these measurements that, even under the conditions of adequate atmospheric turbidity, the scattering phase matrix of atmospheric aerosols is very close to that of spherical particles, i.e., $M_{11} = M_{22}$, $M_{33} = M_{44}$, $M_{34} = -M_{43}$, $M_{12} = M_{21}$ with all the rest elements being negligible. This fact, therefore, justifies the assumptions made in [21] in the interpretation of the experimental data.

The statistical analysis of the measurements of scattering phase matrices led to a classification of atmospheric aerosols. This analysis made it possible to distinguish between several types of scattering phase matrix, as well as between corresponding types of optical weather such as mist, haze, foggy haze, and haze with drizzle. Different patterns of behavior of the elements $M_{11}$, $M_{21}$, $M_{33}$, and $M_{43}$ for each of these formations are observed.

It should be noted, however, that the above results were obtained in one geographical region, so their applicability to other regions cannot be possible without carrying out the appropriate statistically-validated investigations.

The interpretation of experimental measurements of scattering phase matrices made under field conditions should take into account the geographical location and, as a consequence, the origin of atmospheric aerosols, their shapes, chemical composition, size spectrum, and number density.

## 1.3. LIGHT SCATTERING BY CLOUDS AND FOGS

A quantitative measure of the light attenuation caused by clouds and fogs, as well as by other aerosols, is the volume extinction coefficient $\alpha(\lambda)$ [see (1.2.3)]. As seen, the value of $\alpha(\lambda)$ is determined by the particles' number density, size distribution function, and by the extinction coefficient of an individual particle. A description of these characteristics is presented below, along with the results of calculating $\alpha(\lambda)$.

### 1.3.1. Microphysical Parameters of Clouds and Fogs

The processes of formation of clouds and fogs depend on the many variable factors which determine the growth of droplets. Therefore, the attempts to construct a theory for the prediction of size distribution functions has not yet been successful.

The analytical expressions used at present have been derived as approximations of experimentally-derived histograms.

Most of the experimental data obtained by different authors show that size distribution functions of cloud and fog particles from single peak asymmetric curves (see Figure 1.3.1).

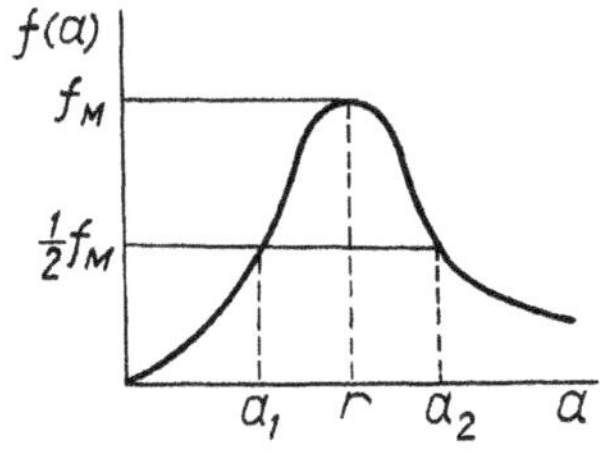


Fig. 1.3.1. Characteristic behavior of a particle size-distribution curve for water clouds and fogs.

The most widely used approximation of the size-distribution function for water clouds and fogs (the so-called gamma distribution function) is written as follows:

$$f(a) = \frac{1}{\Gamma(\mu + 1)} \mu \cdot^{\mu+1} \frac{a^{\mu}}{r^{\mu+1}} \exp(-\mu a/r), \quad (1.3.1)$$

where $\Gamma(\mu + 1)$ is the gamma function and it equals $\mu!$ for integer $\mu$; $r$ and $\mu$ characterize the most probable, or modal, radius of particles and the half-width of the size-distribution, respectively.

In clouds, for example, the mean value of the modal radius is within the range from 3 to 10 μm, while $\mu$ varies from tenths (for broad distributions) to 10 or 12 (for narrow distributions). The most common values of $r$ and $\mu$ are approximately 5 μm and 2, respectively.

For the description of cloud or fog microstructure to be complete, one should know the number density of the particles, $N_0$. This parameter can be found, provided that the function $f(a)$ and water constant $q$ are known. By water content we mean the amount of liquid water contained in droplets occupying 1 $m^3$ of the atmosphere. On average, the various types of clouds (except for strongly-developed cumulus) have a water content $q$ varying from 0.1 to 0.3 $g/m^3$.

For solving some of the problems of laser propagation through clouds or fogs one has to know the relationships between the water content or number density, the meteorological visual range, and the total geometrical cross-section of particles contained in a unit volume. In the case of spherical particles and a gamma size-distribution, the following relationships hold:

$$N_0 = \frac{3.912\mu^2}{S_m F(0.5) \pi r^2 (\mu + 2)(\mu + 1)}; \quad (1.3.2)$$

$$q = \frac{3.912 \times 4r(\mu + 3)\rho_0}{3S_m F(0.5)\mu}; \quad (1.3.3)$$

$$Q = N_0 \int_0^{\infty} \pi a^2 f(a)\, da, \quad (1.3.4)$$

where $\rho_0$ is the density; $S_m$ is the meteorological visual range [$S_m = 3.912/\alpha(0.5)$]; $F(0.5) = \alpha(0.5)/Q$ is the averaged extinction efficiency factor at the wavelength $\lambda = 0.5$ μm; and $\alpha(0.5)$ is the extinction coefficient at this wavelength.

The values of $q$ and the number density $N_0$ calculated using (1.3.2)-(1.3.4) for the case of some typical combinations of the microstructure parameters of clouds and fogs, and with $S_m = 0.2$ km, are presented in Table 1.3.1.

As seen from Table 1.3.1, the parameter of water content, and especially the number density of the droplets, vary widely, while the meteorological visual range remains constant ($S_m = 0.2$ km). Incidentally, this value of $S_m$ is the most probable one for clouds and fogs.

TABLE 1.3.1. Water content and number density of particles in some water droplet clouds.

| Cloud composition | Microstructure parameters | | q, g/ $m^3$ | $N_0$, $cm^{-3}$ |
|---|---|---|---|---|
| | r, μm | μ | | |
| Small droplets | 1 | 2 | 0.031 | 971 |
| | 1 | 10 | 0.015 | 2050 |
| Medium size droplets | 6 | 2 | 0.194 | 28 |
| | 6 | 10 | 0.101 | 65 |
| Large droplets | 10 | 2 | 0.324 | 10 |
| | 10 | 10 | 0.168 | 2.3 |

### 1.3.2. Volume Extinction Coefficients

Quite a comprehensive range of data concerning the theoretical and experimental studies of the extinction properties of clouds and fogs can be found, e.g., in [6].

It can be seen from the calculations of the volume extinction coefficients $\alpha(\lambda)$ of water clouds and fogs that both the absolute values of $\alpha$ and the spectral behaviour $\alpha(\lambda)$ strongly depend upon the microstructure parameters of these formations.

It can easily be seen from Figure 1.3.2 that clouds and fogs create a

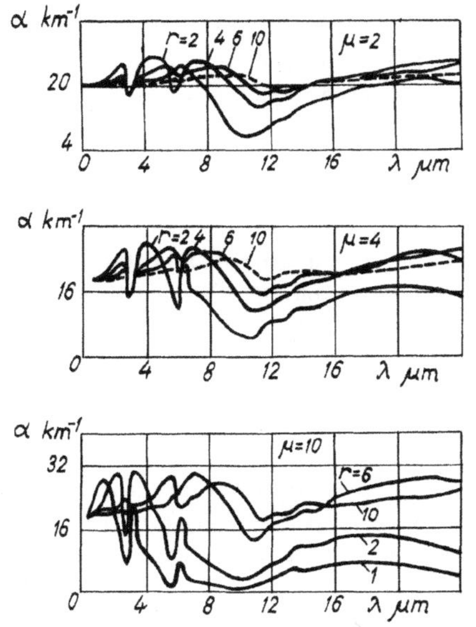


Fig. 1.3.2. Attenuation coefficients of water clouds and fogs within 0.5-25 μm for different values of μ and r.

serious obstacle for laser propagation in the atmosphere, with the exception of those composed of small droplets ($a \leq 1$ μm), whose sizes fall in a narrow range. However, such formations are rarely observed in the atmosphere.

## 1.4. LIGHT SCATTERING BY HAZES

### 1.4.1. Microphysical Parameters of Hazes

Under conditions with a meteorological visual range of 1 to 2 km and further, the turbidity of the atmosphere is caused by the presence of relatively small aerosol particles. These atmospheric aerosols are called hazes. Hazes worsen the visibility in the atmosphere, as compared with those in the purely molecular atmosphere ($S_m \simeq 340$ km).

Numerous investigations of the microphysics of hazes has revealed a great variety of types. The most valuable contributions to these studies have been made by groups headed by K. Ya. Kondratjev and L. S. Ivlev from Leningrad University, U.S.S.R.; G. V. Rosenberg from the Institute of Atmospheric Physics of the U.S.S.R. Academy of Sciences, Moscow, as well as by groups at the University of the Washington State, Seattle, U.S.A. and the University of Wyoming, Laramy, U.S.A. headed by R. D. Charlson, D. D. Hoffman, and D. M. Rosen.

One of the best overviews of the problem was published by Deirmendjian [22], which is quite valuable now. An extensive, profound, and up-to-date analysis of the investigations of the microphysics of hazes was recently made by G. M. Krekov and R. F. Rakhimov in [2]. A summary of the main results obtained at the Institute of Atmospheric Physics, U.S.S.R. Academy of Sciences, is presented in [23]. An extensive bibliography of the investigations of the microphysics of hazes can also be found in [1-7, 22-24]. Here, we shall give only a brief summary of these investigations and present some results.

According to the approach suggested in [23], atmospheric aerosols can be considered to be composed of three distinct fractions. The size spectrum of each fraction can be described by a single peak function that is analogous to the lognormal size distribution. Maxima of the three fraction size spectra are assumed to be at $r = 10^{-2}$, $10^{-1}$, and 1 μm, respectively. In [23] these fractions are called microdispersed, submicron, and coarse fraction, respectively.

Since the microdispersed fraction is not optically active [6], it will not be considered further in our discussions. It is the submicron fraction of atmospheric aerosols which is the basis for the formation of hazes. Finally, the coarse fraction of atmospheric aerosols is the suspension in air of the products of the mechanical fragmentation of the Earth's surface materials.

Now consider the model size distributions of the atmospheric aerosol ensembles suggested by different authors.

The models of clouds, hazes, and precipitation constructed by

TABLE 1.4.1. Parameters of atmospheric aerosol models [22].

| Model | Number density, $cm^{-3}$ | Parameters | | | | | Formation |
|---|---|---|---|---|---|---|---|
| | | $\nu_1$ | $a_m$, μm | $a_{max}$*, μm | $\nu_3$ | $\nu_4$ | |
| Haze M | 100 | $5.3333 \times 10^4$ | 0.05 | 3 | 1 | 1/2 | Marine or coastal aerosol |
| Haze L | 100 | $4.9757 \times 10^6$ | 0.07 | 2 | 2 | 2/1 | Continental aerosol |
| Haze H | 100 | $4.0 \times 10^5$ | 0.1 | 0.6 | 2 | 1 | High altitude or stratospheric aerosol |
| Rain M | 100 ($m^{-3}$) | $5.3333 \times 10^5$ | 0.05 (mm) | 3 (mm) | 1 | 2/1 | Weak to moderate rain |
| Rain | 1000 ($m^{-3}$) | $4.9757 \times 10^7$ | 0.07 (mm) | 2 (mm) | 2 | 2/1 | Heavy rain |
| Hail H | 10 ($m^{-3}$) | $4.0 \times 10^4$ | 1.0 (mm) | 6 (mm) | 2 | 1 | Hail containing a lot of small particles |
| Cumulus $C_1$ | 100 | 2.3730 | 4.00 | 15 | 6 | 1 | Cumulus, stratus, and fogs |
| Cloud $C_2$ | 100 | $1.0851 \times 10^{-2}$ | 4.00 | 7 | 8 | 3 | Clouds with a colored corona |
| Cloud $C_3$ | 100 | 5.5556 | 2.00 | 3.5 | 8 | 3 | 'Mother of pearl' clouds |
| Cloud $C_4$ | 100 | 5.5556 | 4.00 | 5.5 | 8 | 3 | Clouds in which two or three aureoles around the sun are observed |

* $a_{max}$ corresponds to the radius at the level of $N_{max}(a) \times 10^{-3}$.

Deirmendjian [22] are the most widely used. These models are based on the analysis of numerous experimental data. A modified gamma size distribution function

$$N_a(a) = \nu_1 a^{\nu_2} \exp(-\nu_3 a^{\nu_4}), \quad 0 \lesssim a \lesssim \infty \tag{1.4.1}$$

was suggested by Deirmendjian as the general size distribution function of atmospheric aerosol ensembles. Here a is the radius of a particle. $\nu_i$ (i = 1, ..., 4) are empirical parameters. The parameter $\nu_3$ is related to the modal radius $a_m$ as follows:

$$a_m = [2/3(\nu_3)]^{1/3}.$$

The function $N_a(a)$ reaches its maximum value when $a = a_m$. Table 1.4.1 presents the parameters of the models of various atmospheric aerosols according to [22].

Some authors suggest a lognormal size distribution function for describing the haze spectrum, i.e.,

$$f(a) = \frac{1}{\sqrt{2}\,\ln d_s} \exp\left[-\frac{1}{2}\left(\frac{\ln(a/a_m)}{\ln d_s}\right)\right], \tag{1.4.2}$$

where $d_s$ is the standard deviation, and $a_m$ is the modal radius. Whetby [25] suggested a bimodal size distribution, which is the superposition of two expressions similar to (1.4.2) whose parameters are presented in Table 1.4.2.

TABLE 1.4.2. Parameters $a_m$ and $d_s$ of the bimodal size-distribution function of tropospheric aerosols (A denotes the submicron fraction, B the coarse fraction originating from local sources).

| Type of aerosol | $a_m$, μm | | $\ln d_s$ | |
|---|---|---|---|---|
| | A | B | A | B |
| Continental | 0.03 | 0.4 | 0.74 | 0.81 |
| Marine | 0.05 | 0.65 | 0.68 | 0.74 |
| Urban | 0.04 | 0.63 | 0.63 | 0.77 |

For describing the size spectrum of haze particles with radii greater than 0.1 μm, Junge suggested the following as an empirical formula:

$$f(a) = Aa^{-\beta}, \tag{1.4.3}$$

where β is an empirical constant, whose values vary from 2 to 5 depending on the type of aerosol [6].

Junge's formula has become widely used in calculations of aerosol optical characteristics, although it cannot be valid for any particular size spectrum (this limitation applies to the other formulas commonly in use).

Since in the visible, and especially in the IR, region particles with radii of less than 0.1 μm can be considered optically passive [6], the single-peak gamma and lognormal size distributions, as well as Junge's formula, can be used for calculating the optical characteristics of the haze. By varying β in (1.4.3) one can adjust this function to describing corresponding wing of the functions (1.4.1) and (1.4.2).

TABLE 1.4.3. The percentage by weight of the chemical components of continental haze [26].

| | $[SO_4]^{2-}$ | $[NH_4]^+$ | $[NO_3]^-$ | $SiO_2$ | $[CO_3]^2$ | $Fe_2O_3$ | $Al^{3+}$ | Ca | Na | Cℓ | K | $H_2O$ (ice) | Mg and others |
|---|---|---|---|---|---|---|---|---|---|---|---|---|---|
| % by weight | 18 | 4 | 1 | 35 | 8 | 1 | 5 | 9 | 2 | 4 | 1 | 5 | 1 |

In the concluding part of this section we will present the model of the chemical composition of continental haze developed in [26] (see Table 1.4.3). This model is based on the numerous experimental results obtained by a group workung under K. Ya. Kondratjev and L. S. Ivlev at Leningrad University from 1967 to 1972 [2, 4]. These measurements were made in various climatic zones of the U.S.S.R. The statistical analysis of the data did not involve treating data obtained from industrial aerosols or from measurements made in anomalously contaminated atmosphere.

### 1.4.2. Volume Extinction Coefficients

Figure 1.4.1 presents the calculation data of the volume extinction coefficients of hazes, obtained in [6]. The calculations used the Junge size distribution function for spherical water droplets. The values of β were 3, 4, and 5. Minimum and maximum radii of droplets were within the range 0.01 to 10 μm.

One can easily see from Figure 1.4.1 that the microstructure parameters of hazes strongly affect not only the absolute values of the extinction coefficient, but also their spectral behaviour. Thus, for example, the most rapid decrease of the extinction coefficient with increasing wavelength observed in the region from 0.3 to 2.7 μm is also caused by a very rapid falling off of the efficiency factor $K(\rho_M)$ in the visible and near IR, which is characteristic for small particles.

The maxima of the spectral curve of the extinction coefficient, observed

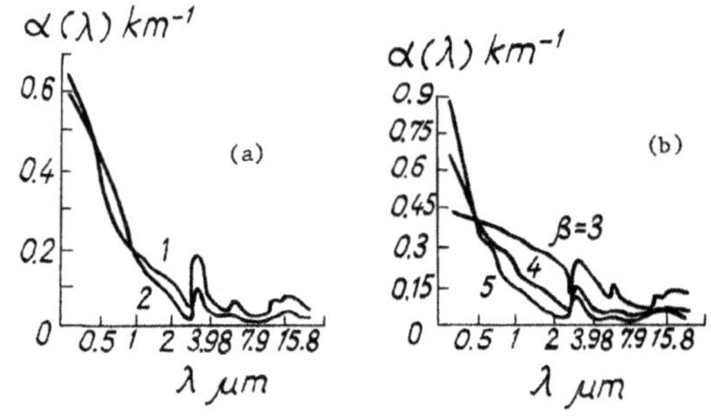


Fig. 1.4.1. Volume coefficients of scattering for hazes in the range 0.3–25 μm with a visual range of 10 km. (a) $\beta = 4$, $a_{min} = 0.1$ μm, $a_{max} = 1.0$ μm (curve 1); $\beta = 4$, $a_{min} = 0.01$ μm, $a_{max} = 10$ μm (curve 2); (b) $a_{min} = 0.05$ μm, $a_{max} = 5.0$ μm.

at 2.9, 6.0, and 17 μm, coincide with the maxima of liquid water absorption bands.

Presented below are the extinction coefficients at specified laser wavelengths, calculated for a haze with the most probable microstructure parameters ($\beta = 4$, $a_{min} = 0.05$; $a_{max} = 5.0$ μm) and assuming that the meteorological visual range is 10 km.

| | | | | | | | | | | |
|---|---|---|---|---|---|---|---|---|---|---|
| λ, μm: | 0.5 | 0.53 | 0.63 | 0.69 | 0.84 | 1.06 | 1.15 | 2.36 | 3.39 | 10.6 |
| α, $km^{-1}$: | 0.40 | 0.38 | 0.32 | 0.29 | 0.24 | 0.18 | 0.17 | 0.07 | 0.08 | 0.01 |

Measurements made near Zvenigorod [27, 28] provided an approximation of the spectrum of the extinction coefficient of hazes for the wavelength range 0.59 to 10 μm:

$$\alpha(\lambda) = \alpha(0.59)(n_0 + n_1\lambda^{-k}) \tag{1.4.4}$$

Here, $n_0$, $n_1$, and k are empirical parameters, which are tabulated for ten different types of weather conditions.

It should be noted, however, that (1.4.4) is only valid for describing the spectral behavior of the extinction coefficient of continental hazes occurring in the Moscow region. The use of (1.4.4) for data obtained in other geographical regions requires additional justification. Moreover, as measurements made in [29, 30] have shown, this expression is not valid for marine hazes. These measurements revealed a significantly slower decrease of the function $\alpha(\lambda)$ with increasing λ than that predicted by (1.4.4). Thus, the analysis of data collected during a summer on the Black Sea coast gave rise to following expression for the extinction coefficient spectrum:

$$\alpha(\lambda) = \{[\alpha(0.59) - C]K\}^n, \tag{1.4.5}$$

where C, K, and n are empirical parameters different for different spectral regions. The validity of this formula is based on a high correlation between $\alpha(\lambda)$ values in the IR and those at $\lambda = 0.59$ μm which was seen in these measurements. Moreover, (1.4.5) is valid for all types of haze observed in this experiment. The values of $\alpha(\lambda)$ calculated for the spectral range of the atmospheric transmission window at 10 μm using both (1.4.4) and (1.4.5) differ significantly. This clearly shows that the microphysical parameters of marine and continental hazes are essentially different.

The experimental study of the spectral behaviour of volume extinction coefficients of continental hazes was undertaken in the vicinity of Tomsk, U.S.S.R. by groups from Leningrad University and the Institute of Atmospheric Optics [31, 32]. The results of this investigation showed quite a complicated dependence of $\alpha(\lambda)$ on aerosol composition. This investigation was possible due to the simultaneous determination of $\alpha(\lambda)$, the size spectrum, and the chemical composition of the haze particles. It was found that the extinction of radiation at wavelengths $\lambda < 2$ μm and $\lambda > 2$ μm is caused by aerosol particles of different origin. One of the interesting facts observed during this study was the existence of maxima in the function $\alpha(\lambda)$ in the region 10 to 12 μm. The origin of these maxima is connected with the large value of the imaginary part of the complex refractive index of the haze particles. It should be noted that the function $\alpha(\lambda)$ of polydispersed water haze has its most significant and broadest minimum just in this region, as shown in Figure 1.4.1. This shows once more that the physical parameters of continental hazes are very varied.

The dependence of the volume extinction coefficient for visible radiation on the relative humidity is another interesting fact revealed in this complex experimental study. It was found that the volume extinction coefficient increases with increasing relative humidity in the range from 40 to 70%, but then it falls, and only when the relative humidity reaches 85% does the volume extinction coefficient increase again. Such behavior was observed during the same two-year period and under conditions when there were no air mass changes. As it was found from the statistical analysis of the experimental data on this dependence, the value of the coefficient of mutual correlation between the relative humidity and the extinction coefficient was approximately 0.6.

In conclusion, we would like to underline once more the fact that there exists a great variety of types of atmospheric haze with different sets of optical parameters. It should be also noted that the experimental data on hazes available to date do not allow a complete classification of hazes.

## 1.5. MICROPHYSICAL AND OPTICAL CHARACTERISTICS OF PRECIPITATION

All natural precipitations are composed of large particles whose Mie parameter $\rho_M \gg 1$. The experimental data on the size spectra of rain droplets available from the literature show that the shape of the size distribution function, in this case, is the same as for clouds and fogs. The microstructure parameters of the rain droplets depend on the rainfall intensity and on the distance from the cloud of origin. The number density of the droplets can vary in the range from 100 to 20,000 $m^{-3}$, the droplet radius varies, in general, from several hundredths of a millimeter to several millimeters, the water content of rains alternates between several hundredths of a gram per cubic meter, and the intensity of the rainfall ranges from several tenths to several tens of millimeters per hour.

Since, for all rain droplets, we have $\rho_M \gg 1$, then the following relationship is valid for all the laser wavelengths:

$$\alpha(\lambda) = N_0 \int_0^\infty \pi a^2 K(\rho_M, m_a) f(a)\, da = 2Q. \qquad (1.5.1)$$

Thus, as seen from (1.5.1), the volume extinction coefficient of precipitation for UV, visible and near IR, and middle IR radiation is independent of wavelength. It is determined quantitatively by the total geometrical cross-sectional area of the droplets occupying a unit volume.

The dependence of $\alpha$ values on the variations of microstructure parameters is insignificant as compared with a very high correlation between the rainfall intensity $I_R$ and $\alpha(\lambda)$. The relationship between $I_R$ mm/h and $\alpha$ $km^{-1}$ is written as follows:

$$\alpha(\lambda) = 0.21 I_R^{0.74}.$$

The value of the coefficient of mutual correlation between $\ell g\ \alpha$ and $\ell g\ I_R$ is approximately $0.95 \pm 0.01$. An even better correlation is observed between $\ell g\ \alpha$ and $\ell g\ q$, where $q$ ($g/m^{-3}$) is water content of the rain. The coefficient of mutual correlation in this case is $0.97 \pm 0.01$. The behavior of $\alpha(\lambda)$ in snowfalls is similar.

A rainfall of moderate intensity - about 10 mm/h ($\alpha(\lambda) \simeq 1$ $km^{-1}$) - removes about 60% of the laser beam energy at a distance of 1 km.

Nonselective spectral behavior of $\alpha(\lambda)$, which is predicted theoretically, is not observed experimentally and the deviation from theoretical estimates is larger for a larger difference between two wavelengths where $\alpha(\lambda)$ is measured.

Physically, this discrepancy between theory and experiment can be explained by the fact that in the case of large particles, the scattered radiation is mainly concentrated in a narrow cone around the direction of light propagation. Since any optical device used for measuring atmospheric transmission has a finite angular aperture, then the forward-scattered radiation will contribute to the total optical flux entering the optical receiver of the device, thus decreasing the attenuation of the light. On the other hand, the cone angle depends on the Mie parameter $\rho_M$, i.e., on $\lambda$, and hence the contribution made by stray light should also have a spectral dependence. Taking this into account, one can easily understand why the experimental extinction coefficient spectra differ from the theoretically nonselective cases described in [33, 35].

## 1.6. SCATTERING PHASE FUNCTIONS OF POLYDISPERSED AEROSOLS

The volume extinction coefficient is the quantitative measure of the energy removed from a beam by a medium due to scattering in all directions.

In many problems of light propagation through the atmosphere, information is needed not only on energetic losses but also on the angular distribution of the energy removed from a beam. The problems of laser sounding of the atmosphere, high-level detection and ranging, and many other related problems, can be mentioned in this connection.

The scattering phase functions of polydispersed ensembles of spherical particles can be easily calculated. A lot of calculated data on this function can be found elsewhere in the literature for various wavelengths and different sets of microstructure parameters.

Figures 1.6.1 and 1.6.2 present some results of calculations for clouds, fogs, and hazes and for the most widely used lasers, viz.: the He-Ne laser

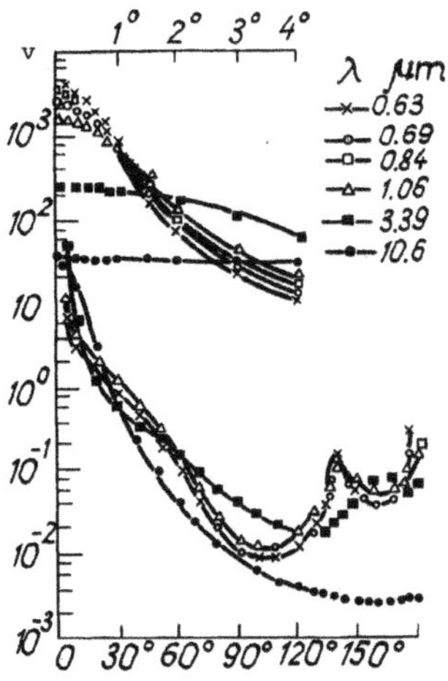


Fig. 1.6.1. Scattering phase function of water clouds and fogs for gamma-distribution parameters r = 5 μm and μ = 2.

(λ = 0.63 μm; 1.15 μm; 3.51 μm); the $CO_2$ laser (λ = 10.6 μm); the ruby laser (λ = 0.69 μm); the neodymium-glass laser (1.06 μm), and the semiconductor laser (λ = 0.84 μm).

Figure 1.6.1 shows the data calculated for fogs and clouds whose size spectra are described by a gamma size-distribution function, while Figure 1.6.2 represents the characteristics of atmospheric hazes composed of spherical particles with a Junge size spectrum.

As seen from these figures, the scattering phase functions of water clouds, fogs, and hazes are quite smooth. Also, fogs and clouds have very

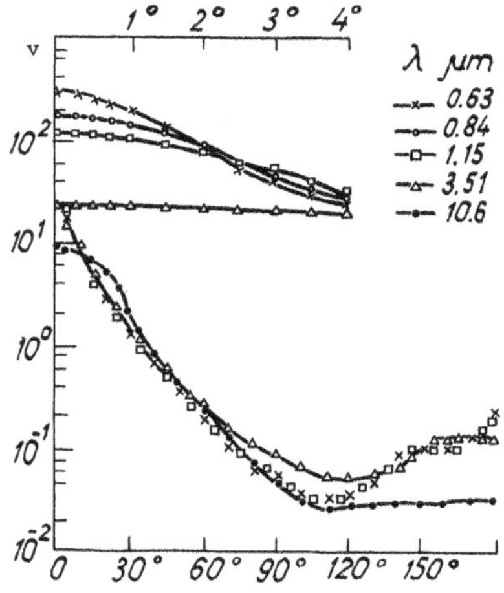


Fig. 1.6.2. Scattering phase function of water haze with a Junge particle size distribution ($\beta = 3$, $a_{min} = 0.05$ μm, $a_{max} = 5.0$ μm).

asymmetric scattering phase functions. The forward-scattered flux, in this case, exceeds the backward-scattered radiation flux by 3 to 5 orders of magnitude. In the case of hazes, this value is about 2 to 3 orders of magnitude.

The monograph [6] presents a description of many experimental results of measuring the scattering phase functions of atmospheric aerosols. These results qualitatively agree with the calculated data. A quantitative comparison is hampered by a lack of data on the microphysical parameters of aerosols. It also should be noted that all the measurements of aerosol scattering phase functions were made within the visible range of electromagnetic radiation.

## REFERENCES: CHAPTER 1

[1] V. E. Zuev: Laser Beams in the Atmosphere (plenum, New York, 1982), in Russian.

[2] G. M. Krekov and P. F. Rakhimov: Opto-Sounding Model of a Continental Aerosol (Nauka, Novosibirsk, 1982), in Russian.

[3] V. E. Zuev: Laser Radiation Propagation in the Atmosphere (Radio i Svyaz, Moscow, 1981), in Russian.

[4] K. Ya. Kondratiev, D. V. Poznyakov: Atmospheric Aerosol Models (Nauka, Moscow, 1981), in Russian.

[5] V. E. Zuev, M. V. Kabanov: Optical Signal Transport (under Noise Conditions) (Sovetskoe Radio, Moscow, 1977), in Russian.

[6] V. E. Zuev: Visible and IR Wave Propagation in the Atmosphere

(Sovetskoe Radio, Moscow, 1970), in Russian.

[7] F. Robinson and R. C. Robbins: Emission Concentrations and Rate of Particulate Atmospheric Pollutants (Amer. Petrol. Inst., Publ. N4076, 1971).

[8] D. Deirmendjian: 'Atmospheric extinction of infrared radiation', Quart. J. Roy. Met. Soc. 86, N369, 371-381 (1960).

[9] G. van de Hulst: Light Scattering by Small Particles (Inostran. Literatura, Moscow, 1961), in Russian.

[10] V. S. Malkova: 'Light scattering by haze particles', Izv. Akad. Nauk SSSR Fiz. Atmos. Okeana 1, N1, 109-113 (1965), in Russian.

[11] R. Eiden: 'The elliptical polarization of light scattered by a volume of atmospheric air', Appl. Opt. 5, N4, 569-576 (1966).

[12] D. Deirmendjian: Electromagnetic Radiation Scattering by Spherical Polydispersed Particles. Translation from the English (Mir, Moscow, 1971), in Russian.

[13] A. P. Prishivalko: 'The Effect of Relative Humidity on the Elements of the Light Scattering Matrix by Systems of Homogeneous and Inhomogeneous Particles of Atmospheric Aerosols', All-Union Symp. on Laser Radiation Propagation in the Atmosphere (Institute of Atmospheric Optics, Siberian Branch, U.S.S.R., 1975) pp. 6-7 (in Russian).

[14] L. S. Ivlev and S. I. Popova: 'Complex refractive index of the matter in dispersive phase of atmospheric aerosol', Izv. Akad. Nauk SSSR Fiz. Atmos. Okeana 9, N10, 1034-1043 (1973), in Russian.

[15] G. I. Gorchakov and G. V. Rozenberg: 'Measurements of the light scattering matrix in the lower layer of the atmosphere', Izv. Akad. Nauk SSSR Fiz. Atmos. Okeana 1, N12, 1279-1288 (1965), in Russian.

[16] G. I. Gorchakov: 'Light scattering matrix on the lower layer of the atmosphere', Izv. Akad. Nauk SSSR Fiz. Atmos. Okeana 2, N6, 595-605 (1966), in Russian.

[17] G. I. Gorchakov and G. V. Rozenberg: 'Correlation associations between optical characteristics of finely dispersed smoke', Izv. Akad. Nauk SSSR Fiz. Atmos. Okeana 3, N6, 611-620 (1967), in Russian.

[18] G. V. Rozenberg and G. I. Gorchakov: 'The degree of polarization ellipticity of the light scattered by atmospheric air as the means for investigating aerosol microstructure', Izv. Akad. Nauk SSSR Fiz. Atmos. Okeana 3, N7, 699-713 (1967), in Russian.

[19] G. V. Rosenberg: 'Optical investigations of atmospheric aerosols', Usp. Fiz. Nauk 95, N1, 159-208 (1968), in Russian.

[20] G. I. Gorchakov: 'The light scattering matrix and types of optical weather', Izv. Akad, Nauk SSSR Fiz. Atmos. Okeana 9, N2, 204-209 (1974), in Russian. And: 'On the choice of characteristic features in classification of light scattering matrixes', Izv. Akad. Nauk SSSR Fiz. Atmos. Okeana 10, N12, 1321-1371(1975), in Russian.

[21] B. S. Pritchard and W. G. Elliott: 'Two instruments for atmospheric optics measurements', J. Opt. Soc. Am. 50, N3, 191-202 (1960).

[22] V. M. Orlov et al.: Elements of Light Scattering Theory and Optical

Sounding, ed. by V. M. Orlov (Nauka, Novosibirsk, 1982), in Russian.

[23] G. V. Rozenberg et al.: 'Optical Parameters of Atmospheric Aerosol', in Physics of the Atmosphere and the Problem of Climate (Nauka, Moscow, 1980), pp. 216-257, in Russian.

[24] S. S. Butcher and R. T. Charlson: An Introduction to Air Chemistry (Acad. Press, New York, 1972).

[25] K. T. Whetby: 'Modeling of Atmospheric Aerosol Particle Size Distribution', Progress Report (Particle Technology Lab., University of Minnesota, U.S.A., 1975), p. 42.

[26] V. E. Zuev et al.: 'Calculation of a stratified model of an atmospheric aerosol for laser sounding at $\lambda = 0.6943$, 1.96, 2.36, and 10.6 μm', Izv. Vyssh. Uchebn. Zaved. **11**, 39-47, in Russian.

[27] Yu. S. Georgievskii: 'On the spectral transmittance of hazes within the range 0.37 to 1.0 μm. Izv. Akad. Nauk SSSR Fiz. Atmos. Okeana **5**, N4, 388-394 (1969), in Russian.

[28] V. L. Filippov and S. O. Mirumyants: 'Aerosol IR radiation attenuation in atmospheric transmission windows: I. winter hazes; II. spring and autumn; III. summer hazes', Izv. Akad. Nauk SSSR Fiz. Atmos. Okeana **8**, N7, 818-819 (1971), in Russian.

[29] M. V. Kabanov et al.: 'Some Peculiarities of optical Radiation Attenuation in Marine Hazes', Proc. £rd All-Union Symp. on Laser Radiation Propagation in the Atmosphere (Institute of Atmospheric Optics, Siberian Branch, U.S.S.R. Acad. Sci., Tomsk, 1975), pp. 19-22, in Russian.

[30] M. V. Kabanov et al.: 'Aerosol Attenuation of Visible and IR Radiation in Marine Coastal Haze', in Problems of Remote Sounding of the Atmosphere (Institute of Atmospheric Optics, Siberian Branch, U.S.S.R. Acad. Sci., Tomsk, 1975), pp. 189-207, in Russian.

[31] S. D. Andreev et al.: 'On some peculiarities of spectral transmission of atmospheric hazes in the visible and IR bands', Izv. Akad. Nauk SSSR Fiz. Atmos. Okeana **8**, N12, 1261-1267 (1972), in Russian.

[32] V. E. Zuev et al.: 'New results of the investigation of atmospheric aerosols', Izv. Akad. Nauk S.S.S.R. Fiz. Atmos. Okeana **9**, N4, 371-385 (1973), in Russian.

[33] V. E. Zuev: Atmospheric Transmission for visible and IR Beams (Sovetskoe Radio, Moscow, 1966), in Russian.

[34] M. V. Kabanov: 'On the effect of experimental conditions on the value of the measured scattering coefficient', in Actinometry and Atmospheric Optics (Nauka, Moscow, 1964), pp. 85-90, in Russian.

[35] M. V. Kabanov and Yu. A. Pkhalagov: 'On the spectral transmission of precipitation for IR waves', in Light Scattering in the Earth's Atmosphere (Astrophysical Inst. Kazakh. Akad. Sci., Alma-Ata, 1972), pp. 177-178, in Russian.

CHAPTER 2

# LOW ENERGY (SUBEXPLOSIVE) EFFECTS OF RADIATION ON INDIVIDUAL PARTICLES

## 2.1. REGULAR REGIMES OF DROPLET VAPORIZATION IN THE RADIATION FIELD

Many papers cited in [1-11] deal with the kinetics of vaporization of small particles heated by radiation.

The approaches used in these papers cover all the realistic situations of the surface vaporization of droplets when the stationary optical field is uniformly distributed over a droplet's volume. When considering the limits of applicability of the formulas obtained one can construct a diagram such as that shown in Figure 2.1.1. This diagram presents data concerning the vaporization of water droplets in the air.

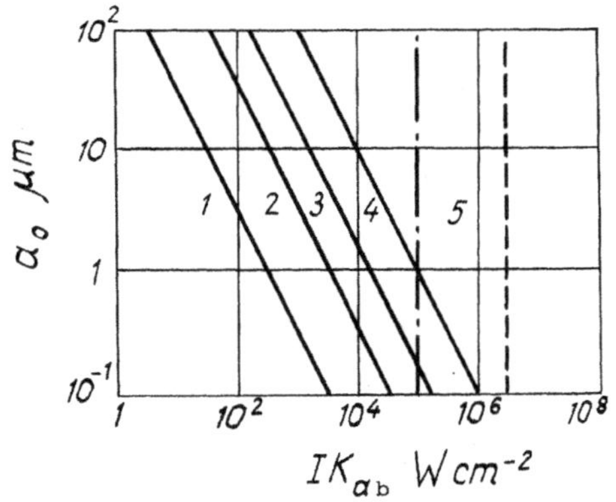


Fig. 2.1.1. Regimes of water drop vaporization in the radiation field.

If I is the radiation power density and $K_{ab}$ is the absorption efficiency factor of a sphere with radius a, then the criteria curve in the regime diagram is described as $IK_{ab}a = c_i$, where $c_i$ are the constants.

In [8], the description of the kinetics of droplet vaporization is based on the linearized stationary equation of energy balance. It can easily be shown that the condition

$$IK_{ab}a \ll \frac{4}{6}\left(\lambda_r + Q_1 D \frac{\mu_\Pi p_0 b}{R_\mu T_0}\right) \tag{2.1.1}$$

should be fulfilled for this equation to be linearized.

Here,

$$Q_1 = Q_e + c_p(T_0 - 273); \qquad b = \frac{Q_e \mu_\Pi}{R_\mu T_0^2},$$

where $\lambda_r$ is the coefficient of heat conductivity of the vapor mixture, $Q_e$ is the specific heat of vaporization, and $c_p$ is the specific heat of vapor under constant pressure. $T_0$ and $p_0$ are the temperature and pressure, respectively, of vapor at an infinite distance. D is the coefficient of diffusion, $\mu_\Pi$ is the molecular weight of the vapor, and $R_\mu$ is the universal gas constant. The product $IK_{ab}a$, obeying (2.1.1), determines the first regime in which the vaporization rate of a droplet is found from the expression

$$\dot{a} = -\frac{IK_{ab}}{4\rho_a\left(Q_1 + \dfrac{\lambda_r R_\mu^2 T_0^3}{D\mu_\Pi^2 p_0 Q_e}\right)}, \tag{2.1.2}$$

where $\rho_a$ is the liquid density.

As shown in [10], a stationary regime can occur, in which the conductive heat transfer from the droplet surface can be neglected.

For such a regime, the following condition must be fulfilled:

$$IK_{ab}a \gg 4\lambda_r(T_b - T_0), \tag{2.1.3}$$

where $T_b$ is the normal boiling point of the liquid. A stationary regime can occur only if

$$IK_{ab}a_0 \ll 4Q_2\rho_a; \qquad Q_2 = Q_e + c_p(T_b - T_0), \tag{2.1.4}$$

where $a_0$ is the initial radius of a droplet.

Conditions (2.1.3) and (2.1.4) determine the regime (3) for water droplets. This regime is shown in Figure 2.1.1 by correspondingly designated band region. The rate of droplet vaporization in this regime is

$$\dot{a} = -\frac{IK_{ab}}{4Q_2\rho_a}. \tag{2.1.5}$$

A stationary regime (2) prevails when conditions (2.1.1) and (2.1.3) are not applicable. The kinetics of droplet vaporization corresponding to this regime are discussed in [11]. Approximation formulas describing the temporal behavior of the radius of a water droplet are derived in this paper. However, it is more convenient if the dependence of the vaporization rate $\dot{a}$ on the problem's parameters is approximated:

$$\dot{a} = -0.496(IK_{ab})^{1.102}. \tag{2.1.6}$$

In the case of any of the regimes, one can find the expression for the droplet radius as a function of time by integrating (2.1.2), (2.1.5), or (2.1.6) taking into account the particular dependence of $K_{ab}$ on a.

In the region located above the boundary of regime 3 one should take into account the nonstationary character of the temperature field inside the droplet. The fourth regime is excluded from the above by the condition under which the droplet explodes:

$$\frac{3IK_{ab}a}{2\pi^2\lambda_a} = \sum_{n=1}^{\infty} \frac{(-1)^n}{n^2}\left[1 - \exp\left(\frac{X_a n^2\pi^2}{a^2} t\right)\right] + T_a = T_{cr}, \tag{2.1.7}$$

where $\lambda_a$ is the coefficient of thermal conductivity of the liquid; $X_a = \lambda_a/c_a\rho_a$; $c_a$ is the specific heat of the liquid; $T_a$ is the temperature of a droplet surface; and $T_{cr}$ is the critical temperature of the liquid. Since the radius of the droplet changes insignificantly during the time interval in which the temperature field reaches its stationary state and, according to [10], $T_a \simeq T_b$, it follows from (2.1.7) that the equation for the upper boundary of regime 4 can be written as follows:

$$IK_{ab}a = 8\lambda_a(T_{cr} - T_b). \tag{2.1.8}$$

The expression for the rate of droplet vaporization in regime 4 obtained in [10] is

$$\dot{a} = \frac{3IK_{ab}}{2Q_2\rho_a\pi^2} \sum_{n=1}^{\infty} \frac{1}{n^2}\left[1 - \exp\left(-\frac{X_a\pi^2 n^2}{a^2} t\right)\right]. \tag{2.1.9}$$

Expression (2.1.9) was obtained under the assumption that the phase boundary is motionless.

The applicability limits for (2.1.9) can be found in the following manner. A stationary temperature field of a droplet irradiated by an homogeneous field is a solution of the Dirichlet boundary value problem for the Poisson equation

$$X_a\,\Delta T_i + J = 0, \qquad J = \frac{3IK_{ab}}{4c_p a\rho_a}, \qquad T_i(a) = T_a, \tag{2.1.10}$$

where $T_i$ is the temperature field inside the particle. The solution of this boundary value problem is well-known:

$$T_i = \frac{a^2}{6X_a} J\left(1 - \frac{r^2}{a^2}\right) + T_a. \tag{2.1.11}$$

According to [10-11], one can show that (2.1.11) correctly describes the process beginning from the moment of time $a^2/6X_a$. Obviously, the amount of heat lost during this time is

$$\Delta q = 4\pi c_a \rho_a \frac{J}{6X_a} \int_0^a r^4\, dr = \frac{2\pi c_a a^5}{15X_a} J. \qquad (2.1.12)$$

On the other hand, $\Delta_q = Q_2\, \Delta M$, where $\Delta M$ is the variation in mass of a droplet. Thus, excluding $\Delta_q$, one can find the value of the relative changes in a droplet's radius:

$$e_1 = \frac{\Delta a}{a} = \frac{IK_{ab} a}{40 Q_2 \rho_a X_a}. \qquad (2.1.13)$$

By integrating (2.1.9), neglecting the movement of the droplet's boundary, and by assuming that $t = a_0^2/6X_a$, one obtains

$$e_2 = \frac{a_0 - a}{a_0} = \frac{IK_{ab} a_0}{24 Q_2 X_a \rho_a} - \frac{3IK_{ab} a_0}{24 Q_2 X_a \rho_a} \sum_{n=1}^{\infty} \frac{1}{n^4} \left[1 - \exp\left(\frac{\pi^2 n^2}{6}\right)\right]. \qquad (2.1.14)$$

Calculations show that $(e_2 - e_1)/e_1 = 7\%$, which, in turn, allows one to neglect the influence of the droplet boundary mobility on the evaporation rate, unless $t \geq a_0^2/6X_a$. For $t > a_0^2/6X_a$, the temperature field of a droplet is described by (2.1.11) and the influence of the droplet boundary mobility is more significant.

Using (2.1.11) one can describe the change of droplet free energy per second by

$$\frac{dF}{dt} = \frac{d}{dt}\left(\frac{\pi}{15} \frac{IK_{ab}}{X_a} a^4\right) = \left(\frac{4\pi}{15} \frac{IK_{ab} a^3}{X_a} + \frac{\pi}{15} \frac{Ia^4 dK_{ab}}{da}\right) \dot{a}. \qquad (2.1.15)$$

The energy balance equation, taking into account the mobility of the droplet's boundary, takes the form

$$-Q_2 \rho_a \dot{a}\, 4\pi a^2 = \pi a^2 IK_{ab} - \frac{dF}{dt}. \qquad (2.1.16)$$

Taking into account (2.1.15), one can find from (2.1.16) that the rate of droplet vaporization is

$$\dot{a} = \frac{IK_{ab}}{4Q_2 \rho_a - \dfrac{4}{15} \dfrac{IK_{ab} a}{X_a} - \dfrac{1}{15} \dfrac{Ia^2}{X_a} \dfrac{dK_{ab}}{da}}. \qquad (2.1.17)$$

Thus, in the fourth regime, droplet vaporization is described as a two-stage process. Up to $t \leq a_0^2/6X_a$, the mobility of droplets boundary can be neglected and the rate of vaporization can be described by (2.1.9). By integrating (2.1.9), one obtains

$$a = \begin{cases} a_0 - \dfrac{3q_R I}{2Q_2\pi^2\rho_a} f(t), & \dfrac{8\pi\kappa_a}{\lambda} a \gg 1, \\ a_0 \exp\left[-\dfrac{12q_R\kappa_a I}{\pi\rho_a\lambda Q_2} f(t)\right], & \dfrac{8\pi\kappa_a}{\lambda} a \ll 1, \end{cases} \tag{2.1.18}$$

$$f(t) = \frac{\pi^2}{6} t - \frac{a_0^2}{X_a\pi^2} \sum_{n=1}^{\infty} \frac{1}{n^4} \left[1 - \exp\left(-\frac{X\pi^2 n^2}{a_0^2} t\right)\right].$$

The expression for the absorption efficiency factor used in the derivation of (2.1.18) was the following:

$$K_{ab} = q_R \left[1 - \exp\left(-\frac{8\pi\kappa_a}{\lambda} a\right)\right],$$
$$q_R = \exp\left[-0.2\left(\sqrt{n_a^2 + \kappa_a^2} - 1\right)\right], \tag{2.1.19}$$

where $(n_a - i\kappa_a)$ is the complex refractive index of the droplet material. For $t > a_0^2/6X_a$ the temperature field of a droplet depends parametrically on time. The rate of droplet vaporization in this case is higher than that in the stationary case, due to the mobility of the droplet boundary. Integrating (2.1.17), one finds that, at this stage of the process, the dependence of the droplet's radius on time can be described by the following:

$$a = (\alpha_1 + r_1) \exp\left(-\frac{t - t_{st}}{\theta}\right) - \alpha_1, \quad \frac{8\pi a\kappa_a}{\lambda} \gg 1,$$
$$\alpha_1 = \frac{15X_a Q_2 \rho_a}{I q_R}; \quad \theta = \frac{60X_a Q_2^2 \rho_a^2}{I^2 q_R^2}; \quad t_{st} = \frac{a_0^2}{6X_a}; \quad r_1 = a(t_{st}); \tag{2.1.20}$$

$$\frac{a^2}{1 + \beta_2 a^2} = \frac{r_1^2}{1 + \beta_2 r_1^2} \exp\left(-\frac{t - t_{st}}{\theta_1}\right), \quad \frac{8\pi a\kappa_a}{\lambda} \ll 1,$$
$$\beta_2 = \frac{2\pi\kappa_a q_R I}{3\rho_a\lambda X_a Q_2}, \quad \theta_1 = \frac{\rho_a\lambda Q_2}{4\pi\kappa_a q_R I}. \tag{2.1.21}$$

The fifth regime is the explosion occurring when the parameters of the liquid in the center of a droplet reach their critical values. The broken line on the regime diagram defines the boundary between regimes where the contribution of the vapor kinetic energy to the energy balance at the droplet surface becomes important. According to [4] one can write

$$(IK_{ab})^2 \simeq 32\rho_{\Pi}^2 Q_2^3 \tag{2.1.22}$$

for this boundary. Here, $\rho_{\Pi}$ is the density of the gas-vapor mixture. The threshold values of the radiation flux $IK_{ab}$ required to initiate the explosion are lower than this boundary and, as a consequence, the regime in which the kinetic energy of vapor affects the vaporization rate does not apply to larger droplets.

The dot-dash line in the regime diagram marks the boundary between the region where the diffusion theory of vapor transfer is applicable and the gas-dynamic region where the vaporization rate is close to that applying in the case of evaporation to a vacuum (i.e., the kinetic evaporation regime). It was assumed, in the analysis of diffuse vapor transfer made in [3], that, taking into account the Stefan flow, the total pressure of the mixture remains constant while, in fact, it must obey the Bernoulli equation

$$\frac{v^2}{2} + \frac{p}{\rho_\Pi} = \text{const}, \tag{2.1.23}$$

where p is the total pressure of the mixture and v is the velocity of the Stefan flow. Since at infinity $v = 0$ and $p = p_0$,

$$p = \rho_\Pi \left(\frac{p_0}{\rho_\Pi} - \frac{v^2}{2}\right). \tag{2.1.24}$$

$p \simeq p_0 \simeq \text{const}$ if the following condition is fulfilled:

$$\frac{v^2}{2} \ll \frac{p_0}{\rho_\Pi} \simeq c_s^2, \tag{2.1.25}$$

where $c_s$ is the speed of sound in the surrounding (ambient) gas. It can be shown that this condition is in good agreement with the definition of the lower limit of the kinetic regime of vaporization given in [11]:

$$IK_{ab} = \frac{4Q_2\mu_\Pi p_0}{R_\mu T_b}\left(\frac{R_\mu T_b}{2\pi\mu_\Pi}\right)^{1/2} \tag{2.1.26}$$

The limit (2.1.26) is only lower than the explosion threshold for droplets whose radii do not exceed 1 μm. It should be noted that, in the regions where the kinetic regime overlaps with the third and fourth regimes, the kinetics of droplet vaporization are described by the formulas valid for these latter regimes.

## 2.2. VAPORIZATION OF HAZE PARTICLES CONSISTING OF A SOLID NUCLEUS AND A SHELL OF SALT IN SOLUTION

Real atmospheric aerosols are known to be formed by the deposition of water at condensation centers. The simplest model of such complex particles is a two-layer spherical particle, at the center of which there is a solid, insoluble nucleus with a shell of salt solution. The salt shell is formed of the soluble part of a condensation center, upon which water is being deposited. It is known that the shapes of condensation centers can differ from a sphere, but, since their sizes do not usually exceed the wavelength of incident radiation, we can simplify the problem and neglect this fact.

The basic system of equations involved in the problem concerning the vaporization of such a two-layer particle includes (1) the equation of vaporization of the liquid shell, (2) the equation describing the variation of the concentration of the solution during particle vaporization, (3) the equations describing the temperature variation for a two-layer particle - air system, and (4) the equation describing the growth of a solid nucleus due to the precipitation of salt after the moment when the concentration of the salt solution exceeds its saturated value.

In general, the problem of solving such a system of equations is complicated, therefore we will use the following approximations [19]:

(1) A particle is spherical.

(2) The process is divided into two stages, separated by the moment in time when the solution near the surface of the solid nucleus reaches its saturation point. Thus, the solid nucleus starts to grow in the second stage of the process.

(3) The optical field inside the particle is homogeneous, valid for small particles $a \ll \lambda$ [1, 7].

(4) A diffusion approximation is used for heat and vapor mass transport, as the diffusion regime is the most probable one for the vaporization of small particles with a wide variation of temperature across their surfaces.

(5) The processes of vaporization and condensation are quasi-stationary, i.e., they occur during a time t which greatly exceeds the characteristic times during which the gradients of the parameters of the droplet material are maintained near the interface $\simeq a_1^2/X_{1(\text{or } 2)}$; $a^2/X_{2(\text{or } 3)}$; $a_1^2/D_2$; $a^2/D_{2(\text{or } 3)}$, where $a_1$, a are the radii of the solid nucleus and the water shell, respectively; $X_1$, $X_2$, $X_3$, and $D_2$, $D_3$ are the coefficients of molecular temperature conductivity and diffusion (molecules of dissolved matter and vapor), respectively. The indices 1, 2, and 3 refer to a solid nucleus, a salt-solution shell, and a vapor-gas mixture, respectively.

### 2.2.1. The Equation describing Particle Vaporization

Let us consider the Knudsen equation [4] for the vapor mass flux from a particle surface:

$$-\rho_a \frac{da}{dt} = \nu_a v_m(\rho_s - \rho_{\text{II}}^a), \tag{2.2.1}$$

where $\rho_{\text{II}}^a$ is the vapor density near the particle surface, $\rho_a$ is the density of the water, $\nu_a$ is the coefficient of vaporization, $v_m$ is the perpendicular component of speed of the vapor molecules with respect to the particle surface, $\rho_s$ is the density of saturated vapor, which depends on the concentration of the salt solution on the particle surface $n_s$ [20] thus:

$$\rho_s = \rho_{so}(T_0/T_a)(1-\phi(n_s))\exp[Q_e(T_a - T_0)/(R_\Pi T_a T_0) + 2\sigma/(\rho_a R_\Pi T_a a)]. \quad (2.2.2)$$

$\rho_\Pi^a$ is found from the condition of vaporization stationarity.

$$-\rho_a \, da/dt = D_3(\rho_\Pi^a - \rho_{\Pi 0})/a, \quad (2.2.3)$$

where $\rho_{so}$ is the density of saturated vapor at temperature $T_0$, $T_0$ is the temperature of the medium at an infinite distance from the particle's center, $T_a$ is the temperature at the particle surface, $\phi(n_s)$ is the function which describes the effect of the vapor pressure descrease over water solutions [20], $Q_e$ is the specific heat of vaporization, $R_\Pi$ is the gas constant of the vapor; $\sigma$ is the coefficient of surface tension of water, and a is the radius of the particle.

Substituting (2.2.2) and (2.2.3) into (2.2.1), we obtain an equation for the rate of particle vaporization, da/dt, during the first stage; this is not written here because of its awkwardness. During the second stage, (2.2.1) is written whilst taking into account the variations of the volume of the vapor due to the growth of the solid particle, i.e., in the left-hand side of (2.2.1) we have: $-\rho_a(da/dt - (a_1^2/a^2)\, da_1/dt)$.

### 2.2.2. The Heat Problem

To solve the problem of heating a two-layer particle we use a system of thermal conductivity with boundary conditions for each layer.

$$C_j \rho_j (\partial T_j/\partial t) = \lambda_j \, \Delta T_j + P_j, \quad (2.2.4)$$

where $j = 1, 2, 3$; $C_j$, $\rho_j$, $\lambda_j$ are the thermal capacity, density, and molecular thermal conductivity, respectively, of the j-th layer matter. $P_j = I\kappa_j V_j$ (but $P_3 = 0$); and $\kappa_j$, $V_j$ are the coefficients of radiation absorption and volumes of the layers.

To define the temperature distribution over the particle radius in the ambient air $T_3(r)$, we use the stationary solution

$$(\partial T_3/\partial R)_{R=+a} = (T_a - T_0)/a; \qquad T_3(\infty, t) = T_0; \quad (2.2.5)$$

$$-\lambda_2(\partial T_2/\partial R)_{R=-a} = -\lambda_3(\partial T_3/\partial R)_{R=+a} - \rho_a Q_e(da/dt); \quad (2.2.6)$$

$$-\lambda_1(\partial T_1/\partial R)_{R=-a_1} = -\lambda_2(\partial T_2/\partial R)_{R=+a_1} + \rho_1 Q_p(da_1/dt), \quad (2.2.7)$$

where $Q_p$ is the specific heat of crystallization of the salt out of solution.

We will assume homogeneous heating of the solid nucleus. Integrating (2.2.1) over the volume, one can obtain the expression for the heat flux balance through the second layer:

$$4\pi[I\kappa_2 V_2 - \lambda_2 a_1(\partial T_2/\partial R)_{R=-a_1} - \lambda_3 a(T_a - T_0) + \\ + \rho_a Q_e a^2(da/dt)] = 4\pi C_2\rho_2 \int_{a_1}^{a} dR\, R^2(\partial T_2/\partial t). \qquad (2.2.8)$$

We now assume that the temperature distribution $T_2$ is close to a stationary one:

$$T_2(R,\ t) = C_0(t) + C_1(t)/R. \qquad (2.2.9)$$

The coefficients $C_0(t)$ and $C_1(t)$ are found from the equations above. Finally, we have:

$$(\partial T_1/\partial t) = C_1\rho_1\{I\kappa_1 + I\kappa_2[(a/a_1)^3 - 1] - \\ - 3\lambda_3(T_a - T_0)/a_1^3 + 3a^2\rho_a Q_e(da/dt)/a_1^3 + \\ + (C_2\rho_a/a)(a/a_1)^2(\lambda_2/\lambda_3 - 1)[1 + (a_1/a)][\lambda_2/\lambda_3 - 1]^{-2} \times \\ \times (T_1 - T_0)(da/dt)[(1 - a_1^3/a^3) - 3(1 - a_1^2/a^2)a_1/(2a)]\} \times \\ \times \{1 - [(C_2\rho_a)/(C_1\rho_1)](a/a_1)^2[1 + (a_1/a)(\lambda_2/\lambda_3 - 1)]^{-1} \times \\ \times [(1 - a_1^3/a^3)(\lambda_2/\lambda_3 - 1) + 3(1 - a_1^2/a^2)/2]\}^{-1}. \qquad (2.2.10)$$

Using a similar procedure, one can write the equation for the second stage of the process during which the precipitation of salt onto the solid nucleus occurs.

### 2.2.3. Variation of Salt Concentration in the Process of Particle Vaporization

To find the distribution of salt over the radius of a particle one should carry out the series expansion $n_s(R,\ t)$ over the running particle radius R. In the first stage, when a solid particle is not yet growing, we will confine ourselves to the first two terms of expansion, $n_s = n_1(t) + n_2(t)R$, i.e., we will use a linear approximation. To find the coefficients $n_1$ and $n_2$ the following two conditions will be used:

(1) There is salt balance in the salt solution:

$$4\pi \int_{a_1}^{a} dR\, R^2 n_s(R,\ t) = 4\pi n_{s0}(a_0^3 - a_{10}^3)/3, \qquad (2.2.11)$$

where $n_{s0}$ is the initial salt concentration in the shell and $a_0$, $a_{10}$ are the initial particle and solid-nucleus sizes, respectively.

(2) There is a condition of quasi-stationarity of the vaporization regime:

$$(dn_s(R,\ t)/dR)_{R=-a} = -n_s(a,\ t)(da/dt)/D_2. \qquad (2.2.12)$$

Finally, we obtain for C(R, t):

$$n_s(R,\ t) = \frac{n_{s0}(a_0^3 - a_{10}^3)(D_1 + a)(1 - R/a)(da/dt)}{(a^3 - a_{10}^3)[D_1 + (da/dt)a] - 3(da/dt)(a^4 - a_{10}^4)/4}. \qquad (2.2.13)$$

Since, with the growth of the solid-particle, the gradient near its surface should be $\simeq 1/R$ for the second stage of the process, we use an expansion of the type

$$n_s(R,\ t) = n_1(t) + n_2(t)R + n_3(t)/R, \qquad (2.2.14)$$

where $n_1$, $n_2$, $n_3$ are the expansion coefficients. The expansion coefficient $n_3$ is found from the additional condition, see (2.2.11) and (2.2.12), that $n_s$ near the surface of the precipitation center $a_1$ is equal to the saturated concentration $-n_s^1$.

## 2.2.4. Growth of the Solid Nucleus

The rate of growth of the solid nucleus is determined by the concentration gradient near its surface

$$(da_1/dt) = (D_1/\rho_1)(\partial n_s/\partial R)_{R=+a_1}. \qquad (2.2.15)$$

Taking into account the expressions for $n_s(R,\ t)$, we determine $\partial n_s/\partial R$. Finally, for the rate of solid particle growth during the second stage, we have:

$$\frac{da_1}{dt} = \frac{D_1C_s'a(da/dt)}{\rho_1a_1^2[D_1/a + (a/a_1 - 1)(da/dt)]}. \qquad (2.2.16)$$

The above relations form a closed system of differential equations which enable one to find variations in size of a two-layer particle and its solid nucleus, the temperature of the nucleus and its shell, the salt concentration in the solution, and other characteristics. The system can be solved easily. In the calculations it was assumed that the solid nucleus consists of graphite and its shell is a NaCl solution. Laser radiation with $\lambda = 10.6$ µm is absorbed by the solid particle and its shell, which have the coefficients $\kappa_1 = 2771$ cm$^{-1}$ and $\kappa_2 = 794$ cm$^{-1}$, respectively ($\kappa_2$ is taken for pure water). $\nu_a = 1$ and $\rho_{\Pi}(\infty) = 0$. The initial concentration of salt by weight was taken as 0.5%.

Figures 2.2.1(a), 2.2.1(b) present the time dependences of the evaporated two-layer particle radius and the growth of the solid nucleus corresponding to different values of laser radiation intensity for two cases: a shell consisting of pure water (a) and a shell consisting of a 5% NaCl solution (b). It is seen that the presence of salt in the solution significantly decreases the rate of particle vaporization. Figure 2.2.2 illustrates

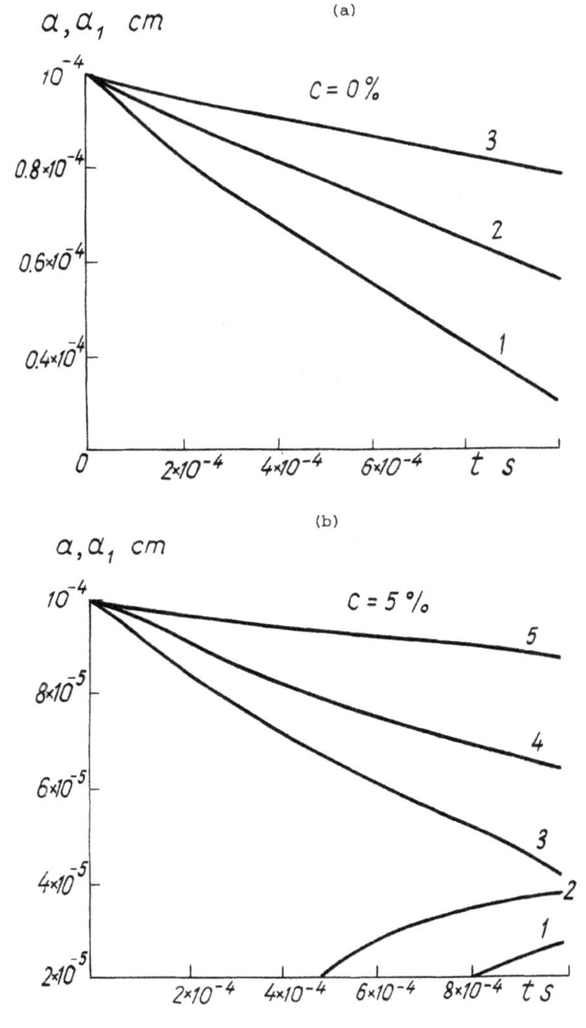


Fig. 2.2.1. Calculated dependences of the variation in radius of a vaporizing drop of water-salt solution on time t. NaCl - solid curve; solid nucleus - dotted curve, measurements taken at different intensities of laser radiation, with $\lambda = 10.6$ μm.
(a) 1; 2; 3 - $I = 10$; 5; 1 $kWcm^{-2}$ - liquid shell.
(b) 1; 2 - $I = 5$; 10 $kWcm^{-2}$ - solid nucleus.
3; 4; 5 - $I = 10$; 5; 1 $kWcm^{-2}$ - liquid shell.

the time dependence of the temperature of the nucleus surface. As follows from the calculations, in the case of a particle with a shell of pure water,

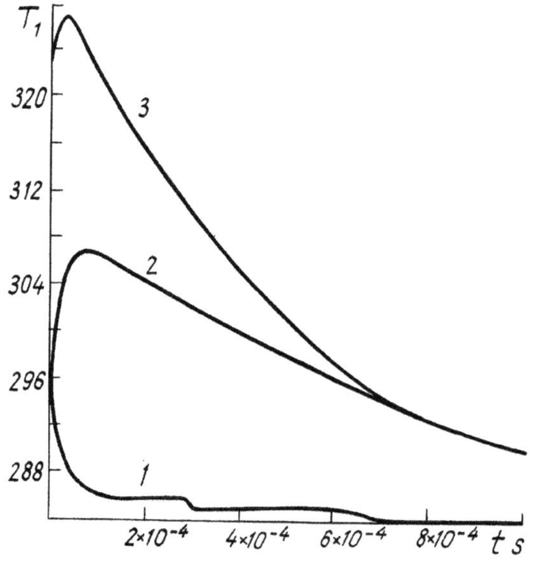


Fig. 2.2.2. The dependence of the temperature of the surface of the solid nucleus of a two-phase particle on the initial concentration of NaCl in water shell, C = 5%, in the process of laser irradiation ($\lambda$ = 10.6 μm). 1; 2; 3 - 1; 5; 10 $kWcm^{-2}$.

the particle is heated and cools more rapidly than one with a shell composed of a salt solution. Moreover, it reaches lower temperatures, i.e., the presence of salt in the solution, even without any consideration of additional heating, results in a heating delay and cooling of the particle due to a decrease in the rate of vaporization.

## 2.3. SOME PECULIARITIES IN THE VAPORIZATION OF SOLID AEROSOL PARTICLES BY HIGH-POWER RADIATION

Dusty hazes composed of solid mineral particles are specific atmospheric phenomena which differ strongly from water droplet aerosols, in both their microphysical and optical parameters of the particulate matter. Consequently, the process of the vaporization of such particles by radiation is characterized by certain peculiarities which do not occur in the vaporization of water aerosols.

The main features of the vaporization process in this case can be summarized as follows: (1) high temperatures or phase transitions that, as a consequence, require that the temperature dependence of the thermophysical parameters of the particles and the surrounding medium have to be taken into account; (2) the vaporization of the solid mineral particles is often accompanied by the thermal dissociation of the molecular complexes forming the dust and (3) strongly metastable states of the overheated substance can be reached in this case, due to the relatively high values of the coefficient of surface tension of the melted matter. The theory of vaporization of solid particles, using a diffusion mechanism for the heat and mass exchange between the particle and the medium, is presented in [12]. This diffusion mechanism works if the partial pressure of saturated vapor $P_s^a$ at the surface of the droplet is less than the atmospheric pressure $P_0$. In situations where $P_s^a \simeq P_0$ corrections are needed to take the Stefan flow into account, this is the case with water droplet aerosols [3]. When $P_s^a \gg P_0$ the hydrodynamic heat and mass transfer takes place as described in [4, 5].

Unfortunately, as yet there have been no systematic experimental studies of the thermal dissociation of solid aerosol particles in the literature. However, some preliminary results of such investigations can be found in [13].

### 2.3.1. The Diffusion Regime of Vaporization of Solid Spherical Particles

Assuming a quasi-stationary character and homogeneity of particle heating, and using the Knudsen relationship for describing the evaporation of vapor from the surface of the particle, one can write the basic system of equations in the following form [12]:

$$-\rho_a(da/dt) = B\exp(-Q_{eff}\mu/R_\mu T_a) - v_m\rho_\Pi^a\nu_a; \tag{2.3.1}$$

$$ac_a\rho_a \frac{4}{3}\frac{\partial T_a}{\partial t} = IK_{ab} + 4\left(\frac{\partial T}{\partial R}\right)_{R=+a} + 4\rho_a Q_{eff}\left(\frac{da}{dt}\right) + 4\sigma_B\varepsilon_B(T_0^4 - T_a^4); \tag{2.3.2}$$

$$-\lambda_T(\partial T/\partial R)_{R=+a} = 2T_0\lambda_0[(T_a/T_0)^{3/2} - 1]/3a; \tag{2.3.3}$$

$$-(\rho_\Pi^a - \rho_\Pi)(da/dt) = D_0[(T_a/T_0)^{3/2} - 1](\rho_\Pi^a - \rho_{\Pi 0})[3a\,\ln(T_a/T_0)/2]^{-1}, \tag{2.3.4}$$

where $\rho_\Pi^a$, $T_a$, and $\rho_\Pi$, T are the vapor density and temperature at the particle surface and of the surrounding medium, respectively; $\mu$, $R_\mu$, $\nu_a$, B, and $v_m$ are the molecular weight, gas constant, vaporization coefficient, and the effective velocity of the molecular (atomic) stream in the backward direction, respectively; and $\lambda_T = \lambda_0(T/T_0)^{1/2}$ and $D_T = D_0(T/T_0)^{3/2}$ are the coefficient of heat conductivity and the diffusion coefficient, respectively, of the vapor-gas mixture in the vicinity of the particle [17]. $Q_{eff}$ is the effective specific heat of the phase transition solid particle-vapor. A great variety of chemical compounds, when heated, undergo decomposition.

The dissociation processes are of a multistage nature and reveal many individual features for the various substances. As an example, below a scheme is given for the thermal decomposition of $N_2CO_3$ particles:

$$Na_2CO_3(s) \rightarrow Na_2O(s) + CO_2 - Q_1, \quad (2.3.5)$$

$$Na_2O(s) \rightarrow Na_2O(\ell) - Q_2 \rightarrow 2Na(g) + \tfrac{1}{2}O_2 - Q_3, \quad (2.3.6)$$

where the letters in parentheses denote the phase: (s) is the solid phase, ($\ell$) is the liquid phase, and (g) is the gas phase; $Q_1$, $Q_2$, $Q_3$ are the specific energies of the decomposition reactions of the original substance ($T_d = 1127$ K), the melting of $Na_2O(s)$ ($T_m = 1190$ K), and the dissociation (boiling) of $Na_2O(\ell)$ ($T_b = 1800$ K), respectively. Generally the quantities $Q_1$ and $Q_2$ are much less than $Q_3$ and, therefore, when numerically assessing the effective heat of vaporization of complex substance according to (2.3.2), one should understand by $Q_{eff}$ the sum $(Q_3)_1 + (Q_3)_2 + \cdots + (Q_3)_i$, neglecting the contributions due to the first and second processes.

Figures 2.3.1 and 2.3.2 illustrate the solution of the system of equations (2.3.1)-(2.3.4) for a quartz ($SiO_2$) particle with an initial radius $a_0 = 1$ µm. In this case, the partial pressure of the vapor $P^a_\Pi$ reaches the value $P_0$ at the temperature $T_a$ of the particle's surface which exceeds the boiling point $T_b$ by a factor of 1.12. This situation occurs because of the compression shock at the particle's surface caused by the small value of the accommodation coefficient of quartz ($\nu_a = 0.022$).

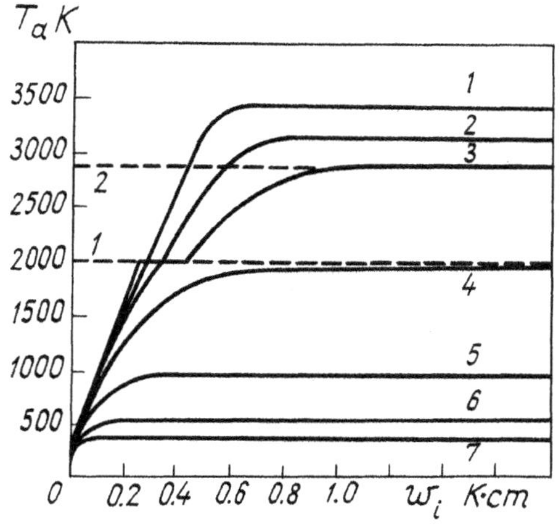


Fig. 2.3.1. The dependence of the temperature of a quartz particle, vaporized under diffuse conditions, on the parameter $w_i = 3tI/(4C_aQ_a)$. Curves 1-7 represent values of $IK_{ab} = 3.16 \times 10^7$; $3.16 \times 10^6$; $10^6$; $3.16 \times 10^5$ Wcm$^{-2}$; $\lambda = 10.6$ µm. Dotted lines 1 and 2 represent the melting and boiling points of the particle material.

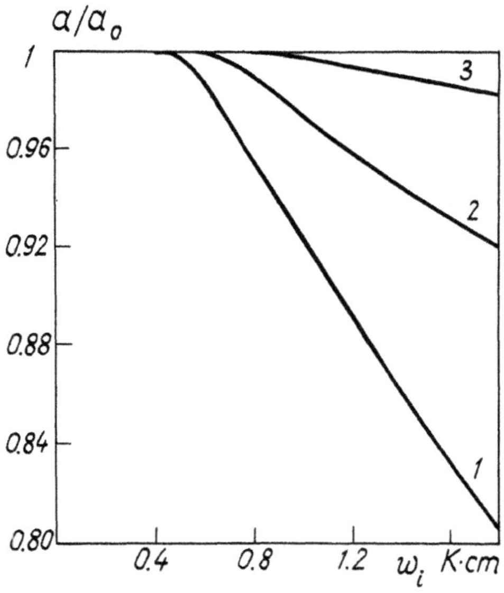


Fig. 2.3.2. Dynamics of vaporization of quartz particle ($a_0 = 1$ μm) under diffuse conditions in a $CO_2$ laser radiation field. The calculated parameters for curves 1-3 are the same as for curves 1-3 in Figure 2.3.1.

## 2.3.2. Pre-Explosion Gas-Dynamic Regime of Vaporization

It is characteristic for metastable overheated states of particles that the probability arises of achieving vapor pressures at the particle surface which exceed atmospheric pressure.

The range of temperature within which the metastable phase exists is from the boiling point $T_b$ up to the temperature at which homogeneous nucleation inside the particle begins during a laser pulse of duration $t_p$: $Jt_pV_a \gg 1$, where $V_a$ is the volume of a particle, and J is the probability density of the homogeneous nucleation of the new phase. The temperature limit of the absolute thermodynamic instability of any substance under normal atmospheric pressure is approximately $0.9T_{cr}$ [4], relative to the critical temperature of a particle $T_{cr}$.

At vapor pressures $P_S$ strongly exceeding the value $P_0$, a gas-dynamic expansion of vapor will take place in the vicinity of the particle, according to the laws of hydrodynamics [4, 5]:

$$\partial\rho_\Pi/\partial t + \mathrm{div}(\rho_\Pi\vec{v}\,) = 0; \tag{2.3.7}$$

$$\partial\vec{v}_\Pi/\partial t + (\vec{v}_\Pi\ \mathrm{grad})\vec{v}_\Pi = -\rho_\Pi^{-1}\ \mathrm{grad}\ P_\Pi; \tag{2.3.8}$$

$$\vec{v}_{\Pi}^2/2 + C_p^{\Pi} T_{\Pi} = \text{const}; \qquad P_0 \cdot \rho_{\Pi}^{-\gamma} = \text{const}, \tag{2.3.9}$$

where $\gamma = C_p^{\Pi}/C_v^{\Pi}$ is the adiabatic exponent. The system (2.3.7)-(2.3.9) is closed by an equation of energy balance of the type (2.3.2).

Boundary conditions are defined, in this case, at infinity and at the boundary of a gas kinetic layer whose thickness h is equal to two to three mean free paths of vapor atoms above the particle surface. In order to find the latter boundary condition, one must solve the Boltzmann kinetic equation which, for a one-dimensional case, has the form [14]

$$v_R \, \partial f/\partial R = (df/dt)_{coll},$$

where $f(v_R)$ is the distribution function of the number density of the vapor particles over the radial velocities $v_R$, and $(df/dt)_{coll}$ is the collision integral.

When $a \gg h$ one can use the known solution of the one-dimensional kinetic equation which, for monatomic vapors and $\nu_a = 1$, is written as follows [14]:

$$T_{\Pi}(R = h) \simeq 0.65 T_a; \qquad \rho_{\Pi}(R = h) \simeq 0.31 \rho_s(T_a); \tag{2.3.10}$$

$$P_{\Pi}(R = h) \simeq 0.2 P_s(T_a); \qquad v_{\Pi}(R = h) = (K_B T_{\Pi} \gamma/\mu)^{1/2}; \tag{2.3.11}$$

$$J_a \simeq 0.82 \rho_s(T_a)(K_B T_a \gamma/2\mu)^{1/2},$$

where the subscript 's' denotes the values characteristic for saturated vapors, and $J_a$ is the density of evaporated substance leaving the particle's surface. The problem of the relationship between the surface temperature of the particle and the vapor pressure is based on the use of the Clausius-Clapeyron equation which is often replaced by empirical relationships for the purpose of making concrete calculations. Such a relationship for oxides takes the form [4]

$$P_s(T_a) = P_b \exp\left[-\frac{Q_\mu}{K_B T_b}\left(\frac{T_b}{T_a} - 1\right)\right], \tag{2.3.12}$$

where $P_b$ is the pressure of the saturated vapor at $T_a = T_b$, and $Q_\mu$ is the work function of one molecule. A corresponding empirical expression for metals can be found in [14].

Since energy losses caused by heat conductivity and thermal emission of radiation are small in comparison with the heat losses due to vaporization, one can derive the following approximate formula for the rate of particle vaporization:

$$\frac{da}{dt} \simeq -\frac{I K_{ab}}{4\rho_a(Q_{eff}(T_a) + 0.35 C_a T_a + 0.65 T_a K_B \gamma/2\mu)}. \tag{2.3.13}$$

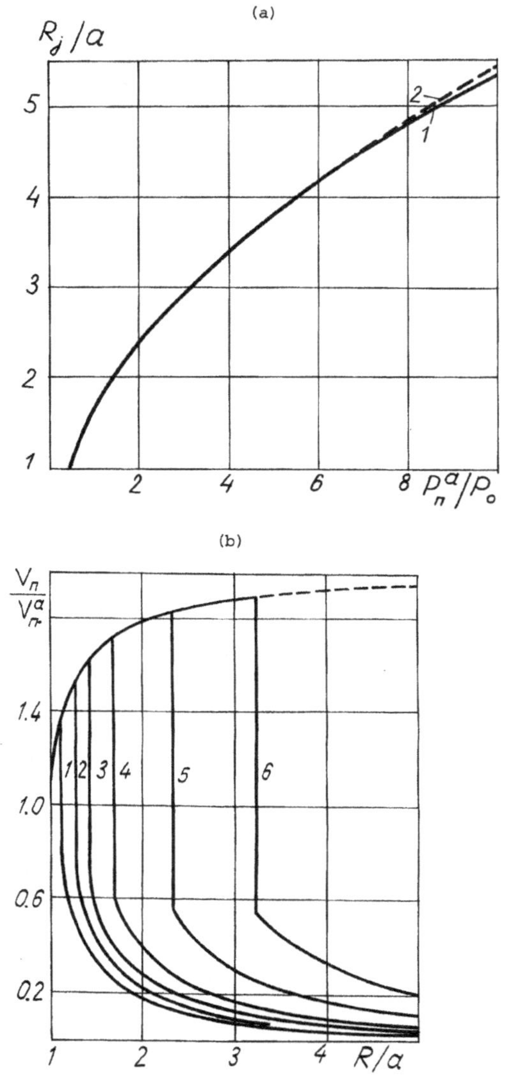


Fig. 2.3.3. Calculated dependence of location of shock jump (a) and vapor flow velocity (b) on the quasi-stationary vaporization of a solid particle under conditions of metastable superheating. Curves 1 and 2 correspond to monatomic ($\gamma = 5/3$) and diatomic ($\gamma = 7/5$) vapor molecules. Curves 1-6 (Figure 2.3.3(b)) correspond to $\gamma = 5/3$.

This expression, in combination with (2.3.11) and (2.3.12), determines the rate of quasi-stationary evaporation and the temperature of metastable overheating of the particle $T_a$.

One specific feature of the quasi-stationary flow of vapor in the gas dynamic regime of vaporization should be noted. In contrast with the case of plane targets irradiated with laser radiation, the counter-pressure generated in the medium surrounding a spherical particle gives rise to the appearance of a shock pressure jump (produced near the droplet's surface in the medium, due to the surface expanding because of laser irradiation). The vapor stream, when passing through this pressure jump, changes the supersonic regime to the subsonic one [4]. On the one hand, the first boundary condition on the surface of the particle gives the solution for the vapor stream velocity, which monotonously increases with an increase of distance between particles, while, on the other hand, the second boundary condition, taking into account the counter-pressure from the medium, demands that the vapor stream velocity at infinity $[v_{\Pi}(R \to \infty) \to 0]$ vanishes. The conditions of energy flux conservation, as well as of momentum and mass, at the shock pressure jump provide the basis for finding the distance between the particle's center and the jump. The following condition for the quasi-stationary vapor flow must be fulfilled for a shock pressure jump to exist:

$$P_{\Pi}(T_a) \gtrsim 5 \left(\frac{2}{\gamma + 1}\right)^{\gamma/(\gamma-1)} P_0 \simeq 2.5P_0. \tag{2.3.14}$$

Figures 2.3.3(a) and (b) present the calculated dependences of the relative distance between the particle center and the shock pressure jump $R_j/a$, and of the relative velocity of the vapor outflow in the vicinity of a particle $V_{\Pi}(R)/V_{\Pi}^a(R \simeq h)$ on the value $P_{\Pi}^a/P_0$, respectively. The contact surface $R_c$ between the vapor and surrounding air exists until $t \ll R_c^2/4D_v$, where $D_{\Pi}$ is the coefficient of molecular diffusion. The value of $R_c$ is estimated using the following equation:

$$R_c \simeq a_0 \left[1 - \frac{W - W_h}{W_b} \left(\frac{\gamma + 1}{10} \frac{K_B T_a \rho_s(T_a)}{\mu P_0} - 1\right)^{1/3}\right], \tag{2.3.15}$$

where $W$, $W_h$, $W_b$ are the energy consumed by the particle during irradiation, the amount of energy required for heating the particle to the temperature necessary for well-developed evaporation, and the energy of vaporization, respectively. As estimates show, the value of $R_c$ is about 20 to 30 times the initial radius of the particle if the particle is completely evaporated in the gas dynamic regime.

## 2.4. BURNING OF CARBON AEROSOL PARTICLES IN A LASER BEAM

The irradiation of aerosol particles made of a thermochemically active substance by a high-power laser can facilitate the combustion of such particles. This, in turn, will cause changes in their optical parameters due to the burning of the aerosol material and the creation of thermal and mass aureoles in the reaction zone. Carbon aerosol particles are of principal interest in the problem of the interaction between radiation and particulate matter, since this type of aerosol is observed in many atmospheric aerosol formations, both of natural and artificial origin. This particular problem is discussed in [13, 15-16].

The process of the combustion of a carbon particle can be described by a system of aerothermochemical equations. Chemical reactions which are considered as fundamental in this process are

$$\begin{aligned} &2C + O_2 \rightarrow 2CO + Q_1; \\ &C + CO_2 \rightarrow 2CO + Q_2; \\ &2CO + O_2 \rightarrow 2CO_2 + Q_3, \end{aligned} \tag{2.4.1}$$

where $Q_k$ $(k = 1, 2, 3)$ is the specific heat of the corresponding reaction. Let the rate of reaction be $A_k$. The system of equation involves the heat equation for a single particle and transfer equations for the mass, momentum, and energy of a gas mixture [13, 16]:

$$\frac{\partial T_a}{\partial t} = \nabla(\lambda_a \cdot \nabla T_a) + \frac{3IK_{ab}}{4C_a\rho_a a}, \quad 0 \leq R \leq a; \tag{2.4.2}$$

$$\mathrm{div}(\rho_k v + j_k) + \frac{M_k A_3}{M_3 b_k} = 0, \quad R > a; \tag{2.4.3}$$

$$\sum_{k=1}^{4} \rho_k M_k = \frac{P}{R_\mu T}; \qquad -\frac{\partial P}{\partial R} = \rho_{mix} v \frac{\partial v}{\partial R}; \tag{2.4.4}$$

$$\mathrm{div}\left(-\lambda_{mix}\frac{\partial T}{\partial R} + T\sum_{k=1}^{4} c_p^k(\rho_k v + j_k)\right) = Q_3 A_3, \quad k = 1, 2, 3, 4. \tag{2.4.5}$$

The boundary conditions for this system are written as follows:

$$-\lambda_{mix}\frac{\partial T}{\partial R} = -\lambda_a \frac{\partial T}{\partial R} - M_a \sum_{k=1}^{2} \frac{A_k Q_k}{M_k} - \sigma_B \varepsilon_B (T_a^4 - T_0^4); \tag{2.4.6}$$

$$\rho_{mix} v = -\rho_a \frac{da}{dt} = \frac{A_1 M_a}{M_2} + \frac{A_2 M_a}{M_2}; \tag{2.4.7}$$

$$\rho_1 v + j_1 = -M_1 A_1/M_3; \qquad \rho_2 v + j_2 = -A_2; \tag{2.4.8}$$

$$\rho_3 v + j_3 = A_1 + 2M_3 A_2/M_2, \tag{2.4.9}$$

where the indices $k = 1, 2, 3, 4$ refer to the characteristics of $O_2$, $CO_2$, $CO$, $N_2$, respectively. The subscript 'a' denotes the characteristics of the particulate matter, and the subscript 'mix' refers to the vapor-gas mixture; the parameters $\lambda_k$, $M_k$, $\rho_k$, $c_p^k$, $j_k$ are the molecular coefficient of thermal conductivity, molecular weight, density, and isobaric specific heat of the $k$th component of the mixture, respectively; $T_a$, $T_0$ are the temperatures of an aerosol particle and the surrounding medium, respectively; $b_k$ is a coefficient introduced into the equations for the purpose of taking into account the stoichiometry of the chemical reactions ($b_1 = 2$, $b_2 = b_3 = 1$, $b_4 = 0$); $R_\mu$ is the universal gas constant, P is the total pressure of the gas mixture; $\sigma_B$ is the Stefan-Boltzmann constant, and $\varepsilon_B$ is the grayness coefficient. The boundary value problem formulated in (2.4.1)-(2.4.9) cannot be solved analytically. An approximate analysis of the problem of carbonic particle combustion is given below, based on the assumption of homogeneous heating of the particulate substance by radiation, this analysis takes into account only the energetically most important heterogeneous reaction $2C + O_2 \rightarrow 2CO + Q_1$. The reactivity of the particulate matter in suspension is defined as follows:

$$g(T_a) = g_0 \exp\left(-\frac{Q_c}{K_B T_a}\right), \tag{2.4.10}$$

where $g_0$ is the pre-exponential factor, $Q_c$ is the activation energy of the combustion reaction for carbon, and $K_B$ is the Boltzmann constant. At relatively low temperatures of particle overheating by radiation, an excess of oxygen molecules at the particle surface is observed, so that the particle burns in a kinetic regime. On the other hand, at high temperatures $T_a$ the reaction rate is controlled by the molecular diffusion of the oxidant.

The expression for the rate of combustion, obtained in [13] following a quasi-stationary approach and taking into account both of the above processes, is written as follows:

$$A_1(a,\ T_a) = \frac{(2M_a P_{10})/(K_B T_a)}{1/g(T_a) + a/D_{eff}(T_a)}, \tag{2.4.11}$$

where $M_a$ is the atomic weight of carbon, $P_{10}$ is the partial pressure of oxygen at infinity, and $D_{eff}$ is the effective coefficient of diffusion:

$$D_{eff} = \frac{1}{\int_1^\infty (dR\ T(R)/T_a)/(R^2 D_1(T(R))}.$$

Here, $D_1$ is the coefficient of molecular diffusion of oxygen, written in a form that indicates the temperature dependence.

The temporal behavior of the radius and temperature of the particle is described by the following approximate equations:

$$da/dt = -A_1(a,\ T_a)/\rho_a; \tag{2.4.12}$$

$$aC_a\rho_a \frac{4}{3}\frac{dT_a}{dt} = IK_{ab} - \frac{4}{a}\int_{T_0}^{T_a} dT'\ \lambda_{mix}(T') +$$

$$+ 4Q_1A_1(a,\ T_a) - 4\sigma_B\varepsilon_B \cdot (T_a^4 - T_0^4). \tag{2.4.13}$$

Equation (2.4.13) has several roots, but only two of them, the smallest and the largest, correspond to a state of steady combustion.

Intermediate roots correspond to a state of unsteady combustion relative to temperature (even roots) or time (odd roots), but such regimes are never observed in practice. Thus, the type of burning regime (i.e., high- or low-temperature), is determined by the history of the process. When the intensity of the incident radiation increases from zero, the reaction takes place at lower temperatures unless the temperature of combustion $T_i$ (the smallest root of (2.4.13)) is reached. After this moment in time a stepwise increase in temperature occurs. In the opposite case, in which the intensity decreases from a sufficiently high level, a monotonic decrease of the combustion temperature occurs, until the extinguishing temperature $T_e$ (the largest root of (2.4.13)) is reached. After that a stepwise temperature decrease takes place, and, consequently, the rate of combustion approaches zero. Temperatures between $T_i$ and $T_e$ are never observed in a quasi-stationary regime. In the case of particles with radii $a \lesssim 10$ µm, this interval of 'forbidden' temperatures practically vanishes. For a spherical particle of a fixed radius a, the temperature of combustion is also fixed, so the expression used for assessing the combustion threshold intensity $I_i$ of the incident radiation, taking into account thermal interaction of the absorbing centers, can be written (according to [13]) as follows:

$$I_i(a, t) \simeq \frac{4\lambda_T(T_i)T_i/(aK_{ab}(a))}{[1 - \exp(-t/t_3)][1 + 4a\pi N_0 X_T t]}, \qquad (2.4.14)$$

where $N_0$ is the number density of the absorbing centers; $X_T = \lambda_T(C_p\rho_0)^{-1}$; $t_3 = a^2 C_a\rho_a/(3\lambda_T(T_i))$; and $\lambda_T(T_i)$ is the coefficient of molecular thermal conductivity of air at temperature $T_i$. The absorption efficiency factor $K_{ab}$ of an individual particle is calculated using the Shifrin approximation formula (2.1.19).

Figure 2.4.1 presents the nomograms for determining the temperature of steady burning of the carbon particle as a function of $IK_{ab}$, calculated using (2.4.13) and the tables [17] of the empirical dependences of $D_1$ and $\lambda_T$ on gas temperature (solid lines), also using a model dependence of the type $\simeq\lambda_T(T_0)(T/T_0)^{1/2}$ (broken curve). As seen from this figure, the empirical curves show a stronger dependence of $D_1$ and $\lambda_T$ of individual particles on temperature than does the model. This leads to qualitatively different results. Thus, for example, the process of combustion in the case of small carbon particles ceases just after the laser is switched off, due to energy losses caused by the molecular thermal conductivity (the heat of reaction does not compensate for these energy losses). However, this conclusion is not valid for a system of carbon particles in which the process of combustion can be sustained through the interaction of temperature fields. This process can also occur in the case of readily-inflammable substances such as oil droplets and alkali metals.

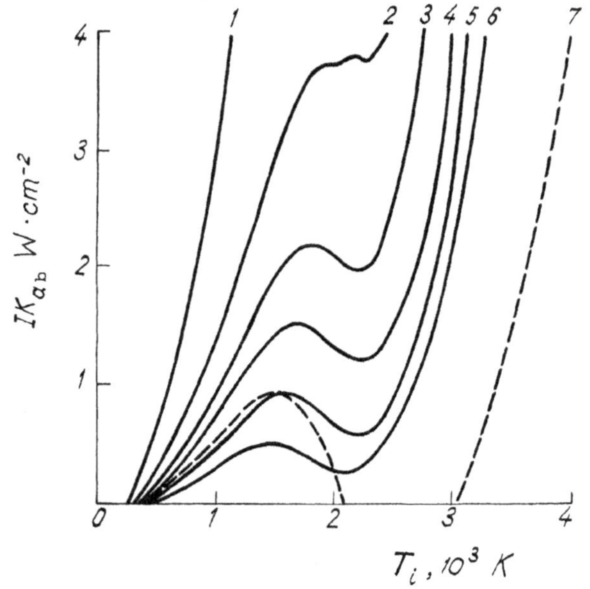


Fig. 2.4.1. Calculated nomograms illustrating the temperature of stationary burning of a carbon particle, depending on the parameter $IK_a$. Curves 1-6 correspond to a = 5, 10, 15, 20, 30, 50 μm, respectively; the dotted curve represents a calculation carried out using the model dependence $\lambda_T = \lambda_0(T_0)(T_a/T_0)^{1/2}$ for a particle with a = 20 μm.

Assuming $T_a$ = const, one can obtain [13] from (2.4.13) the following formulas for estimating the particle's radius as a function of time and characteristic time of particle combustion $t_i$:

$$a(t) \simeq a_0 - \left\{- \frac{A_d}{A_k} + \left[\left(\frac{A_d}{A_k} + 1\right)^2 - \frac{2tA_d}{a_0\rho_a}\right]^{1/2}\right\}; \tag{2.4.15}$$

$$t_i \simeq a_0\rho_a\{A_k^{-1}(a_0) + [2A_d(a_0)]^{-1}\}, \tag{2.4.16}$$

where

$$A_k = \frac{2M_a g P_0}{K_B T_a(a_0,\ IK_{ab})}; \qquad A_d = \frac{2M_a P_0 D_{eff}(a_0,\ T_a)}{K_B a_0^2 T_a(a_0,\ IK_{ab})}.$$

When $A_d/A_k \ll 1$, the process of combustion is controlled by the diffusion of the oxidant molecules to the surface of the particle. This regime dominates at $T_a \gtrsim 1800$ K for $a_0 = 1$ μm and at $T_a \gtrsim 2700$ K for $a_0 = 0.1$ μm.

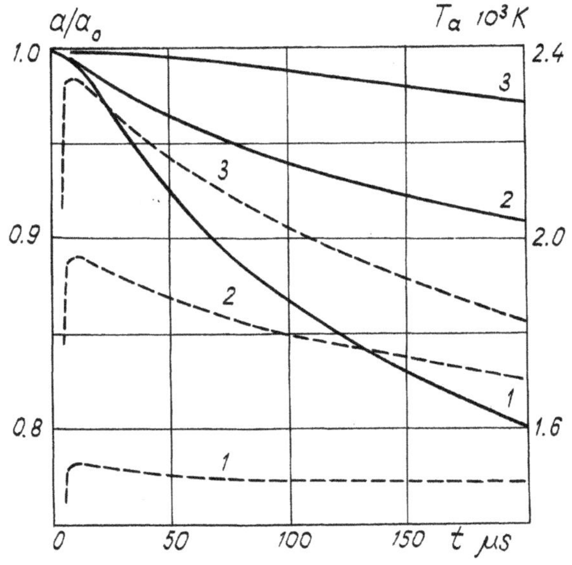


Fig. 2.4.2. Behavior of the relative radius (solid curves) and the temperature at surfaces (broken curves) of burning carbon black particles obtained on the basis of the numerical solution of a total set of aerothermochemistry equations. The parameters of the incident laser radiation are: $\lambda = 10.6$ μm; $I = 5 \times 10^5$ $Wcm^{-2}$. Curves 1-3 correspond to $a_0 = 0.7$; 0.6; 0.5 μm, respectively, with $m_a = 4.3-i3.9$.

A numerical simulation of the full system (2.4.1)-(2.4.9) was made in [16]. Figure 2.4.2 presents the calculated dependences of the relative radii of carbon particles (soot) of submicron size on time when irradiated by laser radiation of wavelength 10.6 μm and power density $I = 0.5$ $MW/cm^2$.

The process of the combustion of aerosol particles in a laser radiation beam has been studied experimentally in [13, 18]. In order to take microphotographs of the combustion process, the particles were mounted on backings made of Al, NaCl, Ag, or quartz fiber 20 to 30 μm in diameter. The time required for a soot particle of about 150 μm diameter, irradiated by $CO_2$ laser radiation ($I = 2.1 \times 10^2$ $W/cm^2$), to catch fire was about 7 msec. as determined from cinegrams [13]. The combustion of the particles took place during a period of 44 msec at a temperature of about 2000 K, and was accompanied by the ejection from the initial particle of small particles (1 μm in diameter) moving at a speed $\simeq 1$ m/sec.

Paper [13] presents the results of an experimental study carried out using a Nd-glass laser ($\lambda = 1.06$ μm) providing a power flux of about 0.5 $MW/cm^2$ in the zone of radiation interaction with soot particles. Micro photographs of soot particle behavior under these conditions are presented in Figure 2.4.5. The initial radius of the particle was 50 μm. As seen from this figure, the burning particle, under the influence of the radiation, accelerates and leaves characteristic tracks on the photograph. The fragmentation of burning particles can also be seen, as well as further fragmentation of these fragments. In the case of coal particles ($a_0 \approx 110$ μm) having a complex chemical composition that includes easily-vaporizable components, when these are irradiated with laser radiation of the same intensity as above, the following stages of the combustion process are observed. During the first ≈70 msec the particle swelled, its radius increased from 110 to 130 μm, then a stage of intensive burning took place until the radius of the particle had been reduced to ≈80 μm. Over a 0.9 sec interval the burning process then changed to the boiling of slag. Two temperature maxima, at 2300 K and 2400 K, were observed during the process of combustion of the coal particle at the first and second stages, respectively. The surface temperature of a burning particle was determined in [13] by measuring the thermal radiation flux at two preselected wavelengths. Figures 2.4.3 and 2.4.4 present the experimental data [13] on the dependence of the temperature of combustion on the intensity of incident radiation, as well as the changes in radius of the soot particles suspended on quartz fibers. The temporal behavior of the radius of a burning particle a(t), measured experimentally, can be approximated by the formula $a(t) = a_0(1 - t/t_i)^{1/2}$, where $t_i \approx 18$ sec, which is in good agreement with the theoretical expression (2.4.15) if $A_d/A_k \ll 1$. In this case, the rate of reaction is limited by the diffusion of the oxidant molecules. As the experimental results obtained in the above-mentioned paper have shown, the influence of the backing was a lowering, by 200 to 220 K, of the temperature of combustion as compared with that for aerosol particles suspended on quartz fibers.

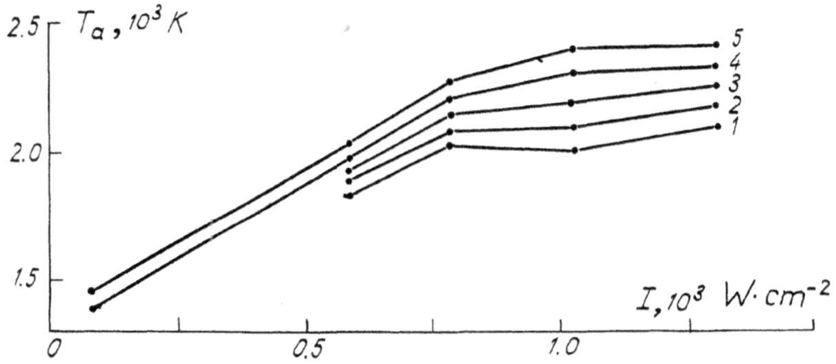


Fig. 2.4.3. Experimental dependences of average temperature of burning of carbon particles on the intensity of incident radiation from a $CO_2$ laser. Curves 1-5 correspond to initial particle radii: $a_0 = 50, 100, 150, 200, 250$ μm, respectively.

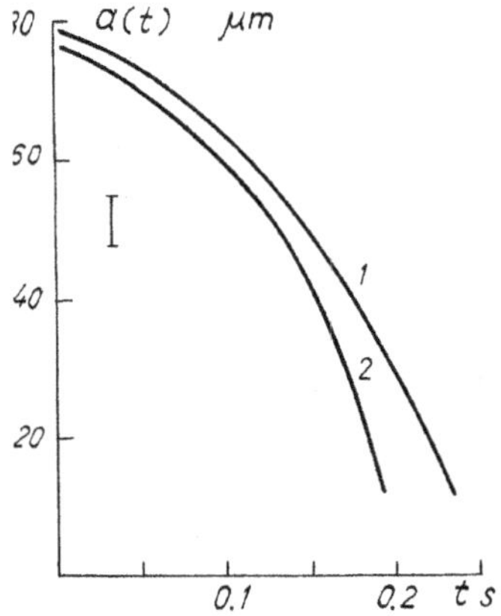


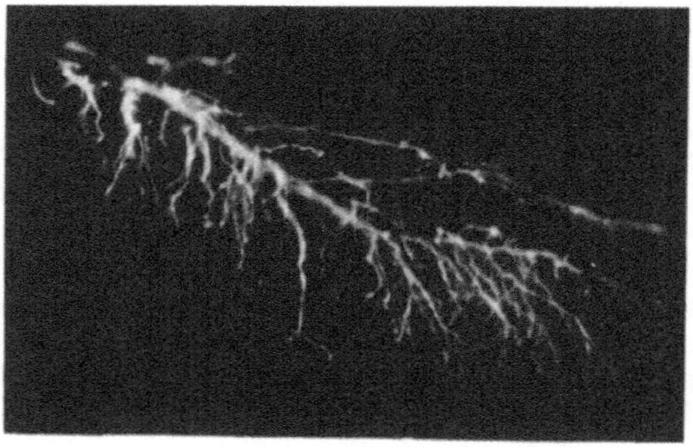


Fig. 2.4.4. (left) Time dependence of the radius of a burning carbon particle fixed to quartz fibers; wavelength of incident radiation $\lambda = 10.6$ μm; intensity $I = 1.02 \times 10^3$ $Wcm^{-2}$. (1) $a_0 = 79$ μm; (2) $a_0 = 77$ μm.

Fig. 2.4.5. (right) Illustration of the effects of fragmentation and acceleration of a burning carbon black particle in the Nd-glass radiation field: $\lambda = 1.06$ μm; $t_p \simeq 1$ ms; $I = 0.5 \times 10^6$ $Wcm^{-2}$.

It should be noted, in conclusion, that these theoretical and experimental results for the combustion of carbon particles in a laser beam cover an intensity range that includes intensities sufficient to overheat the particles to temperatures above their boiling point (for carbon, $T_b =$ $= 4000$ K). At such temperatures the main chemical reactions involving carbon are heterogeneous, i.e., these reactions take place on the surface of a solid phase. However, as follows from § 2.3.2, the metastable overheating of particles to temperatures $T_a \geq T_b$ can take place under high intensities of the incident radiation. In this case, all the chemical reactions occur in the gas phase. An analogous situation can be observed when droplets of oil products or alkali metals, whose boiling point is low, are irradiated by laser radiation. A rough analysis of the combustion process in this case can be found in [15].

## 2.5. INITIATION OF DROPLET SURFACE VIBRATIONS BY LASER RADIATION

The pondermotive forces acting on a droplet placed in the laser beam result in mechanical deformations of it. Various effects of this phenomenon can be observed, depending on the laser beam intensity and the character of the temporal modulation of this intensity, viz. the resonance oscillation and parametric excitation of the surface waves. These effects were first discussed in [21, 22], in which the effects of pondermotive forces on the plane surface of condensed water were studied. These effects were also considered in [23-26, 33], these works studied the interaction between laser radiation and a transparent aerosol. However, attention was focused on investigations into the possibility of the destruction of the transparent particles due to strong deformations occurring in the high-power optical fields [23, 24, 33] and of the effect of Raman light scattering on the oscillating deformations of the droplets as applied to the diagnosis of particle sizes [25, 26]. These effects are not confined to the results of laser action. The destruction of droplets by pondermotive forces in a stationary electrical field is a well-known effect which has been discussed in many papers, see, for example, [27, 28]. Fluctuations in the radar returns from vibrating droplets were considered in [29].

### 2.5.1. Basic Relationships

In general, treatment of the problem of the deformations of a transparent droplet in a high-power light field requires the solution of dynamic equations for viscous incompressible liquids which take into account the action of pondermotive forces [30]:

$$\left(\frac{\partial}{\partial t} + \vec{v}\nabla - \nu\Delta\right)\vec{v} = -\rho_0^{-1}(\nabla p - \vec{f}_E), \quad \operatorname{div}\vec{v} = 0, \tag{2.5.1}$$

where

$$\vec{f}_E = (8\pi)^{-1}\rho_0(\partial\varepsilon_a/\partial\rho)\,\nabla\vec{E}_a^2 \tag{2.5.2}$$

is the volume density of the pondermotive forces in an optically homogeneous medium [31]. $\vec{E}_a$ is the strength of the electric field inside the droplet. The kinematic and dynamic conditions on the free surface of the droplet can be described as follows:

$$\begin{gathered}\left(\frac{\partial}{\partial t} + \vec{v}\nabla\right)F(\vec{r}, t) = 0;\\ \{p - (8\pi)^{-1}\rho_0(\partial\varepsilon_a/\partial\rho)\vec{E}^2 - p_\infty + \sigma(R_1^{-1} + R_2^{-1}) + f\}n_i = \eta\left(\frac{\partial v_i}{\partial x_k} + \frac{\partial v_k}{\partial x_i}\right)n_k,\end{gathered} \tag{2.5.3}$$

where $F(\vec{r}, t) = 0$ is the equation describing the deformed surface; $\vec{r}(x_1, x_2, x_3)$ is the radius vector; $R_1$ and $R_2$ are the principal radii of the curvature of the surface; f is the step in the normal component of the electromagnetic field strength at the surface of the liquid; $\vec{n}$ is the unit vector perpendicular to the outer surface; $\sigma$ is the coefficient of surface tension; and (from [31]):

$$f = \frac{\varepsilon_a - 1}{8\pi}\left[(\varepsilon_a - 1)(\vec{E}_a\vec{n})^2 + \vec{E}_a^2\right]. \tag{2.5.4}$$

The integral form of the initial problem is as follows:

$$\frac{d}{dt}\left\{\frac{\rho_0}{2}\int_{V_d} v^2\, dV' + \sigma S_d\right\} + \frac{\eta}{2}\int_{V_d}\left(\frac{\partial v_i}{\partial x_k} + \frac{\partial v_k}{\partial x_i}\right)^2 dV' = \oint_{S_d} f\vec{v}\vec{n}\, dS, \tag{2.5.5}$$

where $V_d$ and $S_d$ are the volume and surface of the deformed droplet, respectively. Only low-frequency components should be taken into account in (2.5.1)-(2.5.5).

With small perturbations, the linearization of the problem can be accomplished. The solution of such a problem does not differ essentially from that of well-known problems on capillary waves [27, 30, 32].

If $\vec{\xi} = \vec{r} - \vec{r}_0$ is the vector of deformation of the droplet surface, and $r_0$ is the radius vector of the nonperturbed surface, then the weak deformation means that $\xi \ll r_0$, $\xi = |\vec{\xi}|$, $r_0 = |\vec{r}_0|$. The complex amplitude of a slowly-varying electrical field inside a droplet is represented as $\vec{E}_a = \vec{E}_a^0 + \vec{E}_a^\xi$, where $\vec{E}_a^0 = \vec{E}_a$ $(\xi = 0)$, and $\vec{E}_a^\xi$ is the component describing the distortions of the field due to the surface deformation.

All the anticipated effects can be found by analyzing the form of the function f, which contains complete information on the process. In this approach the value $(\xi/r_0) \ll 1$ and, after averaging (2.5.4) over time, one has

$$f = f_0 + f_\xi + O(\xi^2/r_0^2), \tag{2.5.6}$$

where

$$f_0 = \frac{(\varepsilon_a - 1)}{16\pi}|\vec{E}_0|^2\{(\varepsilon_a - 1)(\vec{e}\vec{n}_0)(\vec{e}^*\vec{n}_0) + \vec{e}\vec{e}^*\};$$
$$f_\xi = \frac{(\varepsilon_a - 1)}{8\pi}|\vec{E}_0|^2 \,\mathrm{Re}\{(\varepsilon_a - 1)(\vec{e}\vec{n}_0)(\vec{e}^*\vec{n}_\xi) + \vec{A}\vec{e}^* + (\varepsilon_a - 1)(\vec{A}\vec{n}_0)(\vec{e}^*\vec{n}_0)\}. \tag{2.5.7}$$

Here, $|\vec{E}_0|^2 = \vec{E}_0\vec{E}_0^*$; $\vec{E}_0$ is the complex amplitude of the slowly-varying field in which the particle is placed; $\vec{e} = \vec{E}_a^0/|\vec{E}_0|$; $\vec{A} = |\vec{E}_a|/|\vec{E}_0|$; $\vec{n}_\xi = \vec{n} - \vec{n}_0$; $\vec{n}_0$ is the unit vector normal to the nonperturbed surface. The first, and dominant, term of (2.5.6), i.e., $f_0$, if time-dependent, describes the resonance excitation of the droplet vibrations. The force $f_\xi$ causes the changes in oscillation frequency taking place due to the droplet's surface deformations, thus determining the parametric excitation of the surface waves. If the frequency of the variations in laser beam intensity is not close to the frequency of normal droplet vibrations, then the excitation of capillary waves is due to the parametric build-up of these oscillations.

In this approach, if we have a liquid with a low viscosity (i.e., if the Reynolds number $\mathrm{Re} \simeq \Omega_2 r_0^2 \nu^{-1} \gg 1$, where $\Omega_2 = [(8\sigma)/(\rho_0 r_0^3)]^{1/2}$ is the fundamental frequency of the droplet's normal mode), the liquid flow inside the droplet can be considered to be of a potential character, except for the thin boundary layer. This means that $\vec{v} = \nabla\phi$, where $\phi$ is the potential of velocity field. The pressure inside the droplet, and the vector of surface deformation are described by the following expressions:

$$P = \rho_0 \frac{\partial \phi}{\partial t}, \qquad \vec{\xi} = \int_0^t \nabla\phi \, dt. \tag{2.5.8}$$

The potential $\phi$ is a harmonic function, therefore it can be written

$$\phi(r, \theta, \varphi, t) = \sum_{\ell=0}^{\infty} \sum_{m=-\ell}^{\ell} r^{\ell} \phi_{\ell m}(t) T_{\ell m}(\theta, \varphi), \tag{2.5.9}$$

where $\phi_{\ell m}(t)$ are the coefficients of expansion; $T_{\ell m}$ are the spherical harmonics; and $r$, $\theta$, and $\varphi$ are the spherical coordinates. In the case of liquid with arbitrary viscosity, the problem of small oscillations of the droplet surface can be solved using a series expansion of the hydrodynamic functions over the generalized spherical harmonics [32].

The partial amplitudes $f_{\ell m}$ must satisfy the following equation:

$$\left(\frac{\partial^2}{\partial t^2} + 2t_{v\ell}^{-1} \frac{\partial}{\partial t} + \Omega_\ell^2\right) \phi_{\ell m} = -\rho_0^{-1} \frac{\partial f_{\ell m}}{\partial t}. \tag{2.5.10}$$

This expansion is not the exact corollary of (2.6.3), but has been derived taking into account the energy considerations relevant to situations involving liquids with a low viscosity [34]. The value $\Omega_\ell = (\ell(\ell - 1) \times$ $\times (\ell + 2)(\sigma/\rho_0 r_0^3))^{1/2}$ is the normal mode of the droplet; $t_{v\ell}$ is the time at which the oscillations are damped by the viscosity forces, and $t_{v\ell} =$ $= r_0^2/[\nu(\ell - 1)(2\ell + 1)]$,

$$f_{\ell m} = \int_0^{2\pi} d\varphi \int_0^{\pi} d\theta \sin\theta f(r, \theta, \varphi, t) T_{\ell m}^{*}. \tag{2.5.11}$$

## 2.5.2. Resonance Excitation of the Capillary Waves

Assuming a temporal behavior of the laser radiation such that

$$\frac{\partial f_0}{\partial t} \gg \frac{\partial f}{\partial t} \xi,$$

and setting $f_{\ell m} = f_{\ell m}^0 \cos \Omega t$, one obtains from (2.5.10) the following expression for the amplitude of the stimulated stationary surface oscillations ($t \gg t_{v\ell}$).

$$\xi_{\ell m} = \frac{f_{\ell m}^0 \ell \cos\left[\Omega t + \operatorname{arctg}\left(\dfrac{2t_{v\ell}^{-1}\Omega}{\Omega^2 - \Omega_\ell^2}\right)\right]}{\rho_0 r_0 ((\Omega^2 - \Omega_\ell^2)^2 + 4\Omega^2 t_{v\ell}^{-2})^{1/2}}. \tag{2.5.12}$$

At the resonance $(\Omega = \Omega_\ell)$,

$$\xi_{\ell m} = \frac{f^0_{\ell m}\ell \sin \Omega_\ell t}{2r_0\rho_0\Omega_\ell t^{-1}_{v\ell}}. \tag{2.5.13}$$

The form of the coefficients $f^0_{\ell m}$ is determined by the distribution of the electromagnetic field near the inner surface of the droplet. For droplets with large diffraction parameters, $kr_0 \gg 1$, these coefficients $f^0_{\ell m}$ can only be determined numerically. If we assume a uniform light field inside the droplet, $\vec{E}^0_a = (3/(\varepsilon_a + 2))\vec{E}_0$, and choose the direction of the vector $\vec{E}_0$ so that $(\vec{E}\vec{n}_0) = |\vec{E}_0| \cos \theta$, one can obtain for $f^0_{\ell m}$:

$$f^0_{\ell m} = \frac{I_0}{C\varepsilon_0^{1/2}}\left(\frac{\varepsilon_a - 1}{\varepsilon_a + 2}\right)^2 6\sqrt{\frac{2}{5}}\,\delta_{2\ell}\,\delta_{0m},$$

where $\delta_{\ell m}$ is the Kronecker symbol, and $I_0 = c\varepsilon_0^{1/2}|\vec{E}_0|^2/8\pi$.

Thus, in the case of a uniform optical field inside the droplet, only ellipsoidal oscillations $(\ell = 2,\ m = 0)$ can be excited by resonance. The amplitude of such oscillations of the droplet surface is

$$\xi^{\max}_{20} = 6\sqrt{\frac{2}{5}}\left(\frac{\varepsilon_a - 1}{\varepsilon_a + 2}\right)^2 \frac{I_0 t_{v2}\Omega_2^{-1}}{\rho_0 r_0 \varepsilon_0^{1/2} c}. \tag{2.5.14}$$

### 2.5.3. The Parametric Excitation of the Capillary Waves

If the force $f_0$ is independent of time, then the complex frequency of the surface oscillations can be determined using

$$\Omega^2_{\ell m} - 2it^{-1}_{v\ell}\Omega_{\ell m} - \Omega^2_\ell + B_{\ell m} = 0, \tag{2.5.15}$$

where $B_{\ell m} = (1/\rho_0 r_0)\beta^2 b_{\ell m}$ and $b_{\ell m}$ are dimensionless coefficients of the series expansion of $f_\xi$:

$$f_\xi = \beta^2 \sum_{\ell=0}^{\infty}\sum_{m=-\ell}^{\ell} b_{\ell m}\xi_{\ell m}T_{\ell m}\exp(i\Omega_{\ell m}t), \tag{2.5.16}$$

where $\xi_{\ell m}$ are the coefficients of the series expansion of

$$\xi = \sum_{\ell=0}^{\infty}\sum_{m=-\ell}^{\ell} \xi_{\ell m}T_{\ell m}\exp(i\Omega_{\ell m}t), \text{ and } \beta^2 = (\varepsilon_a - 1)^2\varepsilon_0^{-1/2}I_0 r_0^{-1}c^{-1}.$$

The real oscillation frequency is determined by

$$\Omega^R_{\ell m} = \tilde{\Omega}_\ell(1 - I_0/I_{\ell m})^{1/2}, \tag{2.5.17}$$

where

$$\tilde{\Omega}_\ell = (\Omega_\ell^2 - t_{v\ell}^{-2})^{1/2}, \qquad I_{\ell m} = \frac{c\sqrt{\varepsilon_0}\rho_0 r_0^2 \tilde{\Omega}_\ell^2}{(\varepsilon_a - 1)^2 b_{\ell m}\ell}.$$

It follows from (2.5.17) that a strong build-up of droplet oscillations can take place if $I_0 > I_{\ell m}$ and $t_p > \Omega_\ell^{-1}$. This was discussed in [23] as a basis for the destruction of transparent droplets in a high-power light field.

The approximations made in the assumption of a uniform light field near the droplet's surface show that the value $b_{20} \simeq 7/(\varepsilon_a + 2)^2$. For $\lambda = 0.63$ μm, $\varepsilon_a = 1.7$, and $r_0 = 10^{-2}$ cm, instability of small droplet oscillations appears at $I_0 \simeq 10^8$ W/cm$^2$ and $t_p > \Omega_{20}^{-1} \simeq 4 \times 10^{-5}$ s.

The case of finite droplet deformations occurring in the high-power monochromatic light field was considered [24], based on (2.5.5), for ellipsoidal droplet oscillations. It was shown that strong deformations of a droplet (the ratio of the long ellipsoid semi-axis to the short one $\gamma \gg 1$) can occur only at a significantly high energy of the incident radiation. The value $\gamma \simeq 10$ corresponds to the intensity $I_{cr} > 50C\sqrt{\varepsilon_0}\sigma/(\varepsilon_a - 1)^2 r_0$ and the pulse duration $t_p > 2\Omega_{20}^{-1}$.

The high energy and intensity of incident radiation necessary for the initiation of strong deformation of a transparent droplet show that this process is less effective in the destruction of a droplet than in its optical breakdown [33].

### 2.5.4. Experiments on the Excitation of the Oscillations of Transparent Droplets using Laser Radiation

Experiments [26] aimed at the detection and investigation of the resonance build-up of droplet oscillations as a result of laser irradiation have been carried out in a fog chamber, using a Q-switched ruby laser and an optical receiver connected to a narrow-band tunable amplifier. Radiation from a He-Cd laser with a wavelength $\lambda = 0.44$ μm was used as a sounding beam. Both the high-power beam and the sounding beam were focused using the lens (19 cm focal length) to a spot 0.3 mm in diameter. The caustics of the beams were made to coincide. The light flux scattered at a 30° angle with respect to the beam axis was collected using a lens (60 mm in diameter) located 15 cm from the beam. The narrow-band amplifier could be tuned into the frequency band from 0.7 to 1.3 MHz. A set of interference filters was used in this experiment to suppress the scattered radiation with a ruby laser wavelength. The experimentally-measured value was the level of light scattering at the wavelength of the sounding beam occurring during irradiation of the aerosol with a pulsed ruby laser. Two regimes of laser irradiation were used in the investigation. In the first one, the laser delivered a random series of peaks, while in the second the use of a KDP crystal modulator allowed the generation of regular peaks at a frequency of 1 MHz. This frequency coincides with the principal mode of droplets with a 3 μm radius. The modal radius of the fog droplets in this experiment was approximately 3 to 5 μm. The fog was generated by the evaporation of water. The optical depth of the fog at the prevailing path length of 10 cm (the length as viewed by the collection lens) was about 0.06.

Figure 2.5.1 shows the results of the measurements of scattered radiation, made under conditions of regular 1 MHz laser pulses, as a function of

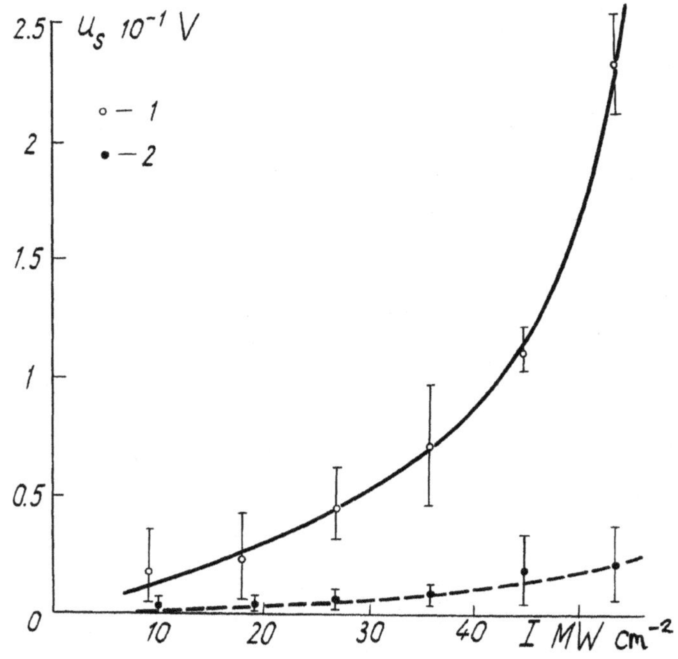


Fig. 2.5.1. The dependence of the intensity of scattered radiation ($\lambda$ = 0.44 μm), modulated at a frequency of 1 MHz, on the intensity of incident radiation (with $\lambda$ = 0.69 μm) in water fog. (1) Modulation frequency and frequency of the light detector are both 1 MHz. (2) Randomly-peaked ruby laser radiation.

the high-power laser beam intensity. The measurements were taken using an oscilloscope. Curves were plotted using the experimental data averaged over 10 to 15 points. Vertical bars represent the scatter of the measurements. As seen from Figure 2.5.1, the scattered light signal at the laser modulation frequency exceeds that from the unmodulated laser beam by more than one order of magnitude, beginning at an initial intensity of incident radiation of 40 MW/$cm^2$. It can also be seen from this figure that, in the case of ruby laser radiation with random intensity peaks, a significant increase in the experimental data scatter (≃30 times) is observed, beginning

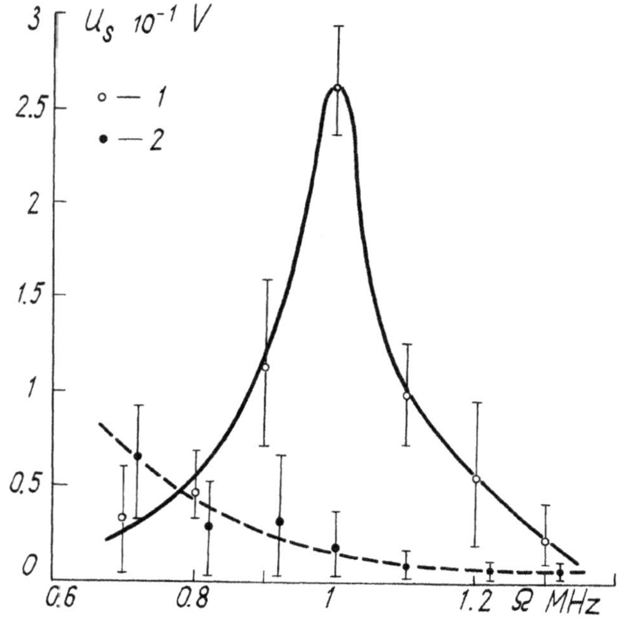


Fig. 2.5.2. Levels of the sounding beam's radiation scattering measured by a frequency-selective, tunable photodetector. (1) Ruby laser radiation modulated at 1 MHz. (2) A regime of randomly-peaked laser generation, $I_0 = 55$ MWcm$^{-2}$.

at the same intensity threshold. This can be explained by the presence of an intense modulation harmonic at 1 MHz in some laser shots. This is confirmed by the data presented in Figure 2.5.2 (broken curve). It can be seen from this figure that the strength of the signal from the detector increases in the low-frequency region of the signal spectrum where the peak of the noise spectrum of the laser radiation with random peaks of intensity occurs.

Figure 2.5.2 clearly illustrates the resonance character of the build-up of fog droplet oscillations at the ruby laser modulation frequency of 1 MHz.

The experimentally-observed mechanism of resonance interaction between modulated laser beam and aerosol can be used in some applications, such as the selective fragmentation of droplets of resonance size, or for remotely measuring the aerosol size distribution function by measuring the amplitudes of sounding beam scattering at different modulation frequencies.

Note, in conclusion, one physically interesting effect of the laser-induced generation of extremely high-frequency radiation (frequencies $\simeq \Omega_\ell$) on particles placed in the external electric field $E_{ext}$ (e.g., in thunder clouds). For $E_{ext} = 1$ MV/m and $I_0 = 10^7$ W/cm$^2$, the power of the extremely high-frequency radiation emitted by an individual particle, where $r_0 = 10$ μm, is $\simeq 10^{-22}$ W.

## REFERENCES: CHAPTER 2


[1] V. E. Zuev: Propagation of Visible and IR Radiation in the Atmosphere (Sovetskoye Radio, Moscow, 1970), in Russian.

[2] V. E. Zuev and A. V. Kuzikovskii: 'Thermal dissipation of water aerosols by laser radiation', Izv. Vyssh. Uchebn. Zaved. Fiz. 11, 106-132 (1970), in Russian.

[3] G. A. Andreev et al.: 'Laser Radiation Propagation in the Atmosphere', in Results of Science and Technology, Radioengineering (Moscow, VINITI, Vol. 11, 1977), pp. 5-148, in Russian.

[4] V. E. Zuev et al.: Nonlinear Optical Effects in Aerosols (Nauka, Novosibirsk, 1980), in Russian.

[5] V. E. Zuev: Laser Radiation Propagation in the Atmosphere (Radio i Svyaz', Moscow, 1981), in Russian.

[6] O. A. Volkovitskii et al.: Propagation of Intense Laser Radiation in Clouds (Gidrometeoizdat, Leningrad, 1982), in Russian.

[7] A. P. Prishivalko: Optical and Thermal Fields inside Light-Scattering Particles (Nauka i Tekhnika, Minsk, 1983), in Russian.

[8] K. S. Shifrin and Zh. K. Zolotova: 'Kinetics of droplet vaporization in a radiation field', Izv. Akad. Nauk SSSR Fiz. Atmos. Okeana 2, N12, 1311-1315 (1966); and 4, N1, 80-84 (1968), in Russian.

[9] F. A. Williams: 'On the vaporization of mist by radiation', J. Heat and Mass Transfer 8, 575-587 (1965).

[10] A. V. Kuzikovskii: 'Dynamics of a spherical particle in a high-power optical field', Izv. Vyssh. Uchebn. Zaved. Fiz. 5, 89-94 (1970), in Russian.

[11] V. E. Zuev et al.: 'Thermal effect of optical radiation on small water droplets', Dokl. Akad. Nauk SSSR 205, N5, 1069-1072 (1972), in Russian.

[12] E. B. Belyaev et al.: 'Laser spectrochemical analysis of aerosols', Kvant. Elektron. 6, 1152-1156 (1978), in Russian.

[13] V. I. Bukaty et al.: 'Combustion of carbon particles initiated by laser radiation', Izv. Vyssh. Uchebn. Zaved. SSSR Fiz. 2, 14-22 (1983), in Russian.

[14] S. I. Anisimov et al.: High-Power Radiation Effect on Metals (Nauka, Moscow, 1970), in Russian.

[15] V. I. Bukaty et al.: 'Combustion of carbon particles in a high-power optical field', Physics of Combustion and Explosion 15, 46-50 (1979), in Russian.

[16] V. S. Loskutov and G. M. Strelkov: 'Laser Radiation Attenuation by a Burning Soot Particle Aerosol', Abstracts 2nd Conf. on Atmospheric Optics (Institute of Atmospheric Optics, Siberian Branch, U.S.S.R. Acad. Sci., Tomsk, 1980), in Russian.

[17] N. B. Vargaftik: Handbook on Thermal Physical Characteristics of Gases and Liquids (Nauka, Moscow, 1972), in Russian.

[18] A. V. Kuzikovskii and V. A. Pogodaev: 'On the combustion of solid particles under the effect of $CO_2$ laser radiation', Physics of Combustion and Explosion 5, 783-787 (1977), in Russian.

[19] Yu. D. Kopytin and G. A. Mal'tseva: 'Laser radiation initiation of heterogeneous photocondensation processes', Izv. Vyssh. Uchebn. Fiz. 3, 95-101 (1978), in Russian.

[20] Yu. S. Sedunov: Physics of Liquid-Droplet Phase Formation in the Atmosphere (Gidrometeoizdat, Leningrad, 1972), in Russian.

[21] A. I. Bozhkov and F. V. Bunkin: 'Optical excitation of surface waves in transparent condensed media', Zh. Eksp. Teor. Fiz. 61, N6, 2279-2286 (1971), in Russian.

[22] V. K. Gavrikov et al.: 'Light scattering stimulated by surface waves', Zh. Eksp. Teor. Fiz. 58, 4, 1318-1331 (1970), in Russian.

[23] A. A. Zemlyanov: 'Stability of small vibrations of a transparent droplet in a high-power light field', Kvant. Elektron. 1, N9, 2085-2088 (1974), in Russian.

[24] A. A. Zemlyanov: 'Deformation and stability of a transparent droplet in a high-power optical field', Izv. Vyssh. Uchebn. Zaved. Fiz. 6, 132-134 (1975), in Russian.

[25] Ya. A. Bykovskii et al.: 'Resonance build-up of droplet surface oscillations due to an electromagnetic field', Kvant. Elektron. 3, N1, 157-162 (1976), in Russian.

[26] Yu. V. Ivanov and Yu. D. Kopytin: 'Selective intersection of laser pulse trains with aerosols', Kvant. Elektron. 12, 1820-1824 (1982), in Russian.

[27] D. V. Strett: Theory of Sound (Gostekhizdat, Moscow, 1955), Vol. 2, in Russian.

[28] P. R. Brazier-Smith: 'The stability of a water drop oscillating with finite amplitude in an electric field', J. Fluid Mech. 50, N3, 417-430 (1971).

[29] M. Brook and J. Latham Don: 'Fluctuating radar echo: modulation by vibrating drops', J. Geogr. Res. 73, N22, 7137-7144 (1968).

[30] L. D. Landau and E. M. Lifshits: Mechanics of Continuous Media (Gostekhizdat, Moscow, 1954), in Russian.

[31] L. D. Landau and E. M. Lifshits: Electrodynamics of Continuous Media (Gostekhizdat, Moscow, 1957), in Russian.

[32] N. D. Kopachevskii and A. D. Myshkis: 'On the free vibrations of a liquid self-gravitating ball taking into account viscous and capillary forces', J. Computational Mathematics and Mathematical Physics 8, N6, 1291-1305 (1968), in Russian.

[33] A. A. Zemlyanov et al.: 'Optical stability of weakly absorbant droplets in intense light fields', Applied Mathematics and Theoretical Physics 4, 33-37 (1977), in Russian.

[34] G. Lamb: Hydrodynamics (Gostekhizdat, Moscow, 1954), in Russian.

CHAPTER 3

# THE FORMATION OF CLEAR ZONES IN CLOUDS AND FOGS DUE TO THE VAPORIZATION OF DROPLETS UNDER REGULAR REGIMES

A unique result of the process of intensive IR beam propagation through water aerosols is the possibility of increasing the transmission of these media. The idea of removing the light scattering property of the aerosol by beam-induced phase transition of the droplets has stimulated the development of nonlinear optics of scattering media, and has provided the basis for the method of beam 'clearing up', or dissipation, of water aerosols. This method is energetically advantageous, since the spatially-selective absorption of light by aerosol particles is the only source of energy for the particles' heat and their eventual vaporization, while the probable resulting increase in the medium's transmission can be several orders of magnitude.

The first works in this field appeared soon after the invention of the laser. However, it can be stated that, in a more general connection, the problem of the effect of radiation on water aerosols had been studied earlier in connection with the role of solar radiation in cloud dynamics [1, 2]. Contemporary knowledge of beam-induced 'clearing' of water aerosols can be considered to be quite complete. The theoretical predictions of both the fundamental and the accompanying effects were later proved experimentally, this included tests in the field. This history is especially true for pre-explosion regimes of vaporization, which were investigated in many original works and reviews [3-5].

This chapter was written to provide a detailed description of the problem. It is natural, therefore, that we will discuss the multiparameter aspects of the problem, including the theories accounting for the effects of recondensation, turbulent transfer of droplets, and refraction distortions of the beam. We hope, however, that readers who are not interested in such a detailed description can make their own choice of reading. It is for this reason that the parameters are introduced gradually.

A basic limitation imposed on any application of the material presented in this chapter concerns only the beam intensity, which should be lower than the droplet explosion threshold. In the case of droplets about 10 μm in diameter, this threshold is $\simeq 3 \times 10^4$ W/cm$^2$ at a wavelength of 10.6 μm. On the other hand, the algorithms developed in this chapter are applicable to the relevant problems, unless the type of nonlinearity characteristic of regular vaporization regimes (accumulating nonlinearity) is changed. Since this type of nonlinearity is also characteristic of weak two-phase explosions, then it is possible to widen the sphere of applicability of these algorithms. Special attention will be paid to this topic in a separate section.

## 3.1. BASIC CHARACTERISTICS OF THE PROCESS OF CLEARING A 'FROZEN' CLOUD

Consider the problem of 'clearing' a polydispersed aerosol using a non-divergent beam under windless conditions, with no gas absorption as assumption one. Under conditions of a known beam intensity at the cloud boundary, $z = 0$, the intensity of a beam must obey the following equation:

$$\partial I/\partial z = -\alpha_N([I, t])I. \tag{3.1.1}$$

The nonlinear properties of the medium are determined by the volume extinction coefficient of the aerosol, $\alpha_N$, which is related to the modifying size spectrum $f(r, [I, t])$ according to the relationship

$$\alpha_N([I, t]) = \pi N_0 \int_0^\infty r^2 f(r, [I, t])K(r)\, dr, \tag{3.1.2}$$

where $N_0$ is the number density of the droplets and $K(r)$ is the extinction efficiency factor of the droplets. The modifying size spectrum should, in turn, satisfy the problem, with initial conditions, posed by the equation

$$\partial f/\partial t + \partial/\partial r(\dot{r}f) = 0. \tag{3.1.3}$$

In this equation the velocity of density f 'movement' along the r-axis is the function which is determined by the droplet's vaporization kinetics. In the discussion below it is presented as follows:

$$\dot{r} = -k_e \frac{K_a(r)I}{4Q_e\rho_L}, \tag{3.1.4}$$

where $k_e$ is the coefficient of heat loss ($k_e \leq 1$), which takes the vaporization regime into account, $K_a(r)$ is the absorption efficiency factor of one droplet, $\rho_L$ is the density of the droplet, and $Q_e$ is the specific heat of vaporization.

The relationships presented above describe a self-contained problem, in which (3.1.1) describes the influence of changes in the medium on the beam intensity. Equations (3.1.2)-(3.1.4) describe the physical aspects of the problem; it should be noted that (3.1.3) is the most important in this part of the problem.

It is evident that the process of 'clearing' is described by (3.1.1), while the other equations furnish additional information concerning the nonlinear extinction coefficient $\alpha_N$ which, apparently, is in this case a function of I and t.

As regards the procedure for arriving at a solution, one should keep in mind that if the physical part of the problem is solved for an arbitrary function $I(\vec{R}, t)$, then in the succeeding investigations into the dynamics of the clearing process only (3.1.1), or its corollaries, is necessary. We will now demonstrate that such a possibility really exists.

For this purpose we will describe the procedure for solving (3.1.3). First, the characteristic equation

$$\frac{dr}{dt} = -k_e \frac{K_a(r)\,I(\vec{R},\, t)}{4\rho_L Q_e} \tag{3.1.5}$$

is integrated, taking into account the condition that $r(0) = r_0$. The result is in the form of function $r_0(r)$, which is then substituted into the initial size spectrum $f_0(r)$:

$$f = f_0[r_0(r)]\ |dr_0/dr|. \tag{3.1.6}$$

The factor $|dr_0/dr|$ is introduced into (3.1.6) to permit the size spectrum to satisfy the condition of full probability in the process of rearrangement, i.e., $\int_0^\infty f(r)\,dr = 1$.

The solution obtained for f can be used for calculating the nonlinear extinction coefficient of the aerosol according to (3.1.2). The integral in (3.1.5) can be represented in the following form:

$$\frac{4\rho_L Q_e}{k_e} \int_r^{r_0} \frac{dr'}{K_a(r')} = \int_0^t I(\vec{R},\, t')\,dt'. \tag{3.1.7}$$

The integral on the right-hand side of (3.1.7) is not modified by the procedure for calculating the nonlinear extinction coefficient of the aerosol, and will be automatically incorporated in the final result. Thus, the nonlinear extinction coefficient will be a function of the following form:

$$\alpha_N([I,\ t]) = \alpha_N(J), \tag{3.1.8}$$

where the integral

$$J = \int_0^t I(\vec{R},\, t')\,dt' \tag{3.1.9}$$

is called the energetic variable. In physical terms, it means the irradiation of the medium at the Lagrangian point, and that, in turn, shows that we are dealing with accumulating nonlinearity.

The procedure itself, and the view of the function $\alpha_N(J)$, both depend on the form of the initial size spectrum $f_0(r)$ and the nature of the parameters $K(r)$ and $K_a(r)$. The details of the calculation procedure can be found in § 3.7. It will be shown in this section that, typically, $\alpha_N(J)$ can be approximated by the exponential functions of the form $\alpha_N(J) = \alpha_0 \exp(-k_e J/J_h)$. Under conditions where the water-content approach is applicable ($2rk_a \ll 1$), i.e., when $K(r) = K_a(r) = (4/3)rk_a$ ($k_a$ is the absorption cross-section of a unit volume of condensed water), the dependence of $\alpha_N$ on J is really exponential, with $\alpha_0 = k_a q/\rho_L$, $J_h = \rho_L Q_e/k_a$ (q is the water content of the cloud). The brief information above concerning nonlinearity is quite sufficient to pave the way for further description of the process of cloud dissipaticn by laser beams.

Now, we would like to note that it is more convenient to use the equation for the energetic variable which is equivalent to (3.1.1):

$$dJ/dz = -\int_0^J \alpha_N(J')\ dJ'. \tag{3.1.10}$$

The solution of this equation in quadratures, under the condition that $J(0) = I_0t = J_0$, is as follows:

$$\int_{I_0t}^J dJ'/\int_0^J \alpha_N(J'')\ dJ'' + z = 0. \tag{3.1.11}$$

It follows from this formula that the velocity of particles at which J (and hence $\alpha_N(J)$ and I) is cobstant is

$$dz/dt = \frac{I_0}{\int_0^{I_0t} \alpha_N(J)\ dJ}. \tag{3.1.12}$$

If $\alpha_N(J)$ is the function vanishing when $J \leq J_c$ ($J_c \leq \infty$), and the integral $\int_0^\infty \alpha_N(J)\ dJ$ converges, then the value dz/dt reaches its stationary limit $U_f$ during the time interval $t_c = J_c/I_0$:

$$U_f = \frac{I_0}{\int_0^\infty \alpha_N(J)\ dJ}. \tag{3.1.13}$$

This formula means that, during the timme interval $t_c$, the profiles of the characteristic values J, I, and $\alpha_N$ are formed and the particle is moving inside the medium at a speed $U_f$. That, in turn, means that the dissipation of the medium is of the wave type. It is clear, at the same time, that there can exist dissipation regimes under which a stationary profile never occurs. For example, the functions $\alpha_N(J)$ decreasing as 1/J can result in such a regime, since in this case $\int_0^\infty \alpha_N(J)\ dJ$ is logarithmically divergent.

We can better consider more realistic situations. In the case of the above exponential form of $\alpha_N(J)$, one has

$$U_f = k_e(I_0/\alpha_0 J_h). \tag{3.1.14}$$

In the water content approach

$$U_f = k_e(I_0/qQ_e).$$

As can be seen from the previous formula, the velocity of the propagation of clearing waves does not depend on the optical properties of the medium. If the linear function $\alpha_N(J) = \alpha_0(1 - k_eJ/J_c)$, we have

$$U_f = 2k_e(I_0/\alpha_0 J_c). \tag{3.1.16}$$

It can be shown that, in order to take into account the finite speed of light, one should use the following formula:

$$U_f = cU_f^{\infty}/(c + U_f^{\infty}), \tag{3.1.17}$$

where $U_f^{\infty}$ is the speed as calculated using (3.1.13) or (3.1.14)-(3.1.16) in corresponding cases.

Now consider the problem of intensity. For this purpose we shall carry out the integrations in (3.1.11) for both linear and exponential dependences of the extinction coefficient of the aerosol on the energetic variable. This procedure results in the derivation of the profile of the energetic variable. The intensity is then obtained by differentiating J with respect to t.

In the case of a linear dependence of $\alpha_N$ on J, it is possible to separate the intermediate zone and the completely clear zone of the beam channel described by the dimensionless variables $T = I/I_0$, $\tau = \alpha_0 z$, $\bar{J} = k_e J/J_c$, and the intensity profile can be presented as

$$T = \frac{4e^{-\tau}}{\bar{J}_0^2\left(\frac{2 - \bar{J}_0}{J_0} + e^{-\tau}\right)^2}, \quad \bar{J}_0 \leq 1;$$

$$T = \begin{cases} 1, & \tau \leq 2(\bar{J}_0 - 1) \\ \dfrac{4\exp - [\tau - 2(J_0 - 1)]}{\bar{J}_0^2\left(\dfrac{2 - \bar{J}_0}{\bar{J}_0} + \exp -[\tau - 2(\bar{J}_0 - 1)]\right)^2}, & \tau \geq 2(J_0 - 1) \end{cases}, \quad \bar{J}_0 \geq 1. \tag{3.1.18}$$

It can be seen from these equations that the steady profile of transmission in the intermediate zone is formed when the boundary value of the dimensionless energetic variable reaches unity. When $\bar{J}_0 > 1$, this profile moves as a whole (without deformation) into the medium, leaving a completely clear zone behind it. The maximum change in optical depth, $\Delta\tau = \tau + \ln T$, observed in the steady intermediate zone in this case is $\ln 4 = 1.386$.

In the case of an exponential dependence of $\alpha_N$ on J, the division of the beam channel into an intermediate zone and a completely clear zone can be made only conventionally, with reference to a pre-defined value of the energetic variable $J_c$. In this case, the dimensionless energetic variable is introduced according to the formula $\bar{J} = k_e J/J_h$, and its profile is

$$\bar{J} = \ln[1 + e^{-\tau}(e^{\bar{J}_0} - 1)]. \tag{3.1.19}$$

The relative intensity is then expressed according to Glickler's formula [6]:

$$T = e^{\bar{J}_0}/(e^{\bar{J}_0} + e^{\tau} - 1). \tag{3.1.20}$$

Figures 3.1.1 and 3.1.2 represent the relative intensity profiles deter-

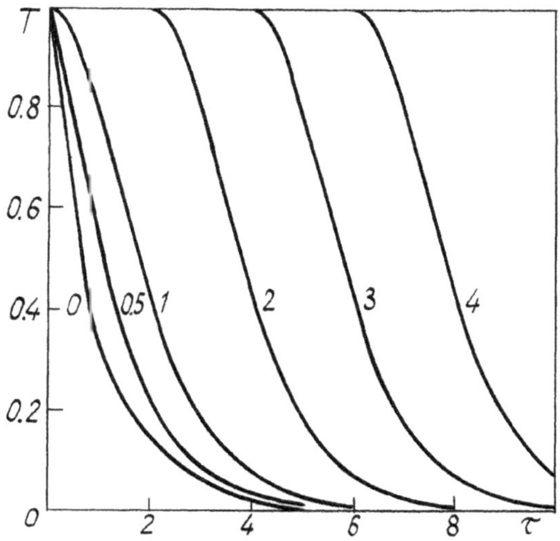


Fig. 3.1.1. Transmission of the cleared zone as a function of optical depth. The extinction coefficient is a linear function of the energetic variable. The numbers on the curves are the boundary values of the dimensionless energetic variable.

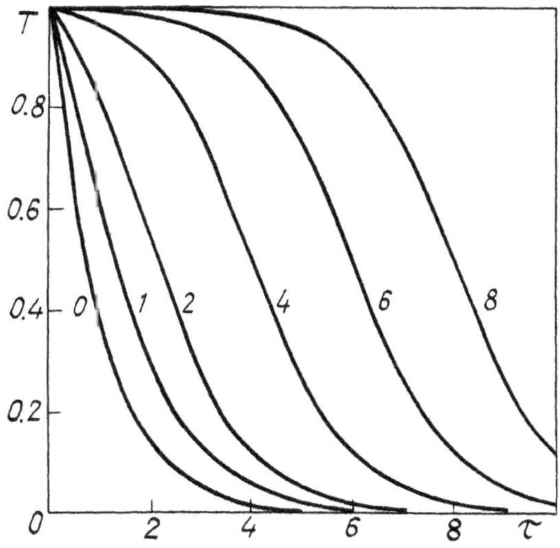


Fig. 3.1.2. Transmission of the cleared zone as a function of optical depth. The extinction coefficient is an exponential function of the energetic variable. The numbers on the curves are the boundary values of the dimensionless energetic variable.

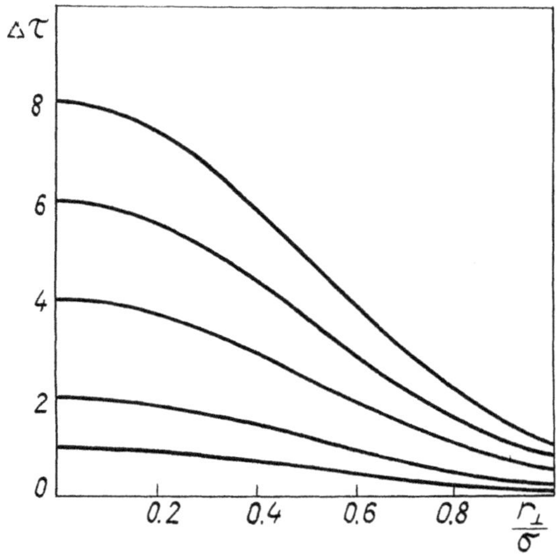


Fig. 3.1.3. Dependence of the changes in optical depth, in the case of a semi-infinite cloud ($\tau \to \infty$), on the distance from the beam's axis. The boundary values of the dimensionless energetic variable along the beam's axis coincide with the value $\Delta\tau$ if $(r_\perp/\sigma) = 0$. The extinction coefficient is an exponential function of the energetic variable.

mined by (3.1.18) and (3.1.20). Figure 3.1.3 shows the dependence of the optical depth of the medium on the distance from the beam axis and on the value of the energetic variable at the axis of a Gaussian beam of intensity

$$I_0 = \frac{2}{\pi}\frac{p}{2}\exp[-2(r_\perp^2/\sigma^2)]. \tag{3.1.21}$$

In cases where only the configuration of the cleared zone is of interest (e.g., the length of cleared zone) it is not necessary to obtain an exact value for the energetic variable. One can use the linearized equation for this purpose. Indeed, by linearizing (3.1.10), one obtains

$$d\bar{J}/d\tau = -\int_0^\infty \bar{\alpha}_N(\bar{J})\ d\bar{J}, \tag{3.1.22}$$

where $\bar{\alpha}_N = \alpha_N/\alpha_0$. The solution of this equation is

$$\bar{J} = \int_0^\infty \bar{\alpha}_N(\bar{J})\ d\bar{J} + \bar{J}_0. \tag{3.1.23}$$

Assuming that $\bar{J} = \bar{J}_c$, where $\bar{J}_c$ is the value of the energetic variable at which $\bar{\alpha}_N(\bar{J})$ vanishes, one obtains from (3.1.23) for the range of the cleared channel

$$\tau = (\bar{J}_0 - \bar{J}_c)/\int_0^\infty \bar{\alpha}_N(\bar{J})\ d\bar{J}. \tag{3.1.24}$$

This is the precise expression. As can be shown by derivation of (3.1.24) with respect to time, this expression is the correct one for the velocity of the clearing wave.

It should be noted here that the first calculations of the nonlinear transmission and velocity of the clearing wave, made by means of the water content approach, were carried out by Lamb and Kinney [7], Glickler [6] then demonstrated the possibility of expanding the limits of applicability of this approach. The case of a linear dependence of the aerosol extinction coefficient on the energetic variable was discussed in [8], in connection with the problem of clearing aerosols consisting of dyes dissolved in water.

## 3.2. STATIONARY CLEARED CHANNELS IN MOVING CLOUDS

Let us modify the problem discussed in § 3.1 introducing the transverse (relative to the beam) transportation of the aerosol at speed $\vec{V}_\perp$. Let us assume that transportation takes place along the X-axis in the positive direction. It is obvious that, under conditions of a stationary beam field, a definite, stationary configuration of aerosol clearing is formed.

The system of equations for the solution of this corresponding self-contained problem is analogous to the system (3.1.1)-(3.1.4). Some peculiarities do appear in the equation for the rearrangement of the size spectrum of the droplets:

$$V_\perp(\partial f/\partial x) + \partial/\partial r(\dot{r}f) = 0. \tag{3.2.1}$$

The boundary condition in this case is the size spectrum in the undisturbed zone, in general at $x = -\infty$. As a matter of fact, the equation is identical to (3.1.3). The solution procedure is also identical to that used for solving (3.1.3). The difference is that, instead of integration over a time variable, we now have integrals over x. The energetic variable is presented as follows:

$$J = \frac{1}{V_\perp}\int_{-\infty}^{x} I(x', y, z)\ dx'. \tag{3.2.2}$$

The equation for the energetic variable has the same form as (3.1.10). All this leads one to the conclusion that the expressions used in the preceding section for describing the configuration of a completely cleared zone and the intensity (in particular, Glickler's formula (3.1.20)) are still valid. It is not necessary to substitute the boundary functions $J_0$ with corresponding stationary values that, consequently, will reveal the dependence of the values sought on the transverse coordinates.

Let us write the expressions for the boundary functions in the case of beams with a uniform intensity distribution over a circular cross-section of radius R, and for a Gaussian beam of the form

$$I_0 = \frac{2}{\pi}\frac{P}{\sigma^2}\exp[-2(x^2 + y^2)/\sigma^2], \qquad (3.2.3)$$

where P is the beam power. In the case of a uniform intensity distribution one has

$$J_0 = \frac{P}{\pi R^2}\frac{x + (R^2 - y^2)^{1/2}}{V_\perp}. \qquad (3.2.4)$$

In the case of a Gaussian beam the boundary function is

$$J_0 = \frac{P}{\sqrt{2\pi}}\frac{1}{\sigma V_\perp}\exp(-2y^2/\sigma^2)\left[1 + \operatorname{erf}\left(\frac{\sqrt{2}}{\sigma}x\right)\right]. \qquad (3.2.5)$$

Let us consider the configuration of a completely clear zone in order to illustrate the above. Using (3.1.24), and assuming the exponential form of the function $\bar{\alpha}_N(\bar{J})$, one obtains $\tau = \bar{J}_0 - \bar{J}_c$. Then let us assume that $\bar{J}_c = 3$. This means that the boundary of the cleared zone is defined at a value $\alpha_0/20$ of the initial extinction coefficient. Taking into account (3.2.4), one can see that, in the case of a uniform distribution of energy over the beam's cross-section, the dependence of the dimensionless range of the cleared zone on the transverse coordinates has the following form

$$\tau = \frac{k_e}{J_h}\frac{P}{\pi R^2}\frac{x + (R^2 - y^2)^{1/2}}{V_\perp} - 3. \qquad (3.2.6)$$

This linear dependence on x describes the characteristic wedge shape of the clear zone. It is obvious that any calculation of the depth of clearing must be made only if $\bar{J}_0 > \bar{J}_c$ when $\tau > 0$. When the incident radiation is weaker, the cleared zone is investigated using Glickler's formula. Incidentally, it follows from this formula that, when $\bar{\alpha}_N(\bar{J}) = e^{-\bar{J}}$, the maximum change in the optical depth of the medium in the intermediate zone is equal to the preset level of complete clearing, $\bar{J}_c$. Note, finally, that the profiles of relative intensity within the completely cleared zone are shown in Figure 3.2.1, and the expression for the depth of the cleared zone, in the case of a Gaussian beam, is as follows:

$$\tau = \frac{k_e}{J_h}\frac{P}{\sqrt{2\pi}}\frac{1}{\sigma V_\perp}\exp(-2y^2/\sigma^2)\left[1 + \operatorname{erf}\left(\frac{\sqrt{2}}{\sigma}x\right)\right] - 3. \qquad (3.2.7)$$

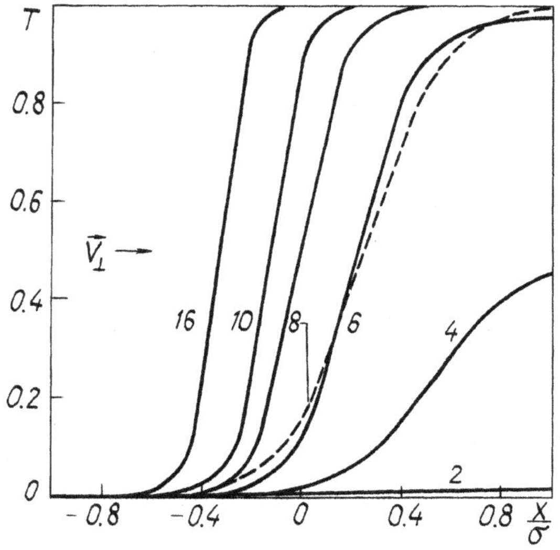


Fig. 3.2.1. Steady value of the optical transmission through a cloud (the initial optical depth being $\tau = 8$) as a function of the transverse coordinate on the plane $y = 0$. The numbers on the curves are the boundary values of the dimensionless energetic variable along the axis of a Gaussian beam. The dashed curve represents the transmission profile in the case of a beam with a uniform intensity distribution, but of constant power and radius ($R = \sigma$), as in the case of a Gaussian beam.

It should be also noted that the full optical depth of the cleared zone (including both the intermediate and the completely cleared zones) is $\bar{J}_0$.

## 3.3. THE UNSTABLE REGIME OF MOVING CLOUD CLEARANCE

First, consider the process of formation of the stable, clear channel. This problem is of particular importance in studying the pulsed regime of medium illumination when the time of interaction is limited, but the transportation of aerosols by wind cannot be neglected.

The equation for the droplet size spectrum in this case is

$$\partial f/\partial t + V_\perp(\partial f/\partial x) + \partial/\partial r(\dot{r}f) = 0. \tag{3.3.1}$$

With the boundary conditions

$$f\Big|_{\substack{t=0 \\ x=-\sqrt{R^2-y^2}}} = f_0(r).$$

Here, R is the full radius of the beam. It is useful to consider this radius to be a constant value, even in the case of a Gaussian beam, in order to correctly describe the stationary and nonstationary stages of the process. Using the coordinate transformation

$$z = \frac{t}{2} + \frac{x + (R^2 - y^2)^{1/2}}{2V_\perp}; \qquad \eta = \frac{t}{2} - \frac{x + (R^2 - y^2)^{1/2}}{2V_\perp}, \tag{3.3.2}$$

one can reduce the problem to the following form:

$$\frac{\partial f}{\partial z} + \frac{\partial}{\partial r}(\dot{r}f) = 0, \qquad f\big|_{z=\mathrm{sgn}\,\eta} = f_0(r). \tag{3.3.3}$$

This means that the problem is formally reduced to the one discussed in § 3.1. The energetic variable is presented as follows:

$$J = \int_{\eta\,\mathrm{sgn}\,\eta}^{z} I(z', \eta, z)\, dz'. \tag{3.3.4}$$

The equation for the energetic variable in this case has the same form as (3.1.10). It can also be stated that all the peculiar features of this problem are associated with the boundary functions, so let us consider them in more detail, filling in the initial coordinates.

For a beam with a uniform power distribution over its cross-section, we have

$$J_0 = \begin{cases} \dfrac{P}{\pi R^2}\,\dfrac{x + \sqrt{R^2 - y^2}}{V_\perp}, & t \geqslant \dfrac{x + \sqrt{R^2 - y^2}}{V_\perp}, \\[2ex] \dfrac{P}{\pi R^2}\, t, & t \leqslant \dfrac{x + \sqrt{R^2 - y^2}}{V_\perp}. \end{cases} \tag{3.3.5}$$

It can be seen from (3.3.5) that the activation zone can be divided into two regions: the stationary region, in which the channel parameters are independent of time, and the nonstationary region, where these parameters do not depend on the transverse coordinates. The boundary between these two regions moves along the X-axis in the positive direction at the wind speed $V_\perp$. It is clear that the stationary region of the cleared channel is formed during the time interval $2R/V_\perp$. It can also be said that the process of clearing the medium occurs in the nonstationary region in the same way as for the 'frozen' cloud situation. The process of forming the wedge-shaped zone of completely cleared atmosphere can be treated, in this case, as the propagation of a plane front of clearing at a speed $u_f$ with shortening of its length along the X-axis taking place at a rate corresponding to the wind speed $V_\perp$ simultaneously. This is illustrated in Figure 3.3.1.

Keeping the above picture in mind, one can easily derive the criteria for efficient beam action on a cloud of finite optical depth. Action of a laser beam on a cloud can be considered to be efficient if the cross-section

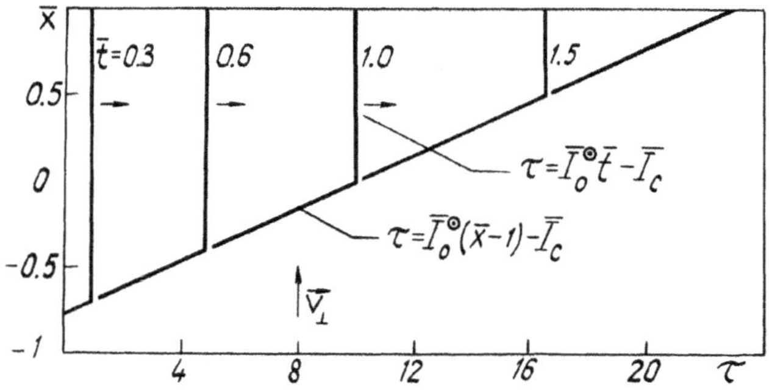


Fig. 3.3.1. Configuration of the completely cleared zone $\tau(\bar{x}, \bar{t})$ in the plane $y = 0$ for the case of an incident beam with a uniform intensity distribution over the beam's cross-section; $\bar{x} = x/R$, $\bar{t} = tV_\perp/R$, $\bar{J}_o^{\circledcirc} = k_e P/\pi R J_h V_\perp$; $\bar{J}_o^{\circledcirc} = 13$, $\bar{J}_c = 3$.

of the completely clear zone is equal to, or only a little less than, the beam cross-section. In other words, the time necessary for the beam-induced clearance of the medium, $t_e = z/U_f$, must be shorter than the time taken for the wind to travel across the beam. Using (3.1.14), one obtains

$$t_e = (J_h \tau)/(k_e \bar{I}_0) \ll 2R/V_\perp. \tag{3.3.6}$$

It is obvious that the time period of effective laser beam action (pulse duration) must not be shorter than $t_e$. If this condition for the effective action of the laser beam is fulfilled, then the cloud can be treated as 'frozen'.

In the case of a Gaussian beam (3.2.3), the boundary value of the energetic variable is

$$J_0 = \frac{P}{\sqrt{2\pi}} \frac{1}{\sigma V_\perp} \exp(-2y^2/\sigma^2) \left\{ \operatorname{erf}\left(\sqrt{2}\,\frac{x}{\sigma}\right) + \right.$$
$$\left. + \operatorname{erf}\left[ \begin{array}{ll} \dfrac{\sqrt{2}}{\sigma}\sqrt{R^2 - y^2}, & t \geq \dfrac{x + \sqrt{R^2 - y^2}}{V_\perp} \\ \dfrac{\sqrt{2}}{\sigma}(V_\perp t - x), & t \leq \dfrac{x + \sqrt{R^2 - y^2}}{V_\perp} \end{array} \right] \right\}. \tag{3.3.7}$$

It can be seen from (3.3.7) that, in the nonstationary region, $J_0$ depends on x and the wind speed $V_\perp$. Since earlier we introduced the finite beam radius R, the time period of the formation of the stationary zone is also finite. However, ideal Gaussian beams have infinite dimensions and hence the time of formation of the stationary situation is also infinite. Therefore, the corresponding expression for $J_0$ can be derived for the asymptotic case of $R \to \infty$. As a result, one obtains

$$J_0 = \frac{P}{\sqrt{2\pi}} \frac{1}{\sigma V_\perp} \exp(-2y^2/\sigma^2) \left\{ \operatorname{erf}\left(\frac{\sqrt{2}}{\sigma} x\right) + \operatorname{erf}\left[\frac{\sqrt{2}}{\sigma} (V_\perp t - x)\right]\right\}. \quad (3.3.8)$$

It is clear that as $t \to \infty$ (3.3.8) takes on the form of (3.2.5) exactly. Figure 3.3.2 illustrates the process of forming the profile of nonlinear transmission by a Gaussian laser beam, as it is described by Glickler's formula and (3.3.8).

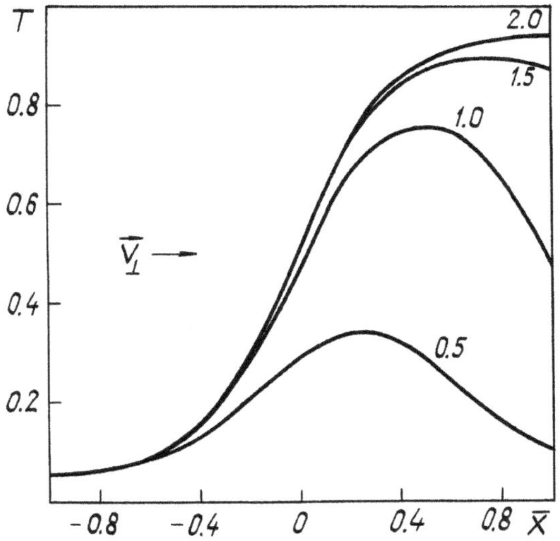


Fig. 3.3.2. Formation of the transmission profile in the intersection zone for the case of a Gaussian beam in windy conditions: $\tau = 3$; $\bar{J}_0^{\otimes} = (k_e/J_h)(P/\sqrt{2\pi})x$; $(1/\sigma V_\perp) = 3$; $y = 0$; $\bar{x} = x/\sigma$. The numbers on the curves are the values of the parameter $\bar{t} = tV_\perp/\sigma$.

## 3.4. THE DETERMINATION OF THE PARAMETERS OF THE CLEARED ZONE TAKING INTO ACCOUNT THE ANGULAR BEAM WIDTH AND WIND SPEED

In spite of the fact that there have been some attempts to generalize Glickler's formula so that it could account for cases involving divergent beams [9, 10], we shall make one more attempt to describe in detail the procedure for obtaining the necessary algorithm. First, consider the stationary problem, which differs from the one considered in § 3.2 by the presence of a beam divergence. Then we shall introduce into the problem the concept of the wind varying along the sounding path, and then consider the process of clearing the 'frozen' cloud using a diverging beam.

The equation for intensity I is

$$\operatorname{div}(\vec{n}I) + \alpha_N(J)I = 0. \tag{3.4.1}$$

For a light beam of angular width $\varphi$ one can obtain from (3.4.1) and (3.2.2), using the small angle approximation, the following equation for the energetic variable:

$$\partial J/\partial z + \vec{n}_\perp \nabla_\perp J + [1/(z + 2\sigma/\varphi)]J + \int_0^J \alpha_N(J)\, dJ = 0, \tag{3.4.2}$$

where $\sigma$ is the radius of the beam 'spot' incident on the cloud, and the vector $\vec{n}_\perp$ has the following components:

$$n_x = \frac{x}{z + 2\sigma/\varphi}, \qquad n_y = \frac{y}{z + 2\sigma/\varphi}.$$

In the case of the exponential formula for $\alpha_N(J)$, if $n_x$ and $n_y$ are taken as the new coordinates, instead of x and y, then (3.4.2) can be reduced to the form

$$\begin{aligned} &d\bar{J}/d\tau + [1/(\tau + \tau_\varphi)]\bar{J} + 1 - e^{-\bar{J}} = 0; \\ &\bar{J} = k_e J/J_h; \qquad \tau = \alpha_0 z; \qquad \tau_\varphi = 2\sigma\alpha_0/\varphi. \end{aligned} \tag{3.4.3}$$

The equation for intensity, written in the coordinate system z, $n_x$, $n_y$, also does not contain partial derivatives

$$d\,\ln I/d\tau + 2/(\tau + \tau_\varphi) + e^{-\bar{J}} = 0. \tag{3.4.4}$$

In order to arrive at a solution, let us consider some particular cases:

(1) Collimated beam $(\tau_\varphi \to \infty)$:

$$\bar{J} = \ln\,[1 + e^{-\tau}(e^{\bar{J}_0} - 1)]. \tag{3.4.5}$$

(2) No thermal effects are observed $(\bar{J} \ll 1)$:

$$\bar{J} = \frac{\bar{J}_0}{a(\tau/\tau_\varphi)}\, e^{-\tau}, \tag{3.4.6}$$

where $a = 1 + \tau/\tau_\varphi = 1 + \varphi z/2\sigma$ is the dimensionless beam width.

(3) Completely cleared zone $(e^{-\bar{J}} \ll 1)$:

$$\bar{J} = \frac{1}{a(\tau/\tau_\varphi)}\left[\bar{J}_0 - \left(\tau + \frac{\tau^2}{2\tau_\varphi}\right)\right]. \tag{3.4.7}$$

As can be shown from the analysis of (3.4.5)-(3.4.7), the generalized approximation which provides for correct asymptotics should have the following form

$$\bar{J} = \ln\left[1 + e^{-\tau}\left(\exp\left(\frac{\bar{J}_0}{a(\tau/\tau_\varphi)}\right) - 1\right)\right]. \tag{3.4.8}$$

The fault inherent in this solution is in the fact that one cannot distinguish, using it, between cases involving a diverging beam and cases involving a collimated one when both have the same power and when the diameter of the collimated beam is equal to the diameter of the diverging beam at the receiver plane. But, nevertheless, (3.4.8) can at least be considered as quite an acceptable initial approach to a better approximation.

Let us now use the Picard algorithm in order to obtain such an approximation. To avoid the incorrect, in the general case, operation of differentiating J with respect to $\bar{J}_0$ (this is due to the fact that the expression for calculating the intensity is as follows: $I = (\partial J/\partial \bar{J}_0)(\partial \bar{J}_0/\partial x)V_\perp$) we shall substitute (3.4.8) not into (3.4.3), but into the equivalent equation (3.4.4). The solution of this equation in the initial coordinate system x, y, z, taking into account the boundary condition $I(x, y, o) = I_0(x, y)$, can be written as follows

$$I(x, y, z) = I_f \exp(-\tau_N), \tag{3.4.9}$$

where the factor $I_f$ describes the propagation of a beam in clear air:

$$I_f = \frac{1}{a^2(z)} I_0\left(\frac{x}{a(z)}, \frac{y}{a(z)}\right); \tag{3.4.10}$$

$\tau_N$ has an obvious meaning: the nonlinear optical depth of the layer

$$\tau_N = \int_0^\tau \frac{d\tau'}{1 + \exp(-\tau')\left[\exp\left(\frac{\bar{J}_0}{a(\tau')}\right) - 1\right]}. \tag{3.4.11}$$

The boundary value of the energetic variable can be calculated using to the following formula:

$$\bar{J}_0 = \frac{k_e}{J_h V_\perp}\int_{-\infty}^{x/a(z)} I_0(x', y/a(z))\, dx'. \tag{3.4.12}$$

This solution generalizes Glickler's formula for the case of stationary 'clearing' regimes caused by diverging beams. For $a = 1$, this expression gives the Glickler formula itself, which can be better written in the form (3.4.9):

$$I(x, y, z) = I_0(x, y) \exp\left(-\ln\frac{e^{\bar{J}_0} + e^\tau - 1}{e^{\bar{J}_0}}\right). \tag{3.4.13}$$

Normally the meaning of a dimensionless beam width a(z) is defined by the following expression:

$$a = 1 + \frac{\varphi}{2\sigma} z = 1 + \frac{\tau}{\tau_\varphi}.$$

However, if the meaning of a(z) in (3.4.10)-(3.4.12) is more general, then these equations can be used for the description of the clearing process caused by any type of beam (both coherent and partially coherent), provided that the geometry can be characterized by the width parameter. Thus, for a Gaussian beam (3.2.3) whose propagation is characterized by a diffraction length $R_d = \pi\sigma^2/\lambda$ and a focal length F, one obtains

$$a = [(1 - z/F)^2 + (z/R_d)^2]^{1/2} = [(1 - \tau/\tau_F)^2 + (\tau/\tau_d)^2]^{1/2}. \tag{3.4.14}$$

Note that the boundary function $\bar{J}_0$, calculated using (3.4.12) for a Gaussian beam, is

$$\bar{J}_0 = \frac{k_e}{J_h} \frac{P}{\sqrt{2\pi}} \frac{1}{\sigma V_\perp} \exp(-2y^2/\sigma^2 a^2(z)) \left[1 + \text{erf}\left(\frac{\sqrt{2}}{\sigma a(z)} x\right)\right], \tag{3.4.15}$$

while the intensity of the beam propagating in clear air is

$$I_f = \frac{2}{\pi} \frac{P}{\sigma^2 a^2(z)} \exp(-2(x^2 + y^2)/\sigma^2 a^2(z)). \tag{3.4.16}$$

In order to illustrate the above, consider the clearing of fog by a 10 kW $CO_2$ laser along a 100 m long path. The beam is of a Gaussian form, with the parameter $\sigma = 2$ cm. Assuming a wind speed of $V_\perp = 1$ m/sec, we shall vary the optical depth of the path (i.e., the fog density, since the path length is fixed). The three situations to be considered are: (1) a beam with diffraction angular divergence, (2) the same beam, but focused on the end of the path, and (3) a beam with an angular beam width of $\varphi = 10^{-3}$ rad, the other parameters being $k_e = 1$ and $J_h = 8$ J/cm$^2$. Under these conditions the value of the energetic variable at the beam axis, according to (3.4.15), is $\bar{J}_0^{\circledast} = 2.4934$. Figure 3.4.1 presents the results of the calculation of the optical depth changes $\Delta\tau = \tau - \tau_N$ along the beam axis, obtained by numerically integrating (3.4.11).

Expression (3.4.11) for the nonlinear optical depth can easily be generalized to apply to the case where the transverse component of the wind speed $V_\perp$ is dependent on z. The corresponding generalized form of (3.4.3) for the energetic variable will take the form

$$d\bar{J}/d\tau + (1/(\tau + \tau_\varphi))\bar{J} + \frac{d \ln V_\perp}{d\tau} \bar{J} + 1 - e^{-\bar{J}} = 0. \tag{3.4.17}$$

The expression for the nonlinear optical depth of a cloud layer, taking into account the variations of the wind speed along the path, is as follows:

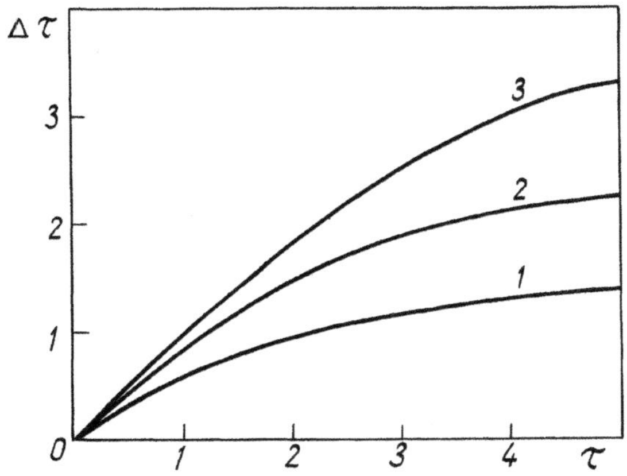


Fig. 3.4.1. Dependence of changes of optical depth along the axis of a Gaussian beam on the fog density. Curve 1 represents data for a beam with an angular width of $\varphi = 10^{-3}$ rad., curve 2 represents data for a collimated beam, and curve 3 for a focused beam; $z = 100$ m, $\sigma = 2$ cm, $V_\perp = 1$ m/sec, $P = 10^4$ W, $k_e = 1$, $J_h = 8$ J/cm$^2$, $F = 100$ m, $\lambda = 10.6$ μm.

$$\tau_N = \int_0^\tau \frac{d\tau'}{1 + e^{-\tau'}\left[\exp\left(\dfrac{\bar{J}_0 V_\perp(0)}{a(\tau')V_\perp(\tau')}\right) - 1\right]}. \tag{3.4.18}$$

As to the intensity, it is described earlier by the general formula (3.4.9). The expression for the boundary function $\bar{J}_0$ should obviously involve the wind speed $V_\perp$ recorded where the beam enters the cloud, i.e., at $z = 0$.

The case of a 'frozen' cloud will also require modifications only in the expressions for $\tau_N$ and the boundary function $\bar{J}_0$. The equation analogous to (3.4.3) in this case has the form

$$d\bar{J}/d\tau + [2/(\tau + \tau_\varphi)]\bar{J} + e^{-\bar{J}} = 0. \tag{3.4.19}$$

The solution procedure gives rise to the following expression:

$$\tau_N = \int_0^\tau \frac{d\tau'}{1 + e^{-\tau'}[\exp(\bar{J}_0/a^2(\tau')) - 1]}, \tag{3.4.20}$$

where

$$\bar{J}_0 = k_e/J_h \int_0^t I_0(x/a(z),\ y/a(z),\ t')\ dt'. \tag{3.4.21}$$

The expressions given in this section for the description of the beam intensity during the process of cloud clearing contain all the information from the models. Probably when $\bar{J}_0 > \bar{J}_c$ only limited information concerning the cleared zone configuration is enough. The value $\bar{J}_c$, here, is (as before) the level of energy at which complete clearing takes place. In the case of $\bar{\alpha}_N(\bar{J}) = e^{-\bar{J}}$, it is expedient to take $\bar{J}_c = 3$. The final expressions for the configuration of the cleared zone in some cases do not involve quadratures making them advantageous as compared with the expressions for nonlinear optical depth.

As shown in § 3.', corresponding numerical investigations can be carried out based on linearized equations for the energetic variable. Let us first consider the linearized variant of (3.4.17):

$$d\bar{J}/d\tau + \left(1/(\tau + \tau_\varphi) + \frac{d\,\ell n\, V_\perp}{d\tau}\right)\bar{J} + 1 = 0. \tag{3.4.22}$$

The solution of this equation taken at the point $\bar{J} = \bar{J}_c$ determines the implicit function $\tau(x, y)$, which describes the front of the completely cleared zone:

$$\frac{V_\perp(0)}{a(\tau)V_\perp(\tau)}\left[\bar{J}_0(x/a(\tau), y/a(\tau)) - \int_0^\tau a(\tau)V_\perp(\tau)/V_\perp(0)\,d\tau\right] - \bar{J}_c = 0, \tag{3.4.23}$$

where $\bar{J}_0(x/a(\tau), y/a(\tau))$ is given by (3.4.12) and the dimensionless beam width is taken in the form of $a(\tau) = 1 + \tau/\tau_\varphi$ for the beams with photometric divergence, or in the form (3.4.14) for the single mode beams.

In the case of a 'frozen' cloud, the surface of the front of the cleared zone is described by the expression obtained from the solution of the linearized equation (3.4.19):

$$1/a^2(\tau)\left[\bar{J}_0(x/a(\tau), y/a(\tau), t) - \int_0^\tau a^2(\tau)\,d\tau\right] - \bar{J}_c = 0, \tag{3.4.24}$$

where $\bar{J}_0(x/a(\tau), y/a(\tau), t)$ is determined by (3.4.21). Consider, for example, the configuration of a completely cleared zone when the beam incident on a 'frozen' cloud has a photometric divergence and uniform power distribution over its cross-section. The function $\bar{J}_0$, in this case, is independent of the transverse coordinates and $\tau$, i.e., $\bar{J}_0 = (k_e/J_h)(Pt/\pi R^2)$. Since $\bar{J}_0$ is a linear function of t, it is useful to write the time taken for clearing as a function of the optical depth of the cleared zone:

$$t = \frac{\bar{J}_c(1 + \tau/\tau_\varphi)^2 + \tau_\varphi/3(1 + \tau/\tau_\varphi)^3 - \tau_\varphi/3}{(k_e/J_h)(P/\pi R^2)}. \tag{3.4.25}$$

In more complicated cases it seems to be advisable to use the numerical integration of the expressions for nonlinear optical depth $\tau_N$, in order to avoid the use of implicit functions.

## 3.5. THE GENERALIZED FORMULA DESCRIBING THE BEAM INTENSITY IN THE PROCESS OF BEAM-INDUCED CLEARING

It can be demonstrated that the expressions in § 3.4 for the beam intensity are corollaries of the general formula

$$I(x,\ y,\ \tau,\ t) = I_f \exp\left[-\int_0^{\tau} \frac{d\tau'}{1 + \exp(-\tau')[\exp(\bar{J}_f(\tau')) - 1]}\right]. \tag{3.5.1}$$

As seen from (3.5.1), the beam intensity in the cleared channel is a function of the beam intensity propagating in clear air. In order to make use of this formula, one should carry out only a relatively simple operation for determining the intensity and the energetic variable during propagation in clear air. If the selective absorption of light by gases is to be taken into account, then one must do this when formulating the above functions relating to beam propagation in clear air. In the case of the nonaberrational propagation of a Gaussian beam, these functions are

$$I_f = \frac{2P}{\pi\sigma^2 a^2(\tau)} \exp\left(-2\,\frac{x^2 + y^2}{\sigma^2 a^2(\tau)} - \frac{\alpha_g}{\alpha_0}\,\tau\right); \tag{3.5.2}$$

$$\bar{J}_f = \frac{k_e}{J_h}\,\frac{P}{\sqrt{2\pi}}\,\frac{1}{\sigma a(\tau) V_\perp(\tau)} \exp\left(-2\,\frac{y^2}{\sigma^2 a^2(\tau)} - \frac{\alpha_g}{\alpha_0}\right) \times$$

$$\times \left\{\mathrm{erf}\left[\frac{\sqrt{2}}{\sigma a(\tau)}\,x\right] + \mathrm{erf}\left[\frac{\sqrt{2}}{\sigma a(\tau)}\,(V_\perp(\tau)t - x)\right]\right\}; \tag{3.5.3}$$

$$a(\tau) = [(1 - \tau/\tau_f)^2 + (\tau/\tau_d)^2]^{1/2},$$

where $\alpha_g$ is the absorption coefficient of the atmospheric gases. Expressions (3.5.1)-(3.5.3) allow one to take into account not only the absorption by atmospheric gases, but also some other factors, for example, variations in wind speed along the beam path, diffraction blurring of the beam, beam focusing, and the nonstationary character of the process. This enables one to make calculations of the parameters of the cleared channel for sufficiently complicated geometries of various types of beam and in a wide range of meteorological conditions. The only situations that are probably outside the limits of applicability of the procedure suggested above are those for which it is impossible to write down the differential equation of the first order for the energetic variable.

## 3.6. THE CLEARED CHANNEL UNDER CONDITIONS OF TURBULENT AEROSOL TRANSPORT

Turbulent air flow is the condition under which the energetic variable does not satisfy the differential equation of the first order, since under these conditions not only the wind speed, but also the direction of the wind, varies along the beam path [11]. Strictly speaking, (3.5.1) is inapplicable under these conditions. However, if the characteristic scale of the wind speed pulsations $\ell_T$ obeys the condition $\ell_T \ll \alpha_0^{-1}$ then, in fact, averaging of the energetic variable over $\tau$ occurs and the problem then requires no account to be taken for rotation in the plane x o y. Thus, by averaging $\bar{J}_f$ over the wind speed pulsations, it is possible to re-establish the applicability of (3.5.1).

Before describing this procedure, we will provide some information on the role of different turbulence scales in the process of aerosol diffusion. The theory of turbulent diffusion uses the following parameters [12]:

$$R_L(z) = \frac{<V(t)V(t+z)>}{<V^2>}; \tag{3.6.1}$$

this is the Lagrangian correlation coefficient, and the Lagrangian turbulence scale

$$T_L = \int_0^\infty R_L(z)\, dz. \tag{3.6.2}$$

Here, V(t) is the pulsation of one of the wind components. The mean square of an aerosol particle's displacement is described by Taylor's formula

$$<x^2(t)> = 2<V^2> \int_0^t dt' \int_0^{t'} R_L(z)\, dz, \tag{3.6.3}$$

from which follows the expression for the coefficient of turbulent diffusion of a particle:

$$K_x = \frac{1}{2}\frac{d}{dt}<x^2(t)> = <V^2> \int_0^t R_L(z)\, dz. \tag{3.6.4}$$

In general, this coefficient is also a function of time, as is the function $<x^2(t)>$. It follows from these formulas that, when $t \ll T_L$,

$$<x^2(t)> \simeq <V^2>t^2; \tag{3.6.5}$$

$$K_x \simeq <V^2>t, \tag{3.6.6}$$

and, when $t \gg T_L$,

$$<x^2(t)> \simeq 2<V^2>T_Lt; \tag{3.6.7}$$

$$K_x \simeq <V^2>T_L. \tag{3.6.8}$$

As seen from these expressions, the coefficient of turbulent diffusion for $t \gg T_L$ is a constant value equal to the coefficient entering the equation of molecular diffusion. Thus, in this case, turbulent diffusion is described in the same manner as molecular diffusion. In terms of spatial scales, the condition $t \gg T_L$ means, in fact, that the turbulence is small-scale turbulence. Applied to a radius R, it can be rewritten as

$$R^2 \gg 2\langle V^2\rangle T_L^2. \tag{3.6.9}$$

Since in real atmospheric conditions the value of $T_L$ is not lower than ~10 sec, the above condition is not met with the beams and their cross-sections used in practice. At the same time (3.6.5) and (3.6.6) are quite useful, since they present the description of the model of stochastic wind.

The energetic variable of a beam propagating in clear air, under conditions in which the wind vector $\vec{V}_\perp$ rotates in a manner dependent on $\tau$, is determined as follows:

$$\begin{aligned} J_f(x, y, \tau) &= \frac{1}{V_\perp(\tau)} \int_{-\infty}^{x'(x,y)} I_f(x(x', y'), y(x', y'), \tau)\, dx'; \\ x'(x, y) &= \frac{V_x}{V_\perp} x + \frac{V_y}{V_\perp} y; \qquad y'(x, y) = -\frac{V_y}{V_\perp} x + \frac{V_x}{V_\perp} y. \end{aligned} \tag{3.6.10}$$

With axially symmetric beams, the required functions $J_f$ are obtained from the ordinary functions like (3.5.3) simply by making the following substitutions:

$$x \to \frac{V_x}{V_\perp} x + \frac{V_y}{V_\perp} y; \qquad y \to -\frac{V_y}{V_\perp} x + \frac{V_x}{V_\perp} y.$$

Along the beam axis, $J_f^{\circledast} \simeq 1/V_\perp(\tau)$ - this makes the averaging procedure much simpler.

Let us consider, for example, the following wind model:

$$V_x = \bar{V}_\perp + \tilde{V}_\perp \cos \omega\tau; \qquad V_y = \tilde{V}_\perp \sin \omega\tau, \tag{3.6.11}$$

where $\omega \gg 1$, according to the condition $\ell_T \ll \alpha_0^{-1}$. Then average $J_f^{\circledast}$ over the period of the pulsating component of rotation of the wind. Note that the averaging could be done with the use of some two-dimensional distribution function for the wind field $f(V_x, V_y)$ but this also requires the use of some wind model. The averaging performed using model (3.6.11) gives rise to

$$\begin{aligned} \langle J_f^{\circledast}\rangle &\simeq \langle 1/V\rangle = 1/\pi \int_0^{\pi} \frac{dx}{\sqrt{\bar{V}_\perp^2 + \tilde{V}_\perp^2 + 2\bar{V}_\perp\tilde{V}_\perp \cos x}} = \\ &= \frac{1}{\pi} \frac{2}{\bar{V}_\perp + \tilde{V}_\perp} K\left(2 \frac{\sqrt{\bar{V}\,\tilde{V}}}{(\bar{V}_\perp + \tilde{V}_\perp)^2}\right), \end{aligned} \tag{3.6.12}$$

where $K(k)$ is a full elliptical integral of the first kind. If $\bar{V}_\perp = 0$, then $\langle J_f^{\circledast}\rangle \sim 1/\tilde{V}_\perp$. It is important that, in this case, the averaged energetic variable $\langle J_f\rangle$ can be calculated not only for the beam axis. It can be shown that, for a beam with a uniform intensity distribution over its cross-section of radius R, one can obtain the following expression:

$$\langle \bar{J}_f(\rho) \rangle = \frac{k_e}{J_h} \frac{P}{\pi R^2} \frac{2E(\rho/R)}{\tilde{V}_\perp}, \tag{3.6.13}$$

where $E(\rho/R)$ is the full elliptical integral of the second kind, and $\rho$ is the distance from the beam axis. In the presence of turbulence, the effect of the clearing process at the beam's periphery can be compared with that at the axis of the laser beam. It can be shown from (3.6.13) that $\langle \bar{J}_f(0) \rangle / \langle \bar{J}_f(R) \rangle = \pi/2$. If the conditions for clearing are uniform relative to $\tau$ then, using (3.5.1) and taking into account (3.6.13) or (3.6.12), one obtains the 'turbulence variant' of Glickler's formula:

$$T = \frac{\exp\langle \bar{J}f \rangle}{\exp\langle \bar{J}_f \rangle + \exp(\tau) - 1}. \tag{3.6.14}$$

Figure 3.6.1 illustrates the case when $\langle \bar{J}_f \rangle$ is determined by (3.6.13). Note, finally, that one should consider the value $\tilde{V}_\perp/\sqrt{2}$ as the rms fluctuation of the wind speed when applying the above equations to the description of a realistic situation.

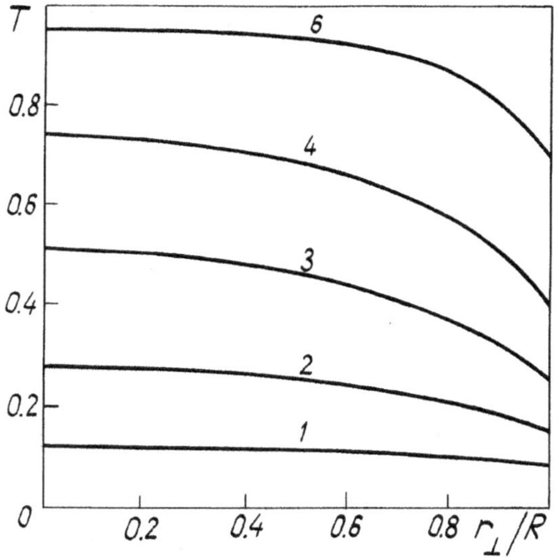


Fig. 3.6.1. Radial profile of cleared channel transmission under conditions of turbulent transportation of the fog. This is the case of a beam with a uniform intensity distribution. The numbers on the curves are the values of the parameter $\langle \bar{J}_f^{\circledast} \rangle = (k_e/J_h) \times (P/\pi R V_\perp)$, $\tau = 3$.

## 3.7. NONLINEAR EXTINCTION COEFFICIENT OF AEROSOLS

Changes of properties of the medium during the process of beam-induced clearing of the medium are described using the nonlinear extinction coefficient of aerosols $\alpha_N$ which, in turn, is the function of energetic variable J. The general relationships determining the principal properties of the extinction coefficient have been given in § 3.1. This section presents a detailed description of the corresponding calculation procedures.

The above-mentioned procedures are described by (3.1.7), (3.1.6), and (3.1.2). It should also be mentioned here that (3.1.7), being the integral of (3.1.5), describes the kinetics of the vaporization of a droplet with initial radius $r_0$. In order to write the explicit form, one should also use the explicit expression for the absorption efficiency factor $K_a(r)$. Works treating the beam-induced clearing process very often use the following approximation formula for this factor (see Shifrin [13]):

$$K_a(r) = (1 - \exp(-2k_a r)) \exp(-0.2\sqrt{n^2 + \kappa^2}), \tag{3.7.1}$$

where $k_a$ is the absorption coefficient of a unit volume of liquid water, and n and $\kappa$ are the components of the complex refractive index.

So, the integration of (3.1.7) results in

$$r = \frac{1}{2k_a} \ln[1 + \exp(-pJ)(\exp(2k_a r_0) - 1)];$$
$$p = k_e \frac{k_a}{2\rho_L Q_e} \exp(-0.2\sqrt{n^2 + \kappa^2}). \tag{3.7.2}$$

Normally, the droplet size spectrum for undisturbed clouds and fogs can be approximated by the following function:

$$f_0(r) = \frac{\mu_0^{\mu_0+1}}{\Gamma(\mu_0 + 1)} \frac{1}{a_0} \left(\frac{r}{a_0}\right)^{\mu_0} \exp(-\mu_0(r/a_0)), \tag{3.7.3}$$

where $a_0$ and $\mu_0$ are parameters of the distribution function ($a_0$ coincides with the modal radius).

Carrying out the inversion of (3.7.2), one obtains

$$r_0 = \frac{1}{2k_a} \ln[1 + \exp(pJ)(\exp(2k_a r) - 1)]. \tag{3.7.4}$$

Therefore, according to (3.1.6), the spectrum dependent on the energetic variable can be written as follows:

$$f(r, J) = \frac{\mu_0^{\mu_0+1}}{\Gamma(\mu_0 + 1)} \frac{1}{a_0} \left(\frac{r_0(r)}{a_0}\right)^{\mu_0} \exp\left[-\mu_0 \left(\frac{r_0(r)}{a_0}\right) + 2k_a r + pJ\right] \times$$

$$\times \frac{1}{1 + \exp(pJ)[\exp(2k_a r) - 1]}. \tag{3.7.5}$$

The dependence of the aerosol extinction coefficient on the energetic variable is determined by averaging the extinction cross-section $K(r)\pi r^2$ over the spectrum $f(r, J)$, see (3.1.2). It can be shown that the description uses the averaging of the extinction cross-section $K(r(r_0, J))\pi r^2(r_0, J)$ over the undisturbed spectrum $f_0(r_0)$. In other words, this means that one can use the following formula instead of (3.1.2):

$$\alpha_N(J) = \pi N_0 \int_0^\infty r^2(r_0, J)K(r(r_0, J))f_0(r_0)\,dr_0, \tag{3.7.6}$$

where $r(r_0, J)$ is a function of the type (3.7.2), or analogous to it. Thus, one can state that the problem of determining the rearranging size-spectrum is avoided, in this case, while the problem of averaging remains for all cases.

Some further simplifications can be made when considering the water content, these were used in the first works on beam-induced clearing [6, 7, 14]. The main results of this approach to the problem can be summarized as follows.

The water content approach works when the condition $2k_a r \ll 1$ is fulfilled, as in the case of droplets with radii $\lesssim 5$ μm and radiation with a wavelength of 10.6 μm. Under these conditions extinction is mainly caused by absorption and $K_a(r) = K(r) = \frac{4}{3}(k_a r)$. The relationship between the radius of a particle and the energetic variable in the water content approach is as follows:

$$r = r_0 \exp\left(-k_e \frac{k_a}{3\rho_L Q_e} J\right). \tag{3.7.7}$$

The water content of the cloud depends on $J$ exponentially, regardless of the size spectrum, and the required dependence of the extinction coefficient on $J$ is

$$\alpha_N(J) = \frac{k_a q}{\alpha_L} \exp\left(-k_e \frac{k_a}{\rho_L Q_e} J\right). \tag{3.7.8}$$

The size spectrum (3.7.3) is assumed to be rearranged so that only the parameter $a_0$ changes according to (3.7.7), while the shape of the spectrum does not change at all.

The last point shows the usefulness of the approximate description of the size spectrum $f(r, J)$ suggested in [11] in which, besides $J$, another parameter $\mu(J)$ is introduced in order to preserve the initial shape of the size spectrum in the case where deviations from the water content regime take place.

This approach permits the use of the results of calculations of polydispersed aerosol extinction coefficients [15] only if the parameters of corresponding formulas are written as functions of the energetic variable. Thus, for so-called soft particles (e.g., water droplets and radiation with $\lambda = 10.6$ μm), one has

$$\alpha_N(J) = \pi N_0 r_\sigma^2 \left(2 - \frac{4}{\mu + 2} \frac{\sin[(\mu + 2)\varphi - \beta][\cos(\varphi + \beta)]^{\mu+3}}{\sin\varphi[\cos\beta]^{\mu+2}} - \right.$$
$$- \frac{4}{(\mu + 2)(\mu + 1)} \frac{\cos[(\mu + 1)\varphi - 2\beta][\cos(\varphi - \beta)]^{\mu+3}}{\sin^2\varphi[\cos\beta]^{\mu+1}} + \quad (3.7.9)$$
$$\left. + \frac{4\cos 2\beta \cos^2(\varphi + \beta)}{(\mu + 2)(\mu + 1)\sin^2\varphi}\right);$$

$$\mu = \frac{a}{\langle r\rangle - a}; \qquad \operatorname{tg}\varphi = \frac{u}{1 + u\operatorname{tg}\beta}; \qquad \operatorname{tg}\beta = \frac{\kappa}{n - 1};$$

$$u = x'/\mu; \qquad x' = 2x(n - 1); \qquad x = 2\pi a/\lambda;$$

$$\left\{\begin{matrix} a \\ \langle r\rangle \\ r_\sigma \end{matrix}\right\} = (1/2k_a)\,\ell n\left[1 + \exp(-pJ)\quad \exp\left(\left\{\begin{matrix} a_0 \\ \langle r\rangle_0 \\ r_{\sigma 0} \end{matrix}\right\} 2k_a\right) - 1\right]; \qquad (3.7.10)$$

$$\langle r\rangle_0 = a_0/\mu_0 + a_0; \qquad r_{\sigma 0} = \sqrt{(\mu_0 + 2)(\mu_0 + 1)a_0^2/\mu_0^2}.$$

As one can see here, we have introduced the third parameter, i.e., rms

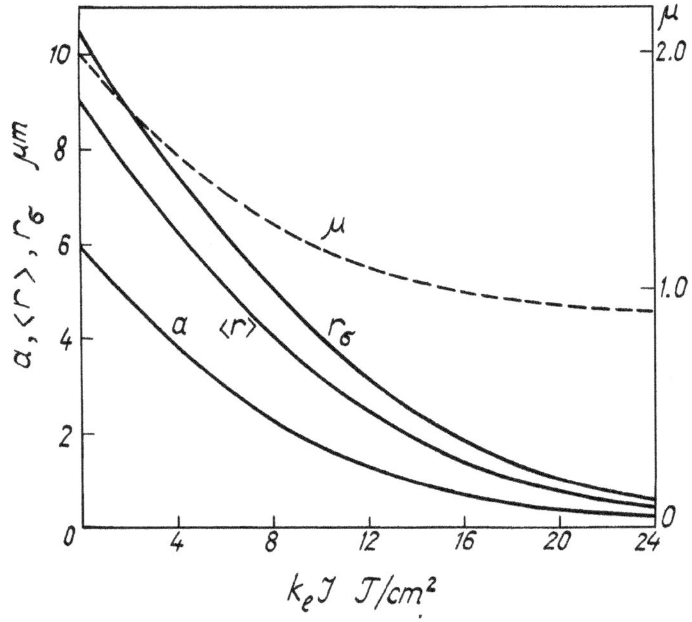


Fig. 3.7.1. Fog droplet microstructure parameters as functions of the energetic variable in the spectral region close to 10.6 μm; m = 1.173 - i 0.083.

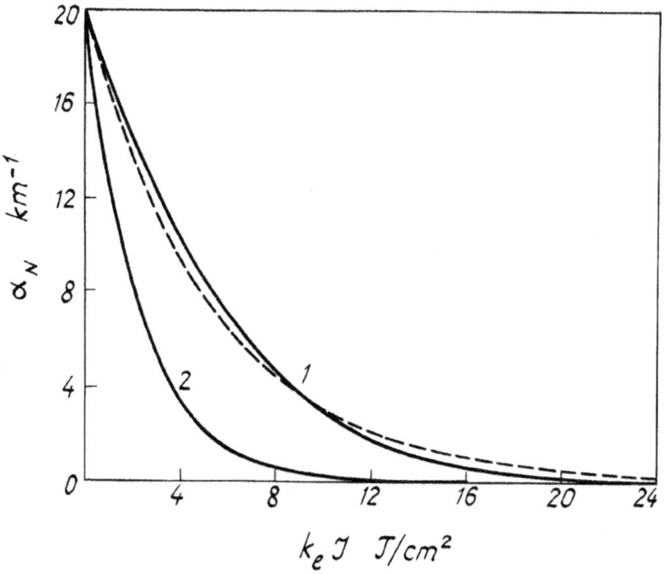


Fig. 3.7.2. Dependence of the nonlinear extinction coefficient of aerosols on the energetic variable in the spectral region close to 10.6 μm. This is the case of a standard fog with $N_0 = 28.8$ cm$^{-3}$, $a_0 = 6$ μm, $\mu_0 = 2$. Curve 1 represents calculations made using (3.7.9) and (3.7.10); the dashed curve represents the exponential approximation at $J_h = 5.4$ J/cm$^2$. Curve 2 represents calculated data following the water content approach.

radius r , which is also assumed to be dependent on J. Figure 3.7.1 presents parameters, calculated as above, of the size spectrum as a function of J for the wavelength 10.6 μm ($n = 1.173$; $\kappa = 0.083$; $k_a = 984$ cm$^{-1}$). Figure 3.7.2 illustrates the dependence of the nonlinear aerosol extinction coefficient on the energetic variable in the case of the most probable parameters of cloud and fog size spectra at the initial number density $N_0 = 28.8$ cm$^{-3}$, which corresponds to the meteorological visual range $S_M = 0.2$ km. In the same figure we present the analogous dependence, calculated using (3.7.8) and following the water content approach.

In experimental studies of the clearing of water aerosols the optical characteristics of the beam channel are monitored, almost exclusively, with the aid of sounding beams in the visual range. Therefore, for a correct interpretation of such experimental data, one should have information concerning the effects of the extinction coefficient in the visible range on the energetic variable of the high-power beam. The necessary relationships can be found by substituting the optical parameters (n, κ, λ) into

(3.7.9) for the visual range, while taking the optical constants entering into (3.7.10) for the case of high-power incident radiation.

The corresponding curve for the sounding beam with $\lambda = 0.63$ μm is shown in Figure 3.7.3. The exponential approximating curve $\alpha_N(J) = \alpha_0 \exp(-k_e J/J_h)$

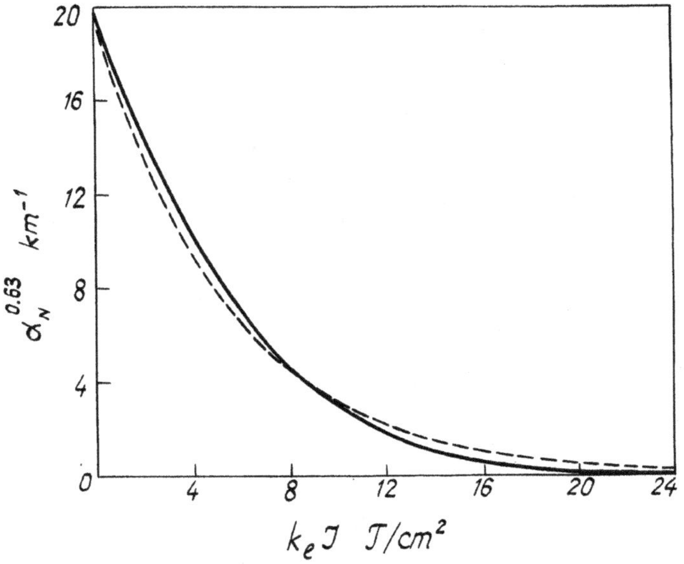


Figure 3.7.3. Dependence of the aerosol extinction coefficient for radiation with $\lambda = 0.63$ μm on the energetic variable. The case of a fog ($N_0 = 28.8$ cm$^{-3}$, $a_0 = 6$ μm, $\mu_0 = 2$) irradiated by a high-power laser beam at $\lambda = 10.6$ μm. The dashed curve represents the exponential approximation at $J_h = 5.4$ J/cm$^2$.

used for describing the cleared channel parameters is also plotted in Figures 3.7.2 and 3.7.3. As seen from these figures, the approximation can be considered to be good enough.

## 3.8. THE INVESTIGATION OF BEAM-INDUCED CLEARING OF NATURAL FOGS

This section presents a discussion of the results obtained when working on the investigation program for studying the influence of the atmosphere on the propagation of high-power c.w. laser beams with $\lambda = 10.6$ μm. The main idea of this experimental program is to make the laser beam parameters as close as possible to the idealized ones, which are used in the theoretical models, trying at the same time to achieve maximum nonlinear effects in the natural atmosphere. Bearing in mind that the influence of the atmosphere is taken into account via generalized parameters, one can state that this approach is designed to obtain highly reliable results using a minimum number of input parameters at a 'large distance' from any interfering factors.

The experimental set-up used for obtaining the field measurements of the nonlinear effects is presented in Figure 3.8.1. The low-pressure $CO_2$

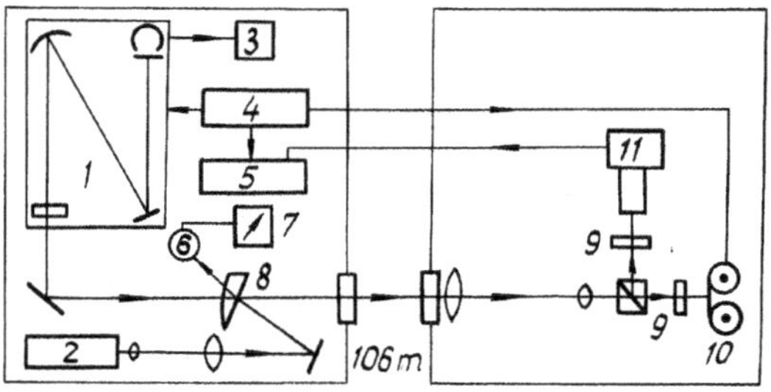


Fig. 3.8.1. Block-diagram of the apparatus used for investigating nonlinear effects under field conditions (1) is the $CO_2$ laser, (2) is the He-Ne laser, (3) is the power meter, (4) is the control unit, (5) is the recording millivoltmeter, (6) is the photodetector, (7) is the microammeter, (8) is the wedge-shaped lense made of KCl, (9) is the neutral attenuating glass, (10) is the camera, and (11) is a PMT.

laser with longitudinal gas mixture circulation ('Photon Sources', model 500) was used in these experiments as the source of high-power radiation. By properly adjusting the resonator made of four reflectors one can arrange for generation in the $TEM_{00}$ mode with a Gaussian amplitude distribution over the beam's cross-section, 9 mm in diameter. The maximum output power of 550 W was achieved in this set-up. With a certain maladjustment of the resonator, the laser can deliver the beam with an angular width of about $5 \times 10^{-3}$ rad and a diameter of 19 mm. The strong 'spike' close to the axis makes an essential difference in this amplitude distribution as compared with a Gaussian one. The output power in this set-up was 410 to 430 W. A wedge-like KCl lens with a focal length of 5 m was used for reducing the beam's divergence and for matching the sounding beam with the high-power one. As a result, a beam was generated whose parameters for clearing the fog at the farthest end of the path were practically the same as in the single-mode regime. Thus, owing to the two beam generation schemes available (for the relevant parameters see Figure 3.8.2), it was possible to study the role of the beam's 'quality'.

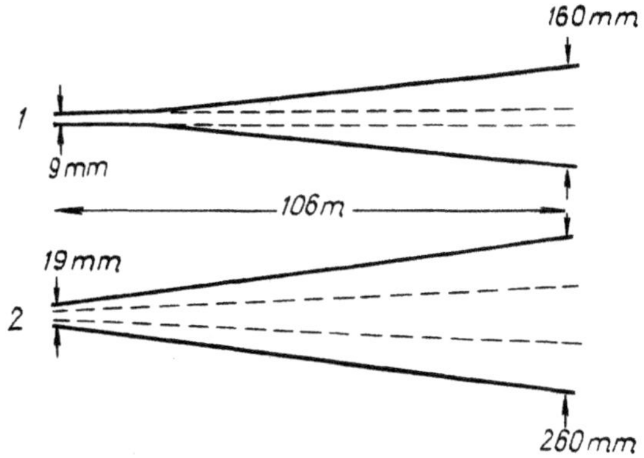


Fig. 3.8.2. Geometrical parameters of the beams. (1) represents the parameters for a single-mode scheme, (2) those for a multimode scheme. Dashed lines denote the beam's boundaries.

The superimposed high-power and sounding beams were directed into the atmosphere through a window made of KCl, which served to avoid the so-called 'pavilion' effect. The influence of thermal self-action in the zone close to the laser source was weakened by the use of artificial wind. At the receiver and along the beam path high-power radiation was rejected by a quartz plate, while the sounding beam (after telescopic transformation in the ratio 1 : 11) was directed into two recording channels. One of the recording channels was equipped with an automatic camera (RPhC-5), while the other was used for photoelectric recording using a PMT. The PMT electric signal, proportional to the beam intensity, was recorded with a millivoltmeter. We used neutral attenuation glass filters to adjust the intensity level in both recording channels. The photographic registration was performed at frequencies of 4 and 8 Hz, with exposure time varying from $0.2 \times 10^{-3}$ sec to $2 \times 10^{-2}$ sec.

The camera and laser operation procedures were monitored with the control unit installed at the transmitter end of the path. The output power of the $CO_2$ laser and the reference signal for the sounding beam were also measured in our experiments. The reference signal was measured with a photocell and a microammeter.

The minimum atmospheric parameters necessary to characterize the process of beam-induced clearing of turbid media are the atmospheric transmission in the visible range (at $\lambda = 0.63$ µm) and the transverse component of the wind speed. The first parameter was inferred from measurements of the sounding beam intensity and the reference beam. The effective value of the second parameter was determined, from an analysis of the nonstationary part of the time behavior of atmospheric transmission, in the form of the ratio of the beam diameter to the time of establishment of the nonlinear transmission of the fog.

The results of measurements obtained using both types of beam are presented in Figures 3.8.8 and 3.8.4. Figure 3.8.3 presents typical examples

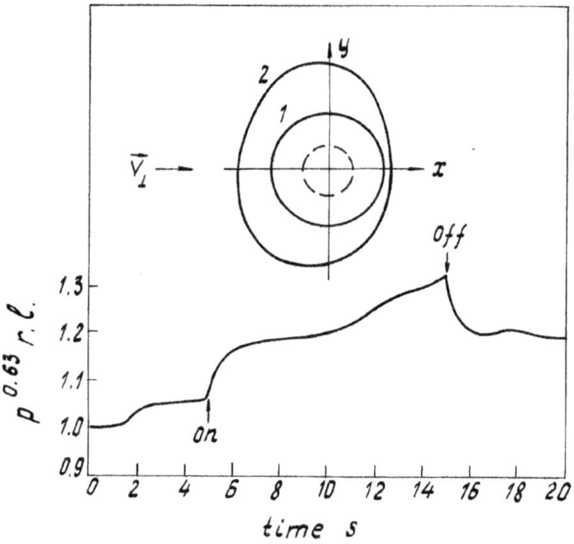


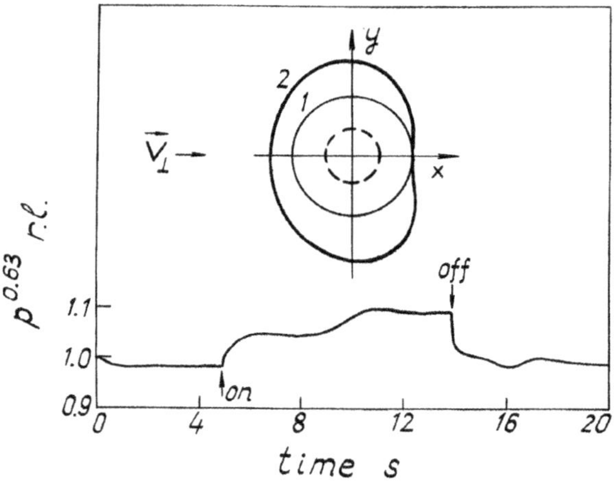


Fig. 3.8.3. Relative power and shape of the sounding pulse at the receiver plane for the case of a natural fog (single-mode scheme). Curve 1 shows the boundary of the beam in undisturbed fog, curve 2 shows the beam's boundary during the stage of steady dissipation of the fog in the high-power beam channel. The dashed line denotes the beam's boundary at the leading edge of the fog formation. (a) $\tau = 1.61$, (b) $\tau = 1.64$, $\bar{J}_0^{\circledast} = 0.5$, $V_\perp \simeq 60$ cm/s.

of the temporal behavior of the sounding beam's intensity and corresponding pictures of the beam's shape (at the level $e^{-2}$) observed with the reference single-mode scheme. In the case of the scheme differing from the reference one (see Fig. 3.8.4), there are no any characteristic peculiarities of the temporal behavior of the beam intensity, while the complicated picture of the intensity redistribution over the beam's cross-section does not allow

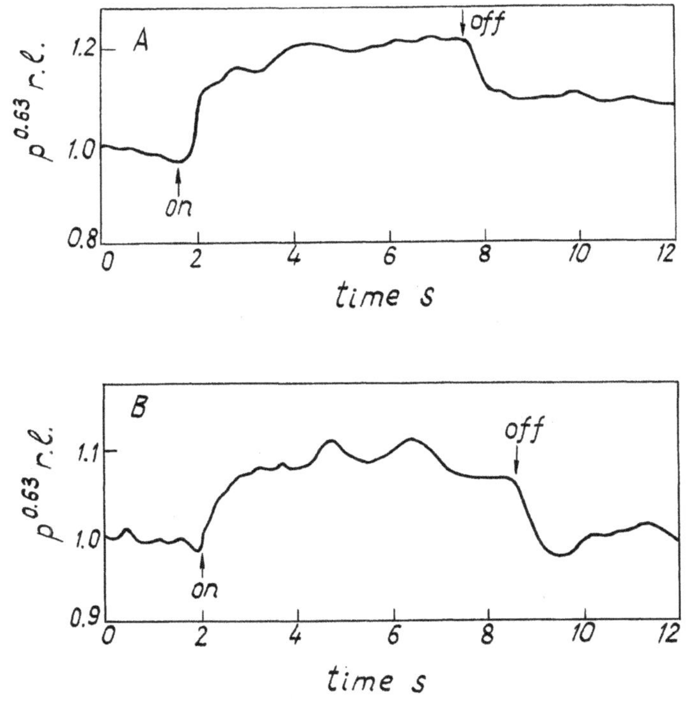


Fig. 3.8.4. Relative intensity of the sounding beam at the receiver plane for the case of a multimode scheme used for clearing the fog. $\bar{J}_0^{\circledast} = 0.5$, $V_\perp \simeq 30$ cm/s; (a) $\tau = 3.78$, (b) $\tau = 3.28$.

one to say anything definite about the beam shape in contrast with the reference scheme. The intensity decrease observed at the partial interception of the beam reliably indicated that the refraction distortions were responsible for the beam's defocusing (in the integral sense) in the non-reference scheme of the experiment. It is just these refraction distortions that are sensitive to the beam's 'quality', while the energetics of transmission are determined quite satisfactorily by such integral characteristics of the beam as full power, angular beam width, and beam diameter.

Now, consider in more detail the comparison of the experimental and calculated results keeping in mind these preliminary conclusions. The field

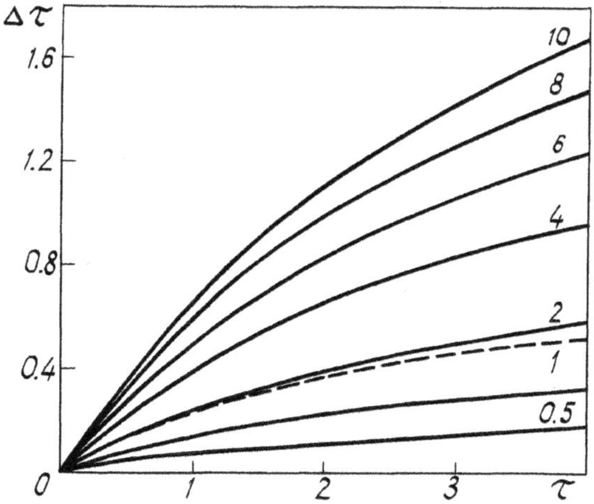


Fig. 3.8.5. Optical depth of the fog as a function of its density along a path of 106 m. The numbers on the curves are the values of the parameter $\bar{J}_0$; $R_d = 6$ m (single-mode scheme) $2\sigma/\varphi = 8.48$ m (multimode scheme). The dashed curve represents data for $2\sigma/\varphi = 25.4$ m.

of expected values of optical depth measured along the beam's axis, calculated using (3.4.11) for various fog densities and various energetic variables of the incident beams, is presented in Figure 3.8.5. As noted above, the curves representing both schemes of beam generation (at one and the same $\bar{J}_0^{\circledast}$) hardly differ from each other (the difference is only 1 or 2 per cent). The increase in fog density is followed by an increase in the transmission ratio $e^{-\tau_N}/e^{-\tau} = e^{\Delta\tau}$, that means that the clearing effect is stronger in the case of denser fogs. This is clearly seen from the data presented in Figures 3.8.3 and 3.8.4. Comparison of curves 2' and 2 in Figure 3.8.5 shows the advantage of using the beam with a diffraction divergence, but with the diameter being the same as in the multimode scheme (i.e., 19 mm). This figure shows that the geometry of a high-power beam in this experimental set up is not optimal. Although the investigation into minimizing the function $\tau_N$ entering into (3.5.1) was not carried out, it seemed intuitively clear that the beam geometry for which the diffraction beam length is close to the path length must be the most advantageous. The thermal self-action of the beam was also studied in these experiments and it was found that the maximum relative value of the beam distortions is reached when the value of the diffraction beam length is related to the path length as follows:

$$R_d = z/\sqrt{3}. \tag{3.8.1}$$

In the case of a Gaussian beam, this relationship is fulfilled if the beam's radius

$$\sigma = (z\lambda/\sqrt{3}\,\pi)^{1/2}. \tag{3.8.2}$$

Such a relationship between the beam parameters and the path length evidently improves the efficiency of the beam-induced clearing of the fog compared with that achievable using the schemes discussed here.

In this case, almost all of the change in the channel's optical depth takes place in the first third of the beam path. This is demonstrated by Figure 3.8.6, which presents the value $\Delta\tau$ as a function of $\tau$ for different positions of the observation point along the beam path. Using this data,

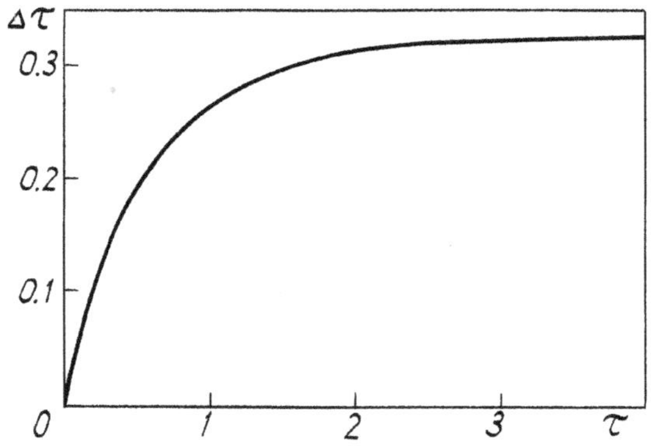


Fig. 3.8.6. Dependence of the changes in the fog's optical depth on the position of the observation point on the propagation path; $\bar{J}_0 = 1$, $\alpha_0 = 3.77 \times 10^{-4}$ cm$^{-1}$.

we determined the effective value of the beam intensity required for making the estimations of the $k_e$ coefficient to be 200 W/cm$^2$. To extract the information on the transverse wind speed we used the analysis of the nonstationary period of the process based on the data collected in the photographic and intensity recordings. Typical values of $V_\perp$ thus obtained ranged from 30 to 100 cm/s. The concrete values of $V_\perp$ and corresponding values of $\bar{J}_0^{\circledast}$ are presented in the captions of Figures 3.8.3 and 3.8.4. The comparison of the calculated and measured values of $\Delta\tau$ (see Fig. 3.8.5) shows a satisfactory agreement between them.

The following information should be noted in connection with the use of an auxiliary sounding beam in the above experiments. Since the calculated value of the beam's intensity along its axis is compared with the total sounding beam intensity measured, the problem arises as to the correctness of such a comparison. The calculations carried out for the beam parameters characteristic of these experiments showed that the above-mentioned substitution results in only a 1.46% relative error in the determination of the transmission of the fog. The calculations of the optical depth of the cloudy layer at the wavelength of the sounding beam should be made according to the following formula:

$$\tau_N^{0.63} = \int_0^L \alpha_0^{0.63}\, dz \Big/ \left[1 + \exp\left(-\int_0^z \alpha_0^{10.6}\, dz'\right)(\exp(\bar{J}_f^{10.6}) - 1)\right] J_h^{10.6}/J_h^{0.63}. \tag{3.8.3}$$

It was assumed, based on the data of § 3.7, that $\alpha_0^{10.6} \approx \alpha_0^{0.63}$, $J_h^{10.6} \approx J_h^{0.63}$ for the situations discussed but, in general, it should be taken into account that (3.8.3) differs from the expressions for $\tau_N$ entering into (3.5.1).

The analysis of the refraction distortions of the beam was carried out only for the single-mode scheme, the beam parameters being the same as in the analysis of intensity.

Figure 3.8.3 presents the results of calculations of the refraction parameters of beams, carried out using the nonaberrational approximation. $a_y$ is the relative beam width along the y-axis, $\Delta x_c$ is the beam drift along the x-axis. The calculations made for the case presented in Figure 3.8.3(a) give $a_y = 3.119$, $\Delta x_c = -0.296$ cm and, for the case presented in Figure 3.8.3(b), $a_y = 3.183$, $\Delta x_c = -0.301$ cm. The comparison of the calculated and the measured parameters shown a good agreement. The broadening of the beam observed along x-axis is caused by aberrations.

## REFERENCES: CHAPTER 3


[1] K. S. Shifrin: 'Optical Investigations of Cloud Particles', in Investigations of Clouds, Precipitation and Thunderstorm Electricity (Gidrometizdat, Moscow, 1957), in Russian.

[2] E. M. Feigelson: Radiation Processes in Stratus (Nauka, Moscow, 1964), in Russian.

[3] O. A. Volkovitsky et al.: High-Power Beam Propagation in Clouds (Gidrometizdat, Leningrad, 1982), in Russian.

[4] V. E. Zuev et al.: Nonlinear Optical Effects in Aerosols (Nauka, Novosibirsk, 1980), in Russian.

[5] V. P. Bisyarin et al.: Radiotekh. 11, 5-148 (1976), in Russian.

[6] S. L. Glickler: Appl. Opt. 10, 644-650 (1971).

[7] G. L. Lamb and R. B. Kinney: J. Appl. Phys. 40, 416-417 (1969).

[8] V. I. Bukaty et al.: Dokl. Akad. Nauk SSSR 217, 52-55 (1974).

[9] R. Kh. Almaev and A. G. Slesarev: Trudy Inst. Eksp. Meteorol. 26 (99), 22-29 (1981).

[10] O. A. Volkovitsky and A. M. Skripkin: ibid., 120-126.

[11] V. E. Zuev and A. V. Kuzikovsky: Izv. Vyssh. Uchebn. Zaved. Fiz. 11, 106-131 (1977).

[12] S. L. Soo: Fluid Dynamics of Multiphase Systems (blaisdell, Massachusetts, 1968).

[13] K. S. Shifrin: Trudy GGO 109, 179-190 (19619, in Russian.
[14] A. P. Sukhorukov et al.: Zh. Tekh. Fiz. Pis'ma Red. 14, 145-150 (1976).
[15] V. E. Zuev: Propagation of Visible and IR Radiation through the Atmosphere (sov. Radio, Moscow, 1970), in Russian.

CHAPTER 4

# SELF-ACTION OF A WAVE BEAM IN A WATER AEROSOL UNDER CONDITIONS OF REGULAR DROPLET VAPORIZATION

The study of the nonlinear propagation of laser beams along a long path through atmosphere contaminated with water aerosol needs to account for various factors affecting the beam's self-action process.

The diffraction distortion of the beam, as well as the fluctuations of the parameters of the medium can, under certain conditions, essentially modify the process of aerosol dissipation by a laser beam (see, e.g., [2] to [7]).

The problem can be investigated most thoroughly and, hence, the dissipation process analyzed in the most detail, using techniques based on the theory of wave propagation in a randomly inhomogeneous media [9, 11, 12].

Based on this approach to solving the problem, the physical approximations providing the description of the beam's behavior, as well as the dynamics of the medium, are discussed below. Let us now discuss the results obtained after solving some problems characteristic of the beam's self-action process in water aerosols when propagating along a long path.

## 4.1. BASIC EQUATIONS OF WAVE BEAM SELF-ACTION IN A DISCRETE SCATTERING MEDIUM

The description of the process of wave beam self-action in a discrete scattering medium, presented below, is based on the methods used in the field of the linear optics of scattering media for studying the energetic parameters of radiation [11, 12].

If the relative perturbation of the medium's dielectric constant (scattering potential) is denoted as $\tilde{\varepsilon} = (\varepsilon - \varepsilon_0)/\varepsilon_0$, then one can write the scalar wave equation for the complex field amplitude as

$$[\Delta + k^2(1 + \tilde{\varepsilon}(\vec{r}, t))]\, E(\vec{r}, t) = 0. \qquad (4.1.1)$$

The boundary condition for this equation is $E(x = 0, \vec{R}, t) = E_0(\vec{R}, t)$.

In problems of aerosol optics where nonlinearity occurs due to thermal perturbations, the parameter $\tilde{\varepsilon}$ entering (4.1.1) can be presented as a superimposition of local perturbations from individual centers. Thus, in the particular problem of the regular vaporization of droplets considered here, the scattering potential of the medium can be represented as

$$\tilde{\varepsilon}(\vec{r}, t) = \sum_{\nu=1}^{N} \varepsilon'_\nu(\vec{r} - \vec{r}_\nu;\ a(a_0;\ [E(\vec{r}_\nu, t)\, E^*(\vec{r}_\nu, t)])), \qquad (4.1.2)$$

where $\varepsilon'_\nu = \varepsilon'_{a\nu} + \varepsilon'_{r\nu}$, $\varepsilon'_{a\nu} = m_a^2 - 1$ is the relative value of the difference between dielectric constants inside the droplet and in the ambient air; $\varepsilon'_{r\nu}$ is the corresponding deviation of the dielectric constant in the region where the perturbations of temperature and water vapor density are observed due to droplet vaporization; and $\vec{r}_\nu$ is the radius vector of the $\nu$-th particle center.

Representation of the scattering potential in the form (4.1.2) shows inhomogeneity and anisotropy in the field of perturbations of the dielectric constants occurring due to the parametric dependence of the perturbation strength on the droplet's redius. In turn, the droplet's radius is a function of the beam intensity at the particle. The other peculiarity of the problem is the random character of the medium. This parameter is a random value because of the random behavior of the values $\vec{r}_\nu$, $a_0$, and N. In the general case, the probability density of $\vec{r}_\nu$ and $a_0$ distributions is also a random function. Thus, the result of averaging over the set of $\vec{r}_\nu$ and $a_0$ values is a random function of external (with respect to scales of local inhomogeneities and the gaps between them) parameters, for example, of the aerosol number density $N_0$, and the initial size-distribution function of the aerosol ensemble b. In addition, $\tilde{\varepsilon}$ can also depend on the transverse component of wind speed $\vec{V}_\perp$, which in turn is a random value.

The main goal of our discussion here is the investigation of the beam self-action, i.e., the mean intensity of the beam propagating through a nonlinear medium is sought. For this purpose one should construct the averaged equations for the energetic parameters of a beam, otherwise the averaging of the corresponding solutions is necessary. Let us formulate the statement based on the information known [13] concerning the scales of the fluctuations of cloud macroparameters (water content, temperature, wind speed, etc.), which is of great importance for further discussion. The scales of local inhomogeneities and the scales of the fluctuations of the macroparameters of the medium are considered to be essentially different. This allows one to carry out the statistical averaging in two steps. First, averaging is carried out over the small-scale fluctuations of $\tilde{\varepsilon}$ caused by the random behavior of $\vec{r}_\nu$, $a_0$, N, and then averaging is carried out over the fluctuations of the macroparameters of the medium.

By so doing one obtains the following equations for the field momenta:

$$\tilde{E} = \langle E(\vec{r}, t)\rangle_{\vec{r}_\nu, a_0, N}; \qquad \tilde{\Gamma}_2 = \langle E(x, \vec{R}_1 t)\, E^*(x, \vec{R}_2, t)\rangle_{\vec{r}_\nu, a_0, N},$$

where the averaging is carried out over the values of random parameters $\vec{r}_\nu$, $a_0$, N. The mean field $\bar{E} = \langle\tilde{E}\rangle$ and the coherence function $\Gamma_2 = \langle\tilde{\Gamma}_2\rangle$ are calculated using $\tilde{E}$ and $\tilde{\Gamma}_2$ by averaging them over the fluctuations of the medium's macroparameters.

Using the diffusion approximation of the random process [12], and assuming $\varepsilon'_\nu$ to be dependent only on the mean taken over the scale of the $\varepsilon'_\nu$ intensity of an incident beam, one can obtain for $\tilde{E}$ and $\tilde{\Gamma}_2$, for the case of a slowly-changing field amplitude $E(x, \vec{R}, t)$ (where $E(x, \vec{R}, t) = E(x, \vec{R}, t)\exp(ikx)$):

$$\partial\tilde{E}/\partial x + 1/2ik\cdot\Delta_{\perp}\tilde{E} - \dot{\Theta}_x\left[\frac{k}{2}\,\delta(\vec{R}' - \vec{R});\ -\frac{k}{2}\,\delta(R' - R)\right]\tilde{E} = 0, \qquad (4.1.3)$$

$$\partial\tilde{\Gamma}_2(x, \vec{R}_1, \vec{R}_2, t)/\partial x + 1/2ik(\Delta_{\perp 1} - \Delta_{\perp 2})\tilde{\Gamma}_2 - \dot{\Theta}_x\left[\frac{k}{2}\,\delta(R' - R_1);\ -\frac{k}{2}\,\delta(\vec{R}' - \vec{R}_2)\right]\tilde{\Gamma}_2 = 0. \qquad (4.1.4)$$

Here, $\delta(\vec{R})$ is the Dirac delta function; $\Delta_{\perp} = \partial^2/\partial y^2 + \partial^2/\partial z^2$; $\dot{\Theta}_x[v_1; v_2] = \partial/\partial x\ \ln \Phi_x^{\tilde{\varepsilon}}[v_1; v_2]$; and $\Phi_x^{\tilde{\varepsilon}}$ is the characteristic function of the field $\tilde{\varepsilon}$:

$$\Phi_x^{\tilde{\varepsilon}}[v_1; v_2] = \left\langle \exp\left\{i\int_0^x dx' \iint_{-\infty}^{\infty} d^2R'[v_1(x', \vec{R}')\,\tilde{\varepsilon}(x', \vec{R}', t) + v_2(x', \vec{R}')\,\tilde{\varepsilon}^*(x', \vec{R}', t)]\right\}\right\rangle_{\vec{r}_\nu, a_0, N}. \qquad (4.1.5)$$

In order to obtain the expression for $\dot{\Theta}_x$ in an explicit form, one should assign certain properties to the randomly inhomogeneous medium beforehand. Let us assume that (1) points $\vec{r}_\nu$ are statistically independent; (2) the probability of finding the center of $\vec{r}_\nu$ on the segment $[\vec{r}_\nu, \vec{r}_\nu + \vec{dr}_\nu]$ depends on $\vec{r}_\nu$ and is determined by the probability density $p_r$ (in the case of a uniform distribution, $p_r = 1/V$, where V is the volume of the particle; (3) the probability of observing N particles in the volume V is described by Poisson's law:

$$P(N) = \bar{N}^N e^{-\bar{N}}/N!,$$

where $\bar{N} = \int_V N_0(\vec{r})\,d\vec{r}$ is the distribution parameter; $N_0$ is the randomly non-homogeneous number density of the aerosol, and $N_0(\vec{r}) = \bar{N}p_r$; (4) the particle radii are described by the initial size-distribution function $f_0(a_0; b(r))$ ($\int_0^\infty da_0 f_0(a_0) = 1$), which depends on the randomly inhomogeneous parameter b; (5) the fluctuations of $N_0(r)$ and $b(r)$ are smooth enough on the scale of a localized inhomogeneity of a particle and its aureole; (6) the inhomogeneity of the dielectric constant field of a macroscale is also smooth on the scale of a localized inhomogeneity.

If conditions (1) to (6) are fulfilled, then one can calculate the function $\dot{\Theta}$ using known Poisson procedures for averaging [10, 12] and obtain the following equations from (4.1.3)-(4.1.5):

$$\frac{\partial E}{\partial x} + \frac{1}{2ik}\Delta_{\perp}\tilde{E} + \frac{k}{2i}\hat{\varepsilon}_e\tilde{E} = 0, \qquad (4.1.6)$$

$$\frac{\partial\tilde{\Gamma}_2(x, \vec{R}, \vec{\rho}, t)}{\partial x} + \frac{1}{ik}\nabla_{\vec{R}}\nabla_{\vec{\rho}}\tilde{\Gamma}_2 + \Bigg[k \cosh\left(\frac{\vec{\rho}\nabla_{\vec{R}}}{2}\right) \times$$

$$\times \operatorname{Im} \hat{\varepsilon}_e(x, \vec{R}, t) - ik \sinh\left(\frac{\vec{\rho}\nabla_{\vec{R}}}{2}\right) \operatorname{Re}(\hat{\varepsilon}_e - 1) - \tag{4.1.7}$$

$$- N_0(x, \vec{R}) \iint_{-\infty}^{\infty} \frac{d\kappa_\perp}{k^2} \exp(i\vec{\kappa}_\perp\vec{\rho}) \left\langle \left| A\left(\frac{\vec{\kappa}_\perp}{k}; a(a_0; \kappa, \vec{R}, t)\right)\right|^2 \right\rangle_{a_0} \Bigg]\tilde{\Gamma}_2 = 0,$$

where $\vec{\rho} = \vec{R}_1 - \vec{R}_2$ is the distance between two points, $\vec{R} = (\vec{R}_1 + \vec{R}_2)/2$ is the coordinate of their center of gravity,

$$\hat{\varepsilon}_e(x, \vec{R}, t) = 1 + (4\pi/k^2)N_0(x, \vec{R}) \int_0^\infty da_0 f_0(a_0, \vec{B}) A(0, a) =$$
$$= \tilde{\varepsilon}_e + i\alpha_a/k \tag{4.1.8}$$

is the complex effective dielectric constant of the medium,

$$\alpha_a = k \operatorname{Im} \hat{\varepsilon}_e$$

is the nonlinear extinction coefficient of a polydispersed medium, and

$$A(\vec{\kappa}_\perp/k; a) = (k/2\pi i) \cdot \iint_{-\infty}^{\infty} d^2R \exp(-i\,\vec{\kappa}_\perp\vec{\rho}) \times$$
$$\times \left[\exp\left( (ik/2) \int_{-\infty}^{\infty} dx'\varepsilon'(x, \vec{R}, a)\right) - 1\right] \tag{4.1.9}$$

is the scattering amplitude caused by a localized inhomogeneity of a particle plus its aureole written following the van de Hulst approach [22].

Let us introduce the ray intensity (or brightness) of the beam

$$I(x, \vec{n}_\perp, \vec{R}, t) = c(\sqrt{\varepsilon_0}/8\pi)(2\pi)^{-2} \iint_{-\infty}^{\infty} d^2\rho \exp(-ik\vec{n}_\perp\vec{\rho})\Gamma_2(x, \vec{R}, \vec{\rho}, t), \tag{4.1.10}$$

where $\vec{n}_\perp = \vec{\kappa}_\perp/k$ is the direction vector.

The transformation (4.1.10) corresponds to the Wigner representation in quantum mechanics [18]. Let us also introduce the limitation $L_\rho < L_R$, where $L_\rho$ is the scale of the function $\tilde{\Gamma}_2$ over the coordinate $\rho$, and $L_R$ is its scale over the R coordinate. This limitation allows one to change the operators in (4.1.7): $\cosh(\vec{\rho}\nabla\vec{R}/2) \to 1$; $\sinh(\rho\nabla\vec{R}/2) \to \rho\nabla_{\vec{R}}/2$. This simplification enables one, in turn, to write a truncated form of (4.1.7). In the representation (4.1.10) it has the form

$$[\partial/\partial x + \vec{n}_\perp\nabla_{\vec{R}} + \tfrac{1}{2}\nabla_{\vec{R}}\tilde{\varepsilon}_e\nabla\vec{n}_\perp + \alpha_a]I_n =$$
$$= N_0 \iint_{-\infty}^{\infty} d^2n'_\perp \langle |A^2(\vec{n}_\perp - \vec{n}'_\perp, a)|\rangle_{a_0} I_n(x, \vec{n}_\perp, \vec{R}, t). \tag{4.1.11}$$

The condition $L_\rho < L_R$ means that (4.1.11) is applicable if the beam is partially coherent. In the case of coherent radiation the range of applicability of (4.1.11) corresponds to the applicability range of the small-angle approach of nonlinear geometrical optics.

Taking into account that the external parameters may be of a random character, (4.1.11) can be considered to be the small-angle approximation of the quasi-nonlinear stochastic transfer equation.

In order to expand the range of applicability of (4.1.11) to the range of applicability of the Markov diffusion approach (under conditions when the scales of localized inhomogeneities are much larger than the light wavelength $\lambda$), one can write, as in linear theory [11], the following equation instead of (4.1.11):

$$[n\nabla_{\vec{R}} + \tfrac{1}{2}\nabla_{\vec{R}}\tilde{\varepsilon}_e\nabla_{\vec{n}_\perp} + \alpha_a]I_n(x, \vec{R}, \vec{n}, t) = \\ = \oint_{4\pi} d\Omega(\vec{n}')G(\vec{n} - \vec{n}'; x, \vec{R})I_n(\vec{r}, \vec{n}', t), \tag{4.1.12}$$

where $\vec{n} = \{n_x, \vec{n}_\perp\}$, $d\Omega(\vec{n})$ is the solid angle element around the $\vec{n}$ direction, and G is the scattering phase function of the medium.

The beam intensity and brightness are related to each other as follows:

$$I(x, \vec{R}, t) = \oint_{4\pi} d\Omega(\vec{n})I_n(x, \vec{R}, \vec{n}, t). \tag{4.1.13}$$

For a direct beam, the law of the conservation of energy is obeyed, i.e.,

$$\operatorname{div} \vec{I} = -\alpha_a I, \tag{4.1.14}$$

where

$$\vec{I} = \oint d\Omega(\vec{n})\vec{n}I_n(\vec{r}, \vec{n}, \vec{t}) \tag{4.1.15}$$

is the Poynting vector of the beam.

In order to apply the above technique to the description of the situation when turbulent fluctuations of the dielectric constant and thermal nonlinearities occur within the beam zone, it is necessary to introduce corresponding components into the real part of the effective complex dielectric constant, viz. the fluctuating one $\tilde{\varepsilon}_f$ and $\tilde{\varepsilon}_g = (\partial\varepsilon/\partial T)\tilde{T}_g$. Here, $\tilde{T}_g$ is the temperature increase due to gaseous absorption. This temperature increase is determined from the thermal conductivity equation:

$$(\partial/\partial t + \vec{v}_\perp\nabla_{\vec{R}} - \chi_g\Delta_\perp)T_g = \alpha_g I/(c_p\rho_g). \tag{4.1.16}$$

The introduction of the fluctuating term into the transfer equation corresponds to the use of a model of a medium with random refraction, which is presented in the form of a randomly-oriented wedge.

Strictly speaking, this approach is applicable only to short paths or to strong fluctuations $\tilde{\varepsilon}_f$, and corresponds to the case when the structure function of the fluctuations $\tilde{\varepsilon}_f$ is described by a quadratic form $D_{\tilde{\varepsilon}_f} \sim \rho^2$.

However, if a correction of the solutions obtained using the quadratic approximation of $D_{\tilde{\varepsilon}_f}$ is made, one can obtain satisfactory approximate solutions, corresponding to realistic paths not only for the linear medium [16], but also for a nonlinear one [17, 26].

In the case of the small-angle approximation of the geometrical optics, (4.1.12) can be derived using the law of conservation of the value $I_n/\tilde{\varepsilon}_e$ along the path of light propagation.

In further discussions we shall assume that the medium has the complex dielectric constant

$$\hat{\varepsilon} = \tilde{\varepsilon} + i\alpha/k, \tag{4.1.17}$$

whose components are

$$\tilde{\varepsilon} = \tilde{\varepsilon}_f + \tilde{\varepsilon}_e + \tilde{\varepsilon}_g; \qquad \alpha = \alpha_a + \alpha_g. \tag{4.1.18}$$

Note that, in the local region occupied by an aerosol, normally $\alpha_a \gg \alpha_g$, so $\alpha = \alpha_a$.

Thus, in the case of a smoothly inhomogeneous nonlinear dispersed media, the laser beam self-action is described by a quasi-nonlinear stochastic parabolic equation in which the effective dielectric constant of the medium is a function of the mean beam intensity over the scale of discreteness, unless the effects connected with thermal aureoles become significant.

Nonlinear light propagation through discrete scattering media can be described by a quasi-nonlinear transfer equation in the cases of partially coherent beams, or when the small-angle approximation of geometrical optics is applicable.

In the case of large-scale fluctuations (as compared with the scales of localized inhomogeneities of the medium's macroparameters), this equation can be considered to be a stochastic one.

## 4.2. THE FIELD OF THE EFFECTIVE COMPLEX DIELECTRIC CONSTANT OF THE AEROSOL WITHIN THE BEAM

Consider the description of the effective complex dielectric constant of an aerosol:

$$\hat{\varepsilon} = 1 + 4\pi k^{-2} N_0(x, \vec{R}) \int_0^\infty da_0 f(a_0; \vec{B}) A(0; a(a_0; [I; \vec{V}_\perp])).$$

Here, $a(a_0; [I; \vec{V}_\perp])$ is the functional presentation of the solution of the kinetic equation for droplet vaporization by light, taking into account the wind velocity field $\vec{V}_\perp$.

$$(\partial/\partial t + \vec{V}_\perp \nabla_{\vec{R}}) a = -\gamma(a, I) I, \tag{4.2.1}$$

where $\gamma(a, I)$ is the function characterizing the regime of droplet vaporization. As follows from (4.2.1), the droplet will be evaporated irregularly when moving along the stochastic trajectory if the velocity field is random, this process causes the appearance of fluctuating components of $\hat{\varepsilon}$.

It is convenient to divide the problem of the field of $\hat{\varepsilon}$ into two sub-problems: (1) the determination of the relationship between $\hat{\varepsilon}$ and the parameters of the beam and the medium; and (2) the description of the statistical characteristic of the field $\hat{\varepsilon}$ when the medium has random parameters.

### 4.2.1. Components of the Effective Complex Dielectric Constant

According to the optical theorem [22], the nonlinear extinction coefficient of an aerosol medium is determined in terms of the imaginary part of the polydispersed scattering amplitude in the forward direction. Under conditions of a regular regime of droplet vaporization, the optical impurities giving rise to light scattering are the droplets themselves surrounded by thermal and mass aureoles. Since the phase shift of the wave in a droplet aureole is small compared with that occurring in the droplet itself, the scattering amplitudes caused by the droplet and its aureole can be considered to be additive, i.e., $A = A_p + A_T$. As a consequence, corresponding extinction coefficients are also additive ($\sigma = \sigma_p + \sigma_T$).

As shown in Chapter 3, the coefficient of nonlinear extinction by droplets of the sizes characteristic of clouds and fogs under regular regimes of evaporation can be written as follows:

$$\alpha_p = \alpha_0 \exp(-\beta J),$$

where $\beta$ is the approximation parameter which depends on the initial beam intensity as well as on the temperature of the surrounding medium and aerosol microstructure. $J$ is the energetic variable characterizing the density of light power incident on the given Lagrangian particle.

Let us calculate the extinction coefficient of the aureole $\sigma_T = 4\pi k^{-1} \operatorname{Im}(A_T(0))$. As follows from the condition of aureole softness,

$$\sigma_T = \pi k^2/2 \iint_{-\infty}^{\infty} d^2q \Phi_{\varepsilon'}(\vec{q}, t), \qquad (4.2.2)$$

where

$$\Phi_{\varepsilon'}(\vec{q}, t) = (2\pi)^3 |F(|\vec{q}|, t)|^2,$$

$$F = (2\pi^2)^{-1} \int_a^{\infty} \varepsilon_T'(\zeta, t, a) \frac{\sin q\zeta}{q} \zeta \, d\zeta$$

is the three-dimensional spectrum of the aureole. Expression (4.2.2) corresponds to the result of the Rayleigh-Gans approach to the theory of light scattering by localized inhomogeneities [22]. Perturbation of the dielectric constant in aureoles is expressed in terms of the perturbations of temperature and water vapor density, according to the Lorentz equation [21] for a binary mixture. Since the density of the mixture $\rho_g = \rho_1 + \rho_2$ and the total pressure $p = p_1 + p_2 = \text{const}$, the relative perturbation of the dielectric constant due to one scattering center is

$$\varepsilon_T' = 1/\varepsilon_0(\partial\varepsilon/\partial T)_p T_\infty T'/T + 1/\varepsilon_0(\partial\varepsilon/\partial\rho_1)\rho_1'. \qquad (4.2.3)$$

The subscripts 1 and 2 denote the vapor and the air, respectively, $T' = T - T_\infty$, $\rho_1 = \rho_1 - \rho_{1\infty}$; T is the temperature of the mixture, and the subscript $\infty$ denotes the undisturbed values. Furthermore,

$$\partial\varepsilon/\partial T = C_2 p\mu_2/R_\mu T_\infty^2, \qquad \partial\varepsilon/\partial\bar{\rho}_1 = (C_1\mu_1 - C_2\mu_2)/\mu_1,$$

where $C_i = 3A_{\mu_i}$, $A_{\mu_i}$, $\mu_i$ are the molecular refraction and molecular weight, respectively, of i-th component. The field T' and $\dot{\rho}_1'$ are found by solving the external nonstationary problem of mass and heat transfer.

Since the extinction cross-section of the aureole is determined mainly by the large-scale components of $F_{\varepsilon'}$, it is quite sufficient to calculate $F_{\varepsilon'}$ using the point source approximation and assuming that $T'T_\infty/T \sim T'$. In this case

$$F_{\varepsilon'}(q, t) = \chi_g/2\pi^2\Big[1/\varepsilon_0(\partial\varepsilon/\partial T)_p \int_0^t dt' a(t')(T_s(t)' - T_\infty) \times$$
$$\times \exp(-q^2\chi_g(t - t')) + 1/\varepsilon_0(\partial\varepsilon/\partial\rho_1)D/\chi_g \times$$
$$\times \int_0^t dt' a(t')(\rho_{1s}(t') - \rho_{1\infty}) \exp(-q^2 D(t - t'))\Big],$$

where the subscript s denotes the values of the corresponding parameters on the droplet's surface. Using a linear approximation, one obtains $\rho_{1s} - \rho_{1\infty} = d\rho^H/dT(T_s - T_\infty)$; $T_s - T_\infty = K_{ab} Ia(1 - \beta_T)/\lambda_T$, where $d\rho^H/dT$ is the derivative of the saturated vapor density with respect to temperature, $\beta_T = Q_e/Q$, Q is the mass energy consumed during the vaporization of a droplet; $Q_e$ is the heat of vaporization; and $\lambda_T$ is the coefficient of air thermal conductivity.

When using the linear approximation $K_{ab} = A \cdot a$, one obtains the following expression for the scattering cross-section of the droplet aureole:

$$\sigma_T(t; \lambda) = \pi k^2 a^2 \chi_g t \ \ln 4[(1/\varepsilon_0)(\partial\varepsilon/\partial T)(T_s - T_\infty) + (1/\varepsilon_0) \times$$
$$\times (\partial\varepsilon/\partial\rho_1)(D_n/\chi_g)(\rho_{1s} - \rho_{1\infty})]^2. \qquad (4.2.4)$$

The estimations of the effective maximum value ($T_s \leq 393$ K, $\rho_{1s} \leq \rho_{2\infty}$) made for the spectral range $\lambda = 0.63$ to 10.6 μm showed that the effect of light scattering by the aureole on the self-action process, as well as on the light propagation in an evaporating droplet, is weak and can be neglected.

Perturbations of the real part of the effective complex dielectric constant of the medium take place due to averaging over the discreteness scales of the fields of temperature and vapor, as well as being due to the evaporating particles themselves. The mean values of the perturbations of temperature and water vapor density in the beam, when the influence of heat and mass diffusion can be neglected, are determined by solving the following equations (see Chapter 3):

$$(\partial/\partial t + \vec{\nabla}_{\perp}\nabla_{\vec{R}})(T_0, \rho_{10}) = 4\pi N_0 \int_0^{\infty} da_0 f_0(a_0, \vec{B}) a^2 (q_s, m_s), \tag{4.2.5}$$

where $q_s$, $m_s$ are the flows of heat and mass of the vapour over the droplet's surface, respectively.

$$(T_0, \rho_{10}) = 4\pi N_0 \int_0^{\infty} da_0 f_0(a_0, \vec{B}) \int_0^{\infty} d\zeta\zeta^2 (T'(\zeta), \rho_1'(\zeta)). \tag{4.2.6}$$

The flows of heat and mass of the vapour over the droplet's surface are

$$q_s = (1 - \beta_T) K_{ab} I/4; \qquad m_s = \beta_T K_{ab} I/4Qe.$$

The real part of the perturbations of the effective dielectric constant of the medium, caused by the perturbations of temperature and water vapor density, can be represented as follows:

$$\tilde{\varepsilon}_{eT} = 1/\varepsilon_0 (\partial\varepsilon/\partial T)_p \tilde{T}_0 + 1/\varepsilon_0(\partial\varepsilon/\partial\rho_1)\tilde{\rho}_{10} = -|\tilde{\varepsilon}_{max}| f_T(J), \tag{4.2.7}$$

where

$$|\tilde{\varepsilon}_{max}| = \bar{f}_T(\infty) \, |1/\varepsilon_0(\partial\varepsilon/\partial T)_p (1 - \beta_T)/C_p \rho_g \beta_T + 1/\varepsilon_0(\partial\varepsilon/\partial\rho_1)| \tag{4.2.8}$$

is the value $\tilde{\varepsilon}_{eT}$ within the zone of complete vaporization of droplets,

$$\bar{f}_T(J) = f_T(J)/f_T(\infty), \qquad f_T = \pi N_0 \beta_T/Q_e \int_0^J dJ' \int_0^{\infty} da K_{ab}(a) f(a),$$

and f(a) is the rearranging size spectrum of evaporating droplets.

The real part of the perturbations of the effective dielectric constant of the medium caused by the aerosol particles is expressed in terms of the amplitude of light scattering by a single particle in the forward direction, but is averaged over the size spectrum. Using the van de Hulst approximation [22] for the scattering amplitude by a single particle in the forward direction, and approximating the size spectrum of evaporating droplets by a gamma size-distribution function, one obtains

$$\tilde{\varepsilon}_{ep} = \frac{N_0}{k^3 (n_a - 1)^2} \left\{ - \frac{\mu + 1}{\mu} \frac{x_m \cos[(\mu + 2)\varphi - g]}{[(x_m/\mu)^2 + [\mathrm{tg}\, g(x_m/\mu) + 1]^2]^{(\mu+2)/2}} + \right.$$
$$\left. + \frac{\cos^2 g \sin[(\mu + 1)\varphi - 2g]}{\left[\left(\frac{x_m}{\mu}\right)^2 + \left[\mathrm{tg}\, g \left(\frac{x_m}{\mu}\right) + 1\right]^2\right]^{(\mu+1)/2}} + \sin 2g \cos^2 g \right\}, \tag{4.2.9}$$

where

$$x = 2ka(n_a - 1); \quad tg\, g = \kappa_a/(n_a - 1); \quad x_m = 2ka_m(n_a - 1);$$

$$\varphi = \arctan[(tg\, g + \mu/x_m)^{-1}],$$

and $a_m$, $\mu$ are the parameters of function f(a).

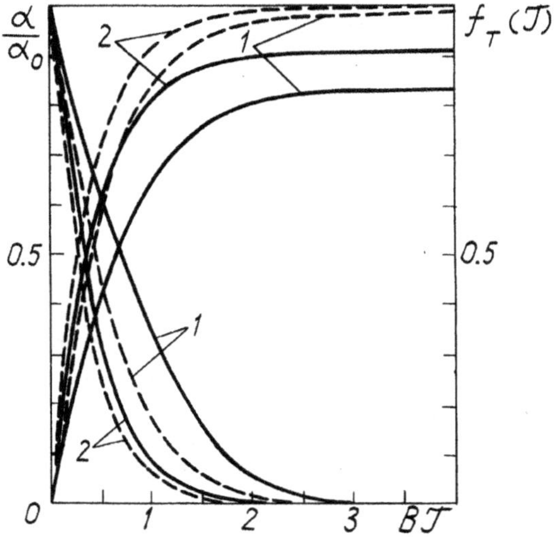


Fig. 4.2.1(a). The dependence of the complex aerosol extinction coefficient and the function $f_T$ on the dimensionless energy density for $\mu_0 = 2$ (solid curves), $\mu_0 = 10$ (broken curves); $a_{m0} = 6$ μm (curve 1) and 2 μm (curve 2).

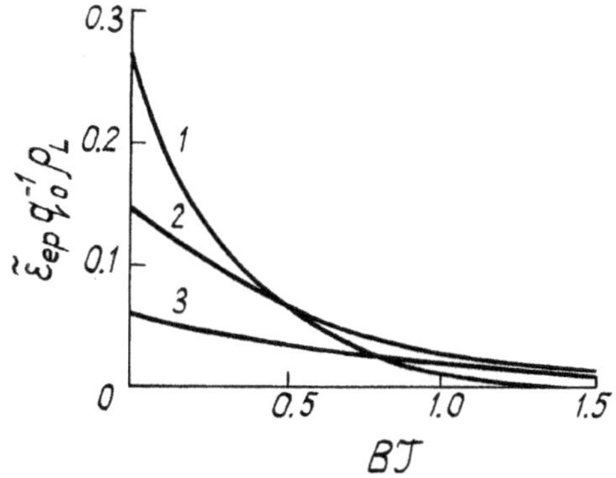


Fig. 4.2.1(b). Real part of the effective dielectric constant of the medium, caused by the presence of aerosol particles, as a function of dimensionless energy density at $a_{m0} = 1$ μm, $\mu_0 = 10$ (curve 1); $a_{m0} = 6$ μm, $\mu_0 = 10$ (curve 2); $a_{m0} = 6$ μm, $\mu_0 = 2$ (curve 3).

Figure 4.2.1 presents the results of calculating the normalized components of the complex effective dielectric constant of the aerosol, namely $\alpha/\alpha_0$, $f_T$ (Fig. 4.2.1(a)), $\tilde{\varepsilon}_{ep}q_0^{-1}\rho_L$ (Fig. 4.2.1(b)), as functions of the dimensionless power density. The calculations used the technique presented in Chapter 3, according to which the rearranging size spectrum of evaporating droplets is approximated by the same function as the initial one, but with the parameters depending on the power density of the incident radiation. The radius of an evaporating droplet was calculated using the following relationship:

$$a = (2k_{ab})^{-1} \ln (1 + \exp(-BJ)(\exp(2k_{ab}a_0) - 1)),$$

where the coefficient B takes into account the different regimes of droplet vaporization:

$$B = \beta_T \exp(-0.2(|m_a| - 1))/2\rho_L Qe.$$

The results calculated refer to the case when the initial size spectrum is set by the gamma size-distribution function with the parameters $a_{m0}$ and $\mu_0$ well approximated by the following functions:

$$\alpha_a = \alpha_0 \exp(-\beta J); \qquad \tilde{\varepsilon}_{eT} = -|\tilde{\varepsilon}_{max}|(1 - \exp(-\beta J)); \qquad \beta = k_1 B,$$

$k_1$ is the parameter of approximation. As seen from Fig. 4.1(b), the relative contribution of $\tilde{\varepsilon}_{ep}$ is significant only at small values of $a_{m0}$ ($a_{m0} \leq 1$ μm) and large values of $\mu_0$ ($\mu_0 \simeq 10$).

Thus, for clouds and fogs having the most probable parameters of microstructure, we have

$$\alpha = \alpha_0 \exp(-\beta J); \qquad \tilde{\varepsilon}_e = -|\tilde{\varepsilon}_{max}| (1 - \exp(-\beta J)). \tag{4.2.10}$$

For the sake of convenience, it is worthwhile to relate the parameters of the complex effective dielectric constant of the medium $\alpha_0$ and $|\tilde{\varepsilon}_{max}|$ to the initial water content $q_0$ of the aerosol, using the coefficients of correspondence

$$\alpha_0 = K_\alpha q_0; \qquad |\tilde{\varepsilon}_{max}| = K_\varepsilon q_0. \tag{4.2.11}$$

These are the most probable parameters of aerosol microstructure: $K_\alpha \simeq 10^8$ km$^{-1}$cm$^3$g$^{-1}$. At $T_\infty = 293$ K and $\beta_T = 0.7$, $K_\varepsilon \simeq 1.6$ g$^{-1}$cm$^3$, $\beta \simeq 0.14$ J$^{-1}$cm$^2$.

### 4.2.2. The Fluctuation Characteristics of the Field of the Complex Effective Dielectric Constant

Let the fluctuations $\tilde{\varepsilon}' = \varepsilon_e - \langle\tilde{\varepsilon}_e\rangle$ and $\alpha' = \alpha - \langle\alpha\rangle$ be of a Gaussian type. In this situation the first and second momenta of the statistical distribution are quite sufficient for a complete description of the statistical properties of the field. Averaging, in our discussion, means averaging over the fluctuations of the external parameters. Let us also assume that $\tilde{\varepsilon}_e = -|\tilde{\varepsilon}_{max}|(1 - \exp(-\beta J))$, $\alpha = \alpha_0 \exp(-\beta J)$, $|\tilde{\varepsilon}_{max}| = K_\varepsilon q_0$, and $\alpha = K_\alpha q_0$. The fluctuations of the coefficients $K_\varepsilon$, $K_\alpha$ and $\beta$ are considered to be weak compared with the fluctuations of the initial water content. These assumptions allow one to consider that the fluctuations $\tilde{\varepsilon}'$ and $\alpha'$ are mainly caused by fluctuations of the initial water content $q_0' = q_0 - \langle q_0 \rangle$ and of the energetic variable $J' = J - \langle J_0 \rangle$, which, in turn, are functions of the transverse wind speed.

Thus, one can obtain the following expressions:

$$\tilde{\varepsilon}' = -K_\varepsilon[q_0'(1 - \exp(-\beta\langle J\rangle)) + \langle q_0\rangle\beta J' \exp(-\beta\langle J\rangle)],$$
$$\alpha' = K_\alpha[q_0' \exp(-\beta\langle J\rangle) - \langle q_0\rangle\beta J' \exp(-\beta\langle J\rangle)], \qquad (4.2.12)$$

which are accurate with respect to the terms of the second order of magnitude. The water content $q \to 0$ in the zone of complete clearing vanishes, so the fluctuations $\tilde{\varepsilon}'$ are determined by the initial level of water content fluctuations, while $\alpha' \to 0$.

Now, consider the influence of the fluctuations of initial water content and the fluctuations caused by the effects of droplet vaporization by laser radiation, under conditions of random wind speed, on the values $\tilde{\varepsilon}'$ and $\alpha'$. The question of water content fluctuations in clouds has been thoroughly discussed in [13] for the inertial range of atmospheric turbulence. Using the results of the semi-empirical theory of atmospheric turbulence [13], one can write the following expression for the structural characteristic of the water content fluctuations:

$$C^2_{q_0'} = C\,\frac{K_{q0}}{K_T}\left[\frac{K_T}{d\langle V_\perp\rangle/dz}\right]^{2/3}\left(\frac{d\Pi}{dz}\right)^2 = C\,\frac{K_{q0}}{K_T}\,L_0^{4/3}\left(\frac{d\Pi}{dz}\right)^2, \qquad (4.2.13)$$

where $C \simeq 2.4$, $K_T$ is the coefficient of turbulent exchange, $K_{q0}$ is the coefficient of turbulent diffusion for the water component of air, $L_0$ is the outer scale of turbulence, z is the altitude, $\Pi = \langle q_0\rangle - (\gamma_a - \gamma_{ba})C_p z/Q_e$, $\gamma_a$ and $\gamma_{ba}$ are the dry and the moist adiabatic gradients. Substance $\Pi$ is considered as a conservative, passive admixture.

Taking this into account, one obtains the following expression for the structural characteristic of the fluctuations:

$$C^2_{\tilde{\varepsilon}'} = K^2_\varepsilon(1 - \exp(-\beta\langle J\rangle))^2 C^2_{q_0'}.$$

The estimation of $C^2_{\tilde{\varepsilon}'}$ in the ground atmospheric layer at $L_0 = 1$ m gives the following results: $C^2_{\tilde{\varepsilon}'} = 1.3 \times K^2_\varepsilon\ (10^{-17}$ to $10^{-15})$ cm$^{-2/3}$. Taking into account that $K_\varepsilon \simeq 1.6$ g$^{-1}$cm$^3$, one can see that the value $C^2_{\tilde{\varepsilon}'}$ is comparable in magnitude with the corresponding value for the case of a turbulent atmosphere with steep water content gradients ($dq_0/dz > 10^{-9}$ gcm$^{-4}$).

Consider the effect of transverse wind speed pulsations on the characteristics of the dielectric constant. For this purpose consider (4.2.1) once more. When $\gamma(a, I) = \gamma(a, I_0)$, its solution has the form

$$-\int_{a_0}^{a} \frac{da}{\gamma(a, I_0)} = J(x, \vec{R}, t). \tag{4.2.14}$$

Here, the function $J(x, \vec{R}, t) = \int_0^t dt' I(x, \vec{R}(t'), t')$ is the energetic variable describing the flow of light energy into a Lagrangian point in the medium, and $\vec{R}(t')$ is the characteristic of (4.2.1) defined by the following equation: $\vec{R}(t') = \vec{R} - \int_{t'}^{t} V_\perp(x, \vec{R}(t''), t'')\, dt''$. The average (over random displacements of the Lagrangian particle of medium) value of $\langle J\rangle$ is as follows:

$$\langle J\rangle = \int_0^t \langle \exp(\vec{R}'(t')\nabla_{\vec{R}})\rangle I(x, \langle\vec{R}(t')\rangle, t')\, dt', \tag{4.2.15}$$

where $\langle\vec{R}(t')\rangle = \vec{R} - \int_{t'}^{t} \vec{V}_\perp(x, \vec{R}(t''), t'')\, dt''$, $\vec{R}' = \vec{R} - \langle\vec{R}\rangle$, $\vec{V}_\perp = \vec{V}'_\perp - \langle\vec{V}_\perp\rangle$. It is also assumed that $I/\langle I\rangle \simeq 1$. It is assumed, when calculating $\langle J\rangle$, that the wind speed fluctuations are Gaussian, stationary, spatially homogeneous and isotropic. This allows one to write the following expression:

$$\langle J\rangle = \int_0^t \exp(\tfrac{1}{2}D(t - t')\Delta_\perp) I(x, \langle\vec{R}(t')\rangle, t')\, dt', \tag{4.2.16}$$

where

$$D(t) = \tfrac{1}{2} \int_0^t\int_0^t dt'\, dt'' \langle \vec{V}'_\perp(x, \vec{R}(t'), t')\vec{V}'_\perp(x, \vec{R}(t''), t'')\rangle, \quad \vec{V}'_\perp = \vec{V}_\perp - \langle\vec{V}_\perp\rangle$$

is the variance of the displacements of a Lagrangian particle of the medium in the field of random wind speed $\vec{V}'_\perp$ [19]. In the limiting cases of $t \ll t_0$ and $t \gg t_0$, the following expressions for $D(t)$ are valid: $D(t) = \tfrac{1}{2}\langle\vec{V}'^2_\perp\rangle t^2$, $(t \ll t_0)$ and $D(t) = 2\chi_t t$, $(t \gg t_0)$, where $t_0$ is the correlation time of the wind speed fluctuations, $\chi_t = D_t(\infty)$, and $D_t(t) = \tfrac{1}{2}dD/dt$ is the coefficient of turbulent diffusion. In some particular cases it is possible to obtain differential equations for $\langle q\rangle$, $\langle T_0\rangle$, $\langle\rho_0\rangle$ using the expression for $\langle J\rangle$. Thus, for example, the equation for the mean water content $\langle q\rangle = \exp(-\beta\langle J\rangle + \tfrac{1}{2}\beta^2\langle J'^2\rangle)$ is

$$\begin{aligned}(\partial/\partial t + \langle\vec{V}_\perp\rangle\nabla_{\vec{R}})\langle q\rangle = {} & -\beta\langle q\rangle I - \langle q\rangle\beta\Delta_\perp \int_0^t dt' D_t(t - t') \times \\ & \times I(x, \langle\vec{R}(t')\rangle, t') + \langle q\rangle\beta^2 \int_0^t\int_0^t dt'\, dt'' D_t(t - t') \times \\ & \times \sum_{i=z,y} \partial/\partial R_i I(x, \langle\vec{R}(t')\rangle, t') \times \\ & \times \partial/\partial R_i I(x, \langle\vec{R}(t'')\rangle, t'').\end{aligned} \tag{4.2.17}$$

The derivation of this equation uses the linearization

$$J' = \int_0^t \vec{R}'(t')\nabla_{\vec{R}} I\, dt'. \tag{4.2.18}$$

The last procedure is applicable if $D(t) \ll L_R^2$, where $L_R$ is the transverse scale of regular inhomogeneity of the dielectric constant. This requirement is equivalent to the conditions $t \ll L_R \langle \vec{V}_\perp'^2 \rangle^{-1/2}$ if $t \ll t_0$, or $t \ll L_R^2/\chi_t$ if $t \gg t_0$. If $t$ and $(\beta I_0)^{-1}$ are much larger than $t_0$, then (4.2.17) becomes the diffusion equation whose diffusion coefficient is $\chi_t$.

Now, calculate the correlation function of the fluctuations taking into account (4.2.18):

$$B_{\tilde{\varepsilon}'} = \langle \tilde{\varepsilon}'(x, \vec{R}_1, t)\tilde{\varepsilon}'(x, \vec{R}_2, t)\rangle =$$
$$= |\tilde{\varepsilon}_{max}|^2 \beta^2 B_{J'}(x, \vec{R}_1, \vec{R}_2, t) \exp[-\beta\langle J(x, \vec{R}_1, t) + J(x, \vec{R}_2, t)\rangle],$$

where $B_{J'} = \langle J'(x, \vec{R}_1, t)J'(x, \vec{R}_2, t)\rangle$.

In the stationary state $t \gg L_R/\langle V_\perp \rangle$, and for $\vec{V}_\perp = \vec{V}_z$ and $\partial I(\vec{R}_y = 0)/\partial R_y = 0$, one can obtain the variance of the fluctuations $\tilde{\varepsilon}'$ in the form

$$\langle \tilde{\varepsilon}' \rangle^2 = \frac{|\tilde{\varepsilon}_{max}|^2 \beta^2}{\langle V_\perp \rangle^2} \exp(-2\beta\langle J(x, \vec{R})\rangle)\langle J(x, \vec{R})\rangle^2 \langle V_\perp'^2 \rangle. \tag{4.2.19}$$

Since we used (4.2.18), (4.2.19) describes the maximum possible effect and, as a consequence, one can assess (using (4.2.19)) the maximum value of the structural constant of the induced fluctuations of the dielectric constant:

$$C_{\tilde{\varepsilon}'}^2 = C_{V_\perp'}^2 K_\varepsilon^2 \langle q_0 \rangle^2 e^{-2}/\langle v_\perp \rangle^2 \simeq L_0^{-2/3} K_\varepsilon^2 \langle q_0 \rangle^2 e^{-2}. \tag{4.2.20}$$

One can see from (4.2.20) that, for $L_0 = 1.25$ m, $q_0 = 10^{-1}$ to 1 gm$^{-3}$, and $C_{\tilde{\varepsilon}}^2 = K_\varepsilon^2 5.41$ ($\times 10^{-17}$ to $10^{-15}$) cm$^{-2/3}$. That means, taking into account that $K_\varepsilon \simeq 1.6$ g$^{-1}$cm$^3$, $K_\varepsilon^2$ exceeds $C_{\tilde{\varepsilon}_f}^2$, only at very high initial water content, i.e., $q_0 > 1$ gm$^{-3}$.

Thus, we have shown that the level of induced fluctuations of the real part of the dielectric constant of the aerosol medium is nearly always lower than that observed in clear atmosphere. This enables us in further discussion to neglect the effect of induced fluctuations on the mean intensity of a laser beam.

In contrast with gaseous media, the beam self-action in aerosol media is determined mainly by the imaginary part of the complex dielectric constant of the aerosol. Therefore, its mean level and fluctuations are very important factors affecting the process of beam self-action. Wind speed fluctuations affect the self-action process strongly. One can obtain, within the framework of the assumptions above,

$$\langle \alpha'^2 \rangle = \langle \alpha_0^2 \rangle \beta^2 \langle J \rangle^2 \exp(-2\beta\langle J \rangle)\langle V_\perp'^2 \rangle / \langle V_\perp \rangle^2,$$

from which it appears that high levels of variance of the extinction coefficient fluctuations $\langle \alpha'^2 \rangle \simeq \langle \alpha_0^2 \rangle$ can occur only at large values of the ratio $\langle V_\perp'^2 \rangle / \langle V_\perp \rangle^2$.

## 4.3. DESCRIPTION OF THE MEAN INTENSITY OF A BEAM

Consider the description of the process of laser radiation transfer in a dispersed media irradiated by high-power laser radiation. The mean intensity of a laser beam in this case can be calculated using a parabolic equation, or the transfer equation and the single scattering approximation.

### 4.3.1. The Method of Transfer Equation

4.3.1.1. Basic Relationships. Let us reduce (4.2.12) to its integral form, neglecting the effects of multiple light scattering. Then, by presenting corresponding functions of the beam intensity as functions of the spatial coordinates and time, one can obtain from (4.2.12) the following quadratic form:

$$I_n(x, \vec{R}, \vec{n}_\perp, t) = I_{n_0}(\vec{R}(0, \vec{R}, \vec{n}_\perp, t), \vec{n}_\perp(0, \vec{R}, \vec{n}_\perp, t)) \times \\ \times \exp\left\{-\int_0^x \alpha(x', \vec{R}(x', \vec{R}, \vec{n}_\perp, t), \\ \vec{n}_\perp(x', \vec{R}, \vec{n}_\perp, t))\, dx'\right\}, \qquad (4.3.1)$$

where $I_{n_0}$ is the boundary value of the ray intensity of the beam at $x = 0$, and $\vec{R}(x', \vec{R}, \vec{n}_\perp, t)$ and $\vec{n}_\perp(x', \vec{R}, \vec{n}_\perp, t)$ are the characteristics of the differential equation (4.1.12), represented in further discussions as $\vec{R}(x')$ and $\vec{n}_\perp(x')$, respectively. The expression (4.3.1) forms the basis for constructing numerical algorithms for solving the problem and seeking approximate relationships.

If the radiation transfer equation is taken in the framework of the approximation of small scattering angles, then $I_{n_0}$ is related to the boundary value of the beam coherence function, $\Gamma_2(x = 0) = \Gamma_2^0$, by a two-dimensional Fourier transform.

$$I_{n_0}(\vec{R}, \vec{n}_\perp, t) = \frac{c\sqrt{\varepsilon_0}}{8\pi}(2\pi)^{-2}\iint_{-\infty}^{\infty} d^2n_\perp \exp(-ik\vec{n}_\perp\vec{\rho})\,\Gamma_2^0(\vec{R}, \vec{\rho}, t). \qquad (4.3.2)$$

Moreover, for a partially coherent beam,

$$\Gamma_2^0(\vec{R}, \vec{\rho}, t) = \langle E_0(\vec{R} + \vec{\rho}/2, t)E_0^*(\vec{R} - \vec{\rho}/2, t)\rangle. \qquad (4.3.3)$$

Averaging is carried out, in this case, over the fluctuations of the source field. Let us assume that the medium is irradiated with a Gaussian, partially coherent beam [16]:

$$\Gamma_2^0(\vec{R}, \vec{\rho}, t) = E_0^2(t)\exp\left\{-\frac{R^2}{R_0^2} - \frac{\rho^2}{4R_0^2}\left(1 + \frac{R_0^2}{\rho_{co}^2}\right) + \frac{ik\vec{\rho}\vec{R}}{F}\right\}, \qquad (4.3.4)$$

where $R_0$ is the initial radius of the beam, $\rho_{co}$ is the initial radius of coherence of the beam, and F is the effective focal length of the beam's wave front. When $F < 0$, the beam is focused and $F > 0$ corresponds to a non-focused beam. The boundary value of the brightness is written

$$I_{n_0}(\vec{R}, \vec{n}_\perp, t) = \frac{k^2 R_0^2 I_0(t)}{\pi\left(1 + \frac{R_0^2}{\rho_{co}^2}\right)} \exp\left\{-\frac{R^2}{R_0^2} - \frac{k^2 R_0^2}{1 + R_0^2/\rho_{co}^2}\left(\vec{n} - \frac{\vec{R}}{F}\right)^2\right\}. \quad (4.3.5)$$

Correspondingly, the mean intensity of the beam is

$$I_n(x, \vec{R}, t) = \iint_{-\infty}^{\infty} d^2\vec{n}_\perp I_n(x, \vec{R}, \vec{n}_\perp, t). \quad (4.3.6)$$

The characteristic $R(x')$ can be found by solving the problem

$$d^2\vec{R}(x')/dx'^2 = \tfrac{1}{2}\nabla_{\vec{R}}\langle\tilde{\varepsilon}(x', \vec{R}(x'), t)\rangle, \quad \vec{R}(x) = \vec{R}, \quad d\vec{R}(x)/dx = \vec{n}_\perp; \quad (4.3.7)$$

while $\vec{n}_\perp(x')$ is determined as follows: $\vec{n}_\perp(x') = d\vec{R}(x')/dx'$. Here, $x'$ is the coordinate on the atmospheric path of length $x$. If the field of $\tilde{\varepsilon}$ values is of a random character, then (4.3.7) describes random shears of the characteristic lines along with the regular displacements.

It is possible, in some particular cases, to reduce the general expression (4.3.6) to finite analytical forms as, for example, in the case of small fluctuations of the aerosol extinction coefficient or a weak influence of refraction on the process of self-action. The former situation is discussed below as an example.

Assuming that the cooperative distribution of fluctuations of the characteristics $\vec{R}'$ and $\vec{n}_\perp'$ is Gaussian, one can write the mean intensity of the beam as

$$\langle I\rangle = \frac{I_0 K^2 R_0^2}{\pi\left(1 + \frac{R_0^2}{\rho_{co}^2}\right)} \iint_{-\infty}^{\infty} \exp\left\{-\left(\sum_{i=z,y}(A_i\langle R_i(0)\rangle^2 + 2C_i\langle R_i(0)\rangle\langle n_{\perp i}(0)\rangle + B_i\langle n_{\perp i}(0)\rangle^2)\right) - \int_0^x \alpha(x', \langle\vec{R}(x')\rangle)\, dx'\right\} d^2 n_\perp. \quad (4.3.8)$$

The integral equation (4.3.8) can then be solved using the equations for the mean values $\langle\vec{R}(x')\rangle$, $\langle\vec{n}_\perp(x')\rangle$ and the corresponding statistical momenta $\langle R'^2(x')\rangle$, $\langle n_\perp'^2(x')\rangle$, and $\langle\vec{R}'(x')\vec{n}_\perp'(x')\rangle$. The analytical forms of the coefficients $A_i$, $B_i$, $C_i$, which depend on the characteristics momenta, are determined by the initial distribution of the beam intensity over its cross-section. It should be noted that integration of (4.3.8) can only be carried out numerically.

4.3.1.2. The approximation of the small angular, nonlinear divergence of a beam. Consider now the approximation of the small angular, nonlinear divergence of a beam, i.e., when $\left(x \max_{i,j=y,z} \frac{1}{2}\frac{\partial^2\langle\tilde{\varepsilon}\rangle}{dR_i dR_j}\right) \ll 1$. Also, assume that, when $R_y = 0$, $\nabla_{\vec{R}}\tilde{\varepsilon} = \nabla_{\vec{R}_z}\tilde{\varepsilon}$. Under these assumptions, one can obtain the following system of approximate equations for the region near the beam axis ($R_y = 0$):

$$I(x, \vec{R}, t) = (I_0(t)/(g_z g_y)^{1/2}) \exp\left\{- \left[R_z^0(x' = 0)u_{1z}(x) + \frac{dR_z^0(x' = 0)}{dx'} u_{2z}(x)\right]^2 / R_0^2 g_z - R_y^2/R_0^2 g_y - \tau_N(x)\right\}, \quad (4.3.9)$$

where $\tau_N = \int_0^x \alpha \, dx'$ is the nonlinear optical depth, and $g^{1/2}$ is the local dimensionless width of the beam along the j-th axis ($j = z, y$), this is written

$$g_j = (u_{2j}^2/k^2R_0^2)(1 + R_0^2/\rho_{co}^2) + (u_{1j} + u_{2j}/F)^2 + R_0^{-2}\Delta_\perp B(0) \int_0^x dx'[u_{1j}(x)u_{2j}(x') - u_{1j}(x')u_{2j}(x)]^2; \quad (4.3.10)$$

$$B(\vec{\rho}) = \int_{-\infty}^{\infty} dx' \langle\tilde{\varepsilon}_f(x', \vec{R} + \vec{\rho}/2)\tilde{\varepsilon}_f(x, \vec{R} - \vec{\rho}/2)\rangle_{\tilde{\varepsilon}_f},$$

$u_{ij}$ is the fundamental system of solutions of the following equation:

$$d^2u_{ij}/dx^2 = (u_{ij}/2)(\partial^2/\partial R^2)\langle\tilde{\varepsilon}(x, \vec{R}^0(x), t)\rangle \quad (4.3.11)$$

with the boundary condition

$$u_{1j}(0) = du_{2j}(0)/dx = 1; \qquad u_{2j}(0) = du_{1j}(0)/dx = 0. \quad (4.3.12)$$

The solution for the characteristic $\vec{R}^0 = \{\vec{R}_z^0, 0\}$ satisfies the problem

$$d^2\vec{R}^0/dx'^2 = \tfrac{1}{2}\nabla_{\vec{R}}\langle\tilde{\varepsilon}(x', \vec{R}^0, t)\rangle; \qquad \vec{R}^0(x) = \vec{R}; \qquad d\vec{R}^0x/dx = 0. \quad (4.3.13)$$

The functions $g_j$ and $\tau_N$ are calculated along the characteristic $\vec{R}^0$. In the case of a symmetric medium and nonstationary self-action, when the influence of regular movement of the medium on the parameters of the waveguide channel can be neglected ($t \ll R_0/V_0$, where $V_0$ is the regular wind speed), (4.3.9) can be reduced (in the vicinity of the axis $R_z = 0$) to the form

$$I(x, \vec{R}, t) = I_0/g \exp\left\{- \frac{R^2}{R_0^2 g} - \tau_N\right\}, \quad (4.3.14)$$

where $g = g_j$. By taking $\vec{R} = 0$ in (4.3.14) and differentiating it with respect to x, one obtains a differential equation for the amplitude of the mean intensity of a beam

$$dI(x)/dx + \alpha I + \gamma_N I = 0, \quad (4.3.15)$$

where

$$\gamma = g^{-1}(dg/dx). \quad (4.3.16)$$

4.3.1.3. The effective beam parameters. Using (4.1.12), one can obtain equations describing the evolution of the effective beam parameters in refractive media that attenuate the radiation non-uniformly. Thus, for the beam's gravity center, one can write

$$\vec{R}_c(x,\ t) = P(x,\ t)^{-1} \iint_{-\infty}^{\infty} d^2R\vec{R}I(x,\ \vec{R},\ t). \tag{4.3.17}$$

The squared value of the full effective beam width is

$$R_e^2(x,\ t) = P(x,\ t)^{-1} \iint_{-\infty}^{\infty} d^2RR^2I(x,\ \vec{R},\ t). \tag{4.3.18}$$

The squared value of the relative effective beam width is

$$R_{e1}^2 = R_e^2 - R_c^2. \tag{4.3.19}$$

Here, $P(x,\ t) = \iint_{-\infty}^{\infty} d^2RI(x,\ \vec{R},\ t)$ is the beam power satisfying the law of conservation

$$dP/dx = -\gamma_e(x,\ t)P, \tag{4.3.20}$$

where

$$\gamma_e(x,\ t) = P^{-1} \iint_{-\infty}^{\infty} d^2R\alpha(x,\ \vec{R},\ t)I(x,\ \vec{R},\ t) \tag{4.3.21}$$

is the integral aerosol extinction coefficient.

In this section we will consider only regular media. The equations for the integral beam parameters $R_c$ and $R_e^2$ are as follows:

$$\begin{aligned} d^2\vec{R}_c/dx^2 = & \left[(d/dx)\left(\vec{R}_c\gamma_e - P^{-1}\iint_{-\infty}^{\infty} d^2R\vec{R}\ dI\right) + \right. \\ & \left. + P^{-1}\left(\gamma_e \iint_{-\infty}^{\infty} d^2R\vec{I}_\perp - \iint_{-\infty}^{\infty} d^2R\alpha\vec{I}_\perp\right)\right] + \\ & + (2P)^{-1} \iint_{-\infty}^{\infty} d^2R\nabla_{\vec{R}}\tilde{\varepsilon}I; \end{aligned} \tag{4.3.22}$$

$$\begin{aligned} dR_e^2/dx = & \left[\gamma_e R_e^2 - P^{-1}\iint_{-\infty}^{\infty} d^2RR^2\alpha I\right] + 2R_n; \\ dR_n/dx = & \left[\gamma_e R_n - P^{-1}\iint_{-\infty}^{\infty} d^2R\vec{R}\alpha\vec{I}_\perp\right] + n_e^2 + \\ & + (2P)^{-1}\iint_{-\infty}^{\infty} d^2R\vec{R}\nabla_{\vec{R}}\tilde{\varepsilon}I; \\ dn_e^2/dx = & \left[\gamma_e n_e^2 - P^{-1}\iint_{-\infty}^{\infty}\iint_{-\infty}^{\infty} d^2R\ d^2n_\perp n_\perp^2\alpha I\right] + \\ & + P^{-1}\iint_{-\infty}^{\infty} d^2R\nabla_{\vec{R}}\tilde{\varepsilon}\vec{I}_\perp, \end{aligned} \tag{4.3.23}$$

where $\vec{I}_\perp = \iint_{-\infty}^{\infty} d^2n_\perp\vec{n}_\perp I_n$ is the transverse component of the Poynting vector $\vec{I} = \oint_{4\pi} d\Omega(\vec{n})\vec{n}I_n$; $R_n = P^{-1} \iint_{-\infty}^{\infty} \iint_{-\infty}^{\infty} d^2R\ d^2n_\perp\vec{R}\vec{n}_\perp I_n$; and

$n_e^2 = P^{-1} \iint_{-\infty}^{\infty} \iint_{-\infty}^{\infty} d^2R\, d^2n_\perp n_\perp^2 I_n$ is the effective angular beam width.

The boundary conditions for (4.3.22) and (4.3.23) in the plane $x = 0$ are determined using the effective parameters and by taking a particular view of a beam. The terms enclosed by brackets in (4.3.22) and (4.3.23) describe the influence of the inhomogeneities of the medium's extinction coefficient on the effective parameters of the beam. As follows from (4.3.22) and (4.3.23), the effective parameters of narrow beams, whose sizes are small compared with the cross-sectional sizes of the inhomogeneities of the medium's extinction coefficient, are controlled only by the phase distortions of the beam's wave front. In the case of infinitely narrow beams ($I \simeq \delta(R - R_c)$, where $\delta$ is the Dirac delta function, (4.3.22) takes the form of the equation for rays in the small-angle approach of geometrical optics, while (4.3.23) describes the full beam width caused by deflection of rays from the beam axis.

When $|\vec{R}_c| < R_0$ and $R_e^2 \simeq R_0^2$, (4.3.22) and (4.3.23) lead to the so-called nonaberrational approximation.

$$d^2R_{cj}/dx^2 = \tfrac{1}{2}(\partial/\partial R_j)\tilde{\varepsilon}(x, \vec{R} = 0, t) + \sum_{i=z,y} \tfrac{1}{2}R_{ci}\partial^2\tilde{\varepsilon}(x, \vec{R} = 0, t)/\partial R_i \partial R_j;$$

$$d^3R_{ej}^2/dx^3 = (dR_{ej}^2/dx)\,\partial^2\tilde{\varepsilon}(x, \vec{R} = 0, t)/\partial R^2 + + d/dx(R_{ej}^2(\partial^2\tilde{\varepsilon}(x, \vec{R} = 0, t)/\partial R_j^2)), \tag{4.3.24}$$

where

$$i, j = z, y, \quad R_c = \{R_{cz}, R_{cy}\}, \quad R_e^2 = R_{ez}^2 + R_{ey}^2.$$

These relationships allow the qualitative and quantitative analysis of the problem. They allow one, in particular, to analytically describe the integral scales of the inhomogeneities of the dielectric constant of the medium, which are the decisive parameters for describing the propagation of beams through such media.

Let us introduce the effective beam intensity:

$$I_e(x, t) = P(x, t)/R_{e1}^2(x, t) = = P_0 \exp\left[- \int_0^x \gamma_e(x', t)\, dx'\right]/(R_e^2(x, t) - R_c^2(x, t)). \tag{4.3.25}$$

The main tendencies of the evolution of the beam in the medium can be determined by analyzing the functions $\vec{R}_c(x)$, $R_e^2(x)$ at small values of $x$:

$$\gamma_e(x) \approx \gamma_0(0); \quad \vec{R}_c(x) \approx \vec{R}_c(0) + (d\vec{R}_c(0)/dx)x + (\tfrac{1}{2}d^2\vec{R}_c(0)/dx^2)x^2;$$

$$R_e^2(x) = R_e^2(0) + (dR_e^2(0)/dx)x + (\tfrac{1}{2}d^2R_e^2(0)/dx^2)x^2. \tag{4.3.26}$$

If the incident beam is symmetrical (as, e.g., that described by (4.3.4)) with a plane wave front ($F = \infty$), then one has

$$\vec{R}_c(x) = \frac{1}{2}\frac{d^2\vec{R}_c(0)}{dx}x^2 = \frac{1}{4P_{(0)}}\iint_{-\infty}^{\infty} d^2R\nabla_{\vec{R}}\tilde{\varepsilon}(0, \vec{R}, t)I(0, \vec{R}, t)x^2;$$

$$R_e^2(x) = R_0^2 + \left[(k^2R_0^2)^{-1}\left(1 + \frac{R_0^2}{\rho_{co}^2}\right) + \right. \quad (4.3.27)$$

$$\left. + \frac{1}{2P_{(0)}}\iint_{-\infty}^{\infty} d^2R\vec{R}\nabla_{\vec{R}}\tilde{\varepsilon}(0, R, t)I(0, R, t)\right] x^2.$$

Such a consideration clearly shows the existence of integral scales which characterize the behavior of a beam in a nonlinear medium. These scales have the dimensionality of length. Since, in the general case, the media are nonstationary, it is advisable to assess these scales using methods that are based on the principle of a maximum contribution of the nonlinear effect to the interaction process. The first scale characterizes the length of a nonlinear interaction in an aerosol medium,

$$L_a = (\max_t \gamma_e(0))^{-1} = \alpha_0^{-1}. \quad (4.3.28)$$

The scale

$$L_c = R_0^{1/2}\left|\max_t \frac{1}{4P(0)}\iint_{-\infty}^{\infty} d^2R\nabla_{\vec{R}}\tilde{\varepsilon}(0, \vec{R}, t)I(0, \vec{R}, t)\right|^{-1/2} \quad (4.3.29)$$

shows that the beam undergoes a noticeable displacement, as a whole, when travelling the distance

$$x \geq L_c. \quad (4.3.30)$$

Finally, the scale

$$L_N = R_0\left|\max_t \frac{1}{2P(0)}\iint_{-\infty}^{\infty} d^2R\vec{R}\nabla_{\vec{R}}\tilde{\varepsilon}(0, \vec{R}, t)I(0, \vec{R}, t)\right|^{-1/2} \quad (4.3.31)$$

characterizes the angular beam divergence leading to beam defocusing or focusing. These effects can be described by the following inequalities:

$$x \geq L_N, \qquad \theta_N = R_0/L_N \geq \theta_0, \quad (4.3.32)$$

where $\theta_0$ is the initial angular width of the beam and $\theta_N$ is the beam divergence after a nonlinear interaction. In the case considered,

$$\theta_0 = (kR_0)^{-1}(1 + R_0^2/\rho_{co}^2)^{1/2}.$$

4.3.2. The Parabolic Equation Method

As is known, the 'exact' solution of the radiation transfer equation does not describe the wave aberrations caused by interference. Therefore, when it is necessary to account for these effects, one should use numerical methods for solving the parabolic equation.

In its generalized form, the problem of laser beam self-action can be formulated as follows:

$$2ik(\partial E/\partial x) + \Delta_{\perp}E + k^2\hat{\varepsilon}[EE^*]E = 0;$$
$$\hat{\varepsilon}[EE^*] = \tilde{\varepsilon}[EE^*] + i\alpha[EE^*]/k; \qquad E(x = 0) = E_0. \tag{4.3.33}$$

This quasi-nonlinear equation can be solved using a number of numerical methods available from the literature, for example, the splitting method [23], the fast Fourier transform method [15], the method of finite elements [24], and others.

The main peculiarity of the problem, in the case of an aerosol medium, is the strong nonlinearity caused by the dependence of the imaginary part of the complex dielectric constant on the wave intensity. The step of integration over x should be chosen so that the phase change and optical depth increase are sufficiently small.

The parabolic equation method can be recommended as an efficient technique for numerically simulating beam propagation through randomly inhomogeneous media, as well as the propagation of beams with a random field structure (i.e., partially coherent beams).

## 4.4. THE INFLUENCE OF THERMAL DISTORTIONS OF WAVE BEAMS AND FLUCTUATIONS OF THE MEDIUM ON THE BEAM-INDUCED DISSIPATION OF WATER AEROSOLS

The thermal distortions of a high-power beam appearing in the medium due to thermal losses, droplet vaporization, and hence an increase in humidity, along with gaseous nonlinearity can, under certain conditions, limit the penetration of the laser beam into the aerosol. The stochastic distortions of a beam caused by dielectric constant fluctuations can also weaken the beam. Finally, the 'smearing' of the laser beam channel by random wind can also result in a reduction of the clarity of the beam channel.

The most important effects limiting the efficiency of the beam-induced dissipation of aerosols are discussed below, as well as the conditions under which these effects are sufficiently weak.

### 4.4.1. The Influence of Nonstationary Thermal Defocusing on the Beam-Induced Dissipation of Water Aerosols

Consider nonstationary laser beam self-action in water aerosols. By this we understand the process in which both the forced and free convections play insignificant roles and the effects of diffusion on the beam scale are negligible ($t \ll L_R/V_0$, $L_R^2/\chi_g$, $L_R^2/D_\Pi$, where $L_R$ is the transverse beam scale and $V_0$ is the velocity of drift of the medium). The presence of the nonstationary self-action process allows the determination of laser beam potentials necessary for dissipating the water aerosols.

As shown in § 4.1 and § 4.2, the components of the complex effective dielectric constant of a medium are related to the energy density of laser beam in a localized volume (energetic variable) as follows:

$$\alpha = \alpha_0 \exp(-\beta J), \quad \tilde{\varepsilon} = \tilde{\varepsilon}_f - |\tilde{\varepsilon}_{max}|(1 - \exp(-\beta J)) - \left|\frac{\partial \varepsilon}{\partial T}\right| \frac{\alpha_g J}{c_p \rho_g}. \tag{4.4.1}$$

In the case of nonstationary self-action, the energetic variable is defined as $J = \int_0^t I(t')\,dt'$.

The factor seriously limiting the possibility of dissipating water aerosols using laser beams is beam defocusing by thermal lenses formed in the beam channel as a result of nonlinear interactions. The integral scale characterizing this effect (according to (4.3.31)) is

$$L_N = R_0 \left(\max_t \left|\frac{1}{2P(0)} \iint_{-\infty}^{\infty} d^2R\vec{R}(\nabla_{\vec{R}}\tilde{\varepsilon}_e(x = 0) + \nabla_{\vec{R}}\tilde{\varepsilon}_g(x = 0))I(x = 0)\right|\right)^{-1/2}. \tag{4.4.2}$$

If $L_{Na}$, $L_{Ng}$ are the scales describing the action of thermal lenses formed by evaporating droplets heating the air and absorption of light by gases, then calculations give

$$L_{Na} \approx R_0\, e^{1/2}/|\tilde{\varepsilon}_{max}|^{1/2}, \quad L_{Ng} = R_0 \left(\frac{c_p \rho_g}{\left|\frac{\partial \varepsilon}{\partial T}\right| \alpha_g J(x = 0, \vec{R} = 0)}\right)^{1/2}. \tag{4.4.3}$$

The action of thermal lenses becomes dominant in the self-action process when $J > 3\beta^{-1}$, i.e., in the regions of the beam where there are no droplets (the completely cleared zone).

The behavior of a high-power beam (partially coherent in the general case) in an evaporating aerosol media whose dielectric constant undergoes turbulent pulsations can be described using a group of scales, each of which is a characteristic of a linear or nonlinear effect affecting the beam's intensity. Of these scales, the scale of the nonlinear interaction of the laser beam on aerosols, $L_a = \alpha_0^{-1}$, is the most important. This scale also characterize the extinction of radiation in a linear medium.

Another group of scales characterizes the beam's behavior in situations where nonlinear effects do not occur. If the beam of incident radiation is of the form (4.3.5), then these scales are: $L_d = kR_0^2(1 + R_0^2/\rho_{co}^2)^{-1/2}$, which is the diffraction length of a partially coherent beam, and the distance of initial focusing (defocusing) of the beam.

The turbulent blooming of the beam in a linear medium can be characterized by the scale $L_t = (\sqrt{3}/2)kR_0\rho_{c\infty}$, where $\rho_{c\infty}$ is the radius of coherence for a plane wave in the turbulent medium ($\rho_{c\infty} = (0.365C_{\tilde{\varepsilon}_f}^2 k^2 x)^{-3/5}$ if if $D_{\tilde{\varepsilon}_f} \sim \rho^{5/3}$, $\rho_{c\infty} = (0.41C_{\tilde{\varepsilon}_f}^2 k^2 \ell_0^{-1/3} x)^{-1/2}$ if $D_{\tilde{\varepsilon}_f} \sim \rho^2$, where $D_{\tilde{\varepsilon}_f}$ is the structural function of the fluctuations of the dielectric constant in the turbulent atmosphere $\tilde{\varepsilon}_f$, $\ell_0$ is the inner scale of turbulence, and $\rho$ is the spatial separation of the two beams). Finally, the scale of beam divergence caused by the nonlinear effects $L_N$ should be added to the above scales.

The ratio of the different scales shows the relative importance of the different effects, thus determining the character of the self-action process. In the case of beams with $R_0 > 1$ cm and dense aerosol formations ($q_0 > 0.1$ gm$^{-3}$), the situation in which $L_a, L_{Na} < L_d, L_t, L_{Ng}$ takes place in the zone of the beam where droplets are evaporating. This indicates that gaseous nonlinearity, initial beam divergence, and turbulent broadening of the beam efficiently decrease the beam intensity in the completely cleared zone.

The above statements will be illustrated below with numerical calculations. Now, consider the role of nonlinear defocusing in the zone of droplet vaporization on the efficiency of aerosol dissipation by a laser beam. One of the most important characteristics of the 'clearing' process is the time of aerosol dissipation at a given point on the beam's path. This time is defined as the time interval during which the aerosol extinction coefficient reaches some preset small value $\alpha^*$ at a beam energy density $J = J_c$ such that $\alpha(J_c) = \alpha^*$.

The nonaberrational variant of the equations describing the intensity of a laser beam (at its axis) propagating through an evaporating water aerosol is, according to § 4.3,

$$\frac{\partial I(x,\ t)}{\partial x} + \alpha(J)I + \gamma_0(x)I + \gamma_N(x,\ J)I = 0, \tag{4.4.4}$$

where

$$\gamma_0 = g_0^{-1}\frac{dg_0}{dx}, \qquad \gamma_N = g_0^{-1}\left(1 + \frac{g_N}{g_0}\right)^{-1}\frac{dg_N}{dx};$$

$$\frac{d^3 g_N}{dx^3} = 2\frac{d}{dx}(g_0 + g_N)\frac{\partial^2\langle\tilde{\varepsilon}(x,\ 0,\ t)\rangle}{\partial R^2} + (g_0 + g_N)\frac{d}{dx}\frac{\partial^2\langle\tilde{\varepsilon}(x,\ 0,\ t)\rangle}{\partial R^2};$$

$$\frac{d^2 g_N(0)}{dx^2} = \frac{dg_N(0)}{dx} = g_N(0) = 0, \tag{4.4.5}$$

where $g_0$ is the squared dimensionless beam width in a linear medium, and $g_N$ is the correction to $g_0$ to compensate for beam defocusing due to nonlinear effects.

In the case of a beam with a form described by (4.3.5), one can write

$$g_0(x) = x^2/L_d^2 + (1 + x/F)^2 + x^2/L_t^2. \tag{4.4.6}$$

One can obtain from (4.4.4) a relationship which indicates that the beam intensity reaches the value $J_c$ (at a given point in the medium) in a time interval $t_{c\ell}(x)$ after the action of the laser beam begins.

$$\int_0^{t_{c\ell}(x)} I_0(t')\ dt' = \exp\left[\int_0^x \gamma_0(x')\ dx'\right] J_c + \\ + \int_0^x \int_0^{J_c} \exp\left[\int_0^{x'} \gamma_0(x')\ dx'\right] (\alpha(J') + \\ + \gamma_N(x', J'))\ dx'\ dJ'. \qquad (4.4.7)$$

If $I_0(t) = \text{const}$, $g_N/g_0 \ll 1$, and $L_{Ng} \gg L_{Na}$ then, using (4.4.6) and taking into account (4.4.1), one obtains from (4.4.7) that the time necessary for dissipating an aerosol layer of optical depth $\tau = \alpha_0 x$ is

$$t_{c\ell}(\tau) = t_{c\ell}^0(\tau) + t_{c\ell}^N(\tau), \qquad (4.4.8)$$

where

$$t_{c\ell}^0(\tau) = \frac{J_c}{I_0} g_0(\tau) + \frac{\tau \int_0^1 g_0(x\xi)\ d\xi}{\alpha_0 V_{c\ell}} \qquad (4.4.9)$$

is the dissipation time during which beam defocusing does not occur, and $V_{c\ell} = I_0/\int_0^\infty \alpha(J)\ dJ$ is the aerosol dissipation rate at the beam's axis (no defocusing is observed).

$$t_{c\ell}^N(\tau) = \frac{2\xi_1^2}{(\alpha_0 V_{c\ell})} \int_0^\tau \int_0^{\tau'} d\tau'\ d\tau''(\ell n\ g_0(\tau') + 1) \qquad (4.4.10)$$

is the time lag caused by the decrease in dissipation rate due to defocusing. The parameter $\xi_1 = (|\tilde{\varepsilon}_{max}|/(R_0^2\alpha_0^2))^{1/2}$ is the ratio of the nonlinear interaction to the length of thermal defocusing in the zone of droplet vaporization. It is obvious that the condition $\xi_1^2\tau < 1$ means that the influence of the defocusing effect on the self-action process in the zone of droplet vaporization is weak.

Consider a quantitative description of the process using the nonaberrational approach [(4.3.15)-(4.3.10)-(4.3.12)]. In the case of a fixed optical depth of the medium $\tau = \alpha_0 x$, the problem can be characterized by the following parameters: $\bar{J}_0 = \beta I_0 t$; $\Omega = kR_0^2\alpha_0$; $\Omega_0 = k\rho_{c\infty}^2\alpha_0$ (Fresnel's numbers of the scales $R_0$ and $\rho_{c\infty}$); $\bar{F} = \alpha_0 F$; $\xi = (|\tilde{\varepsilon}_{max}|/e)^{1/2}(1/(R_0\alpha_0))$; and $\xi_g$ is the reciprocal dimensionless length of nonlinear interaction due to light absorption by gases: $\xi_g = (|\partial\varepsilon/\partial T|\alpha_g/(c_p\rho_g\beta))^{1/2}(\alpha_0 R_0)^{-1}$; $R_0/\rho_{co}$.

The results obtained by solving this problem numerically are presented in Figures 4.4.1 to 4.4.4. Figure 4.4.1 illustrates the self-action process for a collimated Gaussian beam under conditions when the gaseous nonlinearity and stochastic beam broadening can be neglected. In this case, the self-action process can be thoroughly described using three parameters: $\bar{J}_0$, $\xi$, and $\Omega$. For radiation of $\lambda = 10.6$ μm, the situation in which the gaseous nonlinearity can be neglected is fog dissipation over short distances (the limitation on the optical depth is not imposed). This figure clearly demonstrates the influence of beam diffraction and the thermal

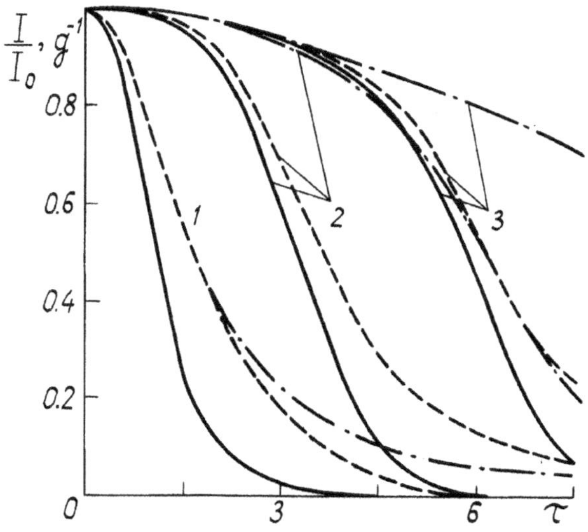


Fig. 4.4.1. Intensity of a collimated Gaussian beam (on its axis) (solid curve) and $g^{-1}$ (dashed curve) as functions of the initial optical depth of the medium in the case of a non-stationary beam's self-action. The parameters of the process are: $\bar{J}_0 = 1.5$ (curve 1); 7.5 (curve 2); 15 (curve 3). $\xi = 0.545$, $\xi_g = \Omega_0^{-1} = 0$; $\Omega = 11.85$. The dot-dash line represents the solution for $I/I_0$ at $\xi = 0$.

defocusing of the beam taking place in the zone of droplet vaporization at the time of aerosol dissipation. The first conclusion that one arrives at is that aerosol dissipation under conditions when diffraction and thermal distortions of the beam are observed within the zone of droplet vaporization is of the wave type. During the course of aerosol dissipation ('clearing') the intensity profile reaches the diffraction level (the junction of curves I and $g^{-1}$).

Now, consider the role of gaseous nonlinearity in the dynamics of transmission within the high-power beam channel. Figure 4.4.2 presents the results for the case when $\alpha_0 = 20$ km$^{-1}$, $\alpha_g = 0.1$ km$^{-1}$, $\beta_T = 0.7$, $\beta = 0.14$, $R_0 = 10$ cm (Figure 4.4.2(a)) and $R_0 = 50$ cm (Figure 4.4.2(b)). As follows from the figures, the defocusing effect caused by gaseous nonlinearity becomes particularly important when $\xi_g\sqrt{\bar{J}_0}\,\tau < 1$. In the case of beams with a small cross-sectional radius, this effect additionally prolongs the dissipation process in water droplet aerosols, while for broad beams it is of significance only in the completely cleared zone, decreasing the mean intensity of the beam in this zone. The broken curves in Figures 4.4.2(b) and 4.4.1 present the function $g^{-1}$. It is seen that the region where $g^{-1}$ and I coincide increases with an increase in $\bar{J}_0$. At the same time one can

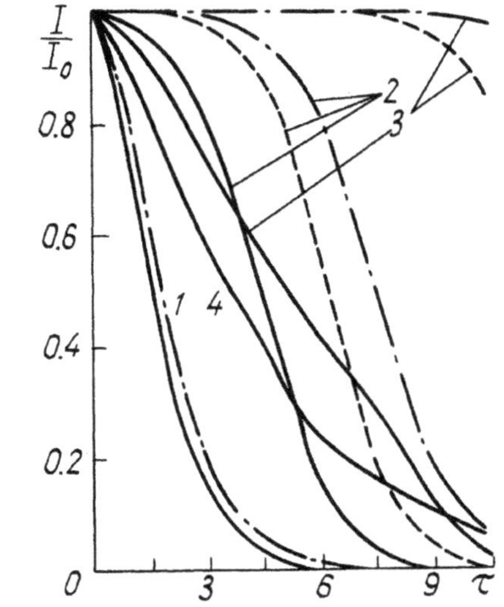


(a)

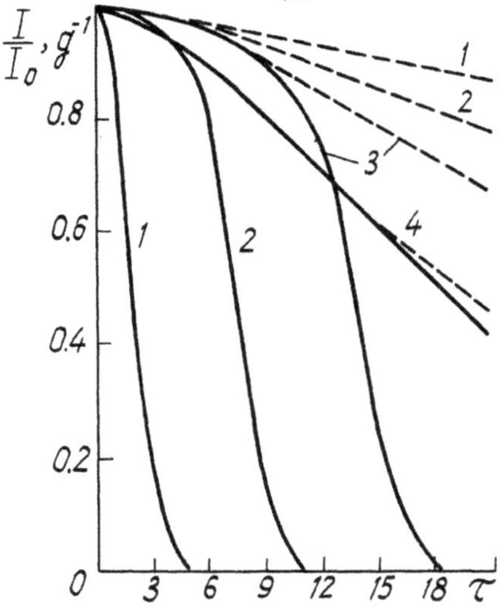


(b)

Fig. 4.4.2. Intensity along the axis of a collimated Gaussian beam as a function of initial optical depth of the medium in the case of nonstationary beam self-action under the conditions: $\bar{J}_0 = 1.5$ (curve 1); 7.5 (curve 2); 15 (curve 3); 30 (curve 4); $\Omega_0^{-1} = 0$.
(a) $\Omega = 118.5$; $\xi = 0.173$; $\xi_g = 0.052$ (solid lines); $\xi_g = 0$ (dashed line); $\xi = \xi_g = 0$ (dot-dash line); (b) $\Omega = 2.95\times10^3$; $\xi = 0.034$; $\xi_g = 0.01$; the dashed line represents the function $g^{-1}$.

see that, in the completely cleared zone, both functions monotonously decrease due to the effect of thermal gaseous defocusing. On the other hand, beam defocusing caused by droplet vaporization is insignificant for such a beam size. As seen from Figure 4.4.2(b), the time interval during which the transmission of the atmosphere at the beam axis reaches the value of 0.9 $g^{-1}$ is practically the same as in the case when no gaseous nonlinearity is observed. This is valid for $\bar{J}_0$ values up to 15. The influence of random broadening of the beam on the beam self-action process is too weak to be significant in this case.

As calculations showed, the defocusing effect caused by droplet vaporization can be weakened by increasing the $R_0\alpha_0^{1/2}$ product. In the case of the most probable parameters of aerosol microstructure, and when $T_\infty = 293$ K, $I_0 \approx 10^2$ W/cm$^2$, $R_0\alpha_0^{1/2} > 10^{-1}$ cm$^{1/2}$, the defocusing effect is too weak to be significant.

For $R_0\alpha_0^{1/2}$ values greater than $10^{-1}$ cm$^{1/2}$, the main factor limiting the extent of 'clearing' is the gaseous thermal nonlinearity, whose role, however, decreases with an increase in the beam's optical radius $R_0\alpha_0$. For $R_0\alpha_0$ values greater than $10^{-2}$, gaseous nonlinearity does not affect the 'clearing' process at all up to $\tau \approx 10$.

Let us consider the self-action process in the case of partially coherent Gaussian beams focused on the far end of the path ($x/F = -1$).

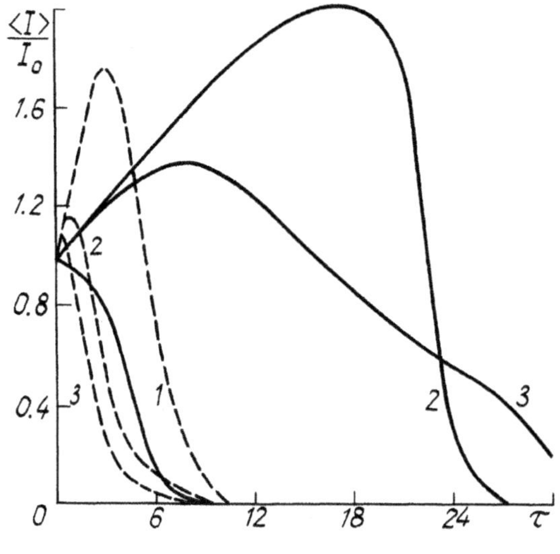


Fig. 4.4.3. Intensity along the axis of a semi-coherent Gaussian beam focused at the farthest end of its atmospheric path as a function of initial optical depth for the case of non-stationary beam self-action, with $\bar{J}_0 = 3$ (curve 1), 15 (curve 2), 30 (curve 3); $\Omega_0 = 6.8\times10^2$; $R_0/\rho_{co} = 10$; $\xi = 0.0345$; $\xi_g = 0.01$; $\bar{F} = -30$; $\Omega = 2.95\times10^3$ (solid lines); and $\xi = 0.109$; $\xi_g = 0.1$; $\bar{F} = -6$; $\Omega = 295$ (dashed lines).

In order to correctly describe the self-action process along a long atmospheric path one should take into account beam defocusing due to the absorption of light by gases. The two types of curves presented in Figure 4.4.3 characterize two cases of beam self-action, viz., in an optically dense ($\alpha_0 = 20$ km$^{-1}$) aerosol medium (the beam is focused at a distance x = 1.5 km), and a slightly turbid ($\alpha = 2$ km$^{-1}$) atmosphere (the beam is focused at a distance x = 3 km). As was to be expected, the defocusing effect due to gaseous nonlinearity deforms the beam, in the case of a slightly turbid atmosphere, so that this leads to a shortening of the time that the medium is in a state of maximum transmission.

Figure 4.4.4 illustrates the influence of the random broadening of a beam on its self-action. The calculations confirm the conclusion that the

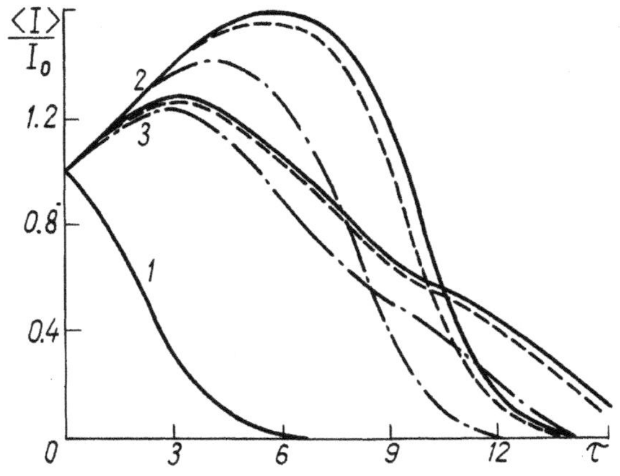


Fig. 4.4.4. The effect of random broadening on the nonstationary self-action of a semi-coherent Gaussian beam with $\bar{J}_0 = 1.5$ (curve 1), 7.5 (curve 2), 15 (curve 3); $\xi = 0.109$; $\xi_g = 0.033$; $\bar{F} = -15$; $\Omega = 295$; $R_0/\rho_{co} = 10$; $\Omega_0 = 6.8\times10^2$ (solid lines); 68 (dashed lines) and $\Omega_0 = 6.8$ (dot-dash lines).

stochasticity of the medium makes a noticeable effect on the beam energetics under the following conditions: (1) the beam self-action of a focused beam takes place in a slightly turbid aerosol at a great distance, when the gaseous absorption of light can be neglected ($\alpha_g \simeq 10^{-2}$ km$^{-1}$), and (2) the fluctuations of the medium's dielectric constant are strong. If these two conditions are fulfilled, then the effect of random beam broadening can result in a significant decrease in the beam's intensity. However, during the course of beam self-action, the random broadening becomes less important compared with the effect of thermal defocusing caused by the absorption of light by gases. Thus, in the case of the range of beam sizes found in practice, the effect of thermal nonlinearity becomes significant at shorter distances rather than effect of random beam broadening.

### 4.4.2. The Influence of Stationary Thermal Distortions of the Beam on the Process of Water Aerosol Dissipation

At times $t \gg L_R/V_0$ from the beginning of the beam self-action process the smearing of the beam channel by the wind becomes the decisive factor in the determination of the process. The motion of medium due to the wind results in the formation of stationary configurations of cleared zones in clouds and fogs (see, e.g., Chapter 3). The nonlinear refraction of the beam in the region of droplet vaporization and the gaseous thermal nonlinearity in the completely cleared zone are the effects which can limit the penetration of a laser beam into the aerosol medium.

Now, consider the main features of laser beam self-action in a water droplet medium moving at a fixed wind speed $V_0$, paying special attention to the influence of thermal distortions of the beam on the self-action process. Let the medium be moving along the Z-axis. The form of the energetic variable which determines the components of the medium's dielectric constants at $t \gg L_R/V_0$ can be described by the following expression:

$$J(x, y, z) = \frac{1}{V_0} \int_{-\infty}^{Z} I(x, y, z')\, dz'. \qquad (4.4.11)$$

If no nonlinear refraction is observed, then the intensity of a beam with a plane wave front can be described by the Glickler formula:

$$I(x, \vec{R}) = I_0(\vec{R})[(e^{\tau} - 1)\, e^{-\beta J_0} + 1]^{-1}; \qquad J_0 = 1/V_0 \int_{-\infty}^{z} I_0(y, z')\, dz'. \qquad (4.4.12)$$

According to this approach, $J = \beta^{-1} \ln[(e^{\beta J_0} - 1)\, e^{-\tau} + 1]$.

The behavior of the laser beam during stationary beam self-action is characterized by the same group of scales as in the nonstationary case. In the stationary case the length of nonlinear refraction for a beam of the form $I_0(R) = I_0 \exp(-R^2/R_0^2)$ is

$$L_{ca}/2 = L_{Na} \approx R_0 (e/|\tilde{\varepsilon}_{max}|)^{1/2};$$
$$L_{cg}/2 = L_{Ng} = R_0^{1/2} \left( \frac{\rho_g c_p V_0}{\left|\frac{\partial \varepsilon}{\partial T}\right| I_0 \alpha_g} \right)^{1/2}. \qquad (4.4.13)$$

As estimations show, the influence of beam divergence and thermal gaseous nonlinearity on the stationary self-action process is significant only in completely cleared zones, i.e., as in the nonstationary case. In the zone of droplet vaporization the essential factor affecting the interaction process is nonlinear refraction on the thermal gaseous lens formed by evaporating droplets ($L_{ca,Na} < L_{cg,Ng}$, $L_d$, $L_t$). In order to elucidate the conditions under which this effect can take place, and to study its influence on aerosol dissipation, it is necessary to find the expression describing the beam intensity. An approximate solution of the problem of the beam channel transmission, taking into account thermal distortions of the beam in the zone of droplet vaporization, can be obtained using (4.3.9)-(4.3.13). Assuming the beam to be only slightly divergent (linear divergence), one can obtain the following formula for the intensity of a beam distorted by weak nonlinear refraction:

$$I(x, y, z) = I_0/(g_z(x, z)g_y(x, z))^{1/2} \times \times \exp\left\{-\frac{y^2}{R_0^2 g_y(x, z)} - \frac{(z - R_{cz})^2}{R_0^2 g_z(x, z)} - \tau_G(x, y, z)\right\}, \quad (4.4.14)$$

where $\tau_G$ is the nonlinear optical depth of the medium according to Glicker's approximation; $R_{cz}(x, z) = 0.5 \int_0^x \int_0^{x'} dx'\, dx''(\partial\tilde{\varepsilon}(x'', 0, z)/\partial z)$ is the displacement of the beam in the vicinity of a point $\{y = 0, z\}$ caused by the wind leading to an asymmetry of the beam channel; and $g_j(x, z) = 1 + \int_0^x \int_0^{x'} dx'\, dx''(\partial^2\tilde{\varepsilon}(x'', o, z)/\partial R_j^2)$ is the local dimensionless beam width along the axis $j = y, z$.

Using the Glickler approximation for the energetic variable, one can obtain

$$R_{cz}(x, z) = R_0(\xi_1^2 Q_0 \exp(-z^2/R_0^2)/2)(\tau \exp(-\beta J_0(0, z)) + + \ln I_G(x, 0, z))/[1 - \exp(-\beta J_0(0, z))]; \quad (4.4.15)$$

$$g_z(x, z) - 1 = \xi_1^2 Q_0 \exp(-z^2/R_0^2)\ (Q_0 \exp(-z^2/R_0^2) - Z/R_0) \times \times (\tau \exp(-\beta J_0(0, z)) + \ln I_G(x, 0, z))/(1 - - \exp(-\beta J_0(0, z)) + Q_0 \exp(-z^2/R_0^2)[\tau + \beta(J(x, 0, z) - - J_0(0, z)) + \exp[-\beta J(x, 0, z)] - \exp[-\beta J_0(0, z)]]\}. \quad (4.4.16)$$

$$g_y(x, z) - 1 = \begin{cases} 2\xi_1^2 \beta J_0(0, z) \left[\dfrac{J(x, 0, z)}{J_0(0, z)} + \ln(J(x, 0, z)/J_0(0, z)) - 1\right], & \beta J_0 < 1 \\ 2\xi_1^2[\beta J(x, 0, z) - \pi^2/6 \ln \beta J(x, 0, z)], & 1 < \beta J_0 < \tau, \end{cases} \quad (4.4.17)$$

where $I_G = [(e^\tau - 1)\exp(-\beta J_0) + 1]^{-1}$, $Q_0 = \beta I_0 R_0/V_0$, and $\xi_1 = (|\tilde{\varepsilon}_{max}/R_0^2/\alpha_0^2)^{1/2}$ is the value characterizing the ratio of the length of the nonlinear interaction to the refraction length. As follows from (4.4.15), in the case of stationary self-action the thermal nonlinearity occurring in the zone of droplet vaporization is efficient if $\xi_1\sqrt{Q_0}\,\tau > 1$. The nonlinear distortions of a beam in the zone of droplet vaporization do not produce any effect on the process of aerosol dissipation if

$$\xi_1 < (\tau_{max} Q_0^{1/2})^{-1}, \tag{4.4.18}$$

where $\tau_{max} = \ln[( T_{tr}^{*-1} - 1) \exp(\sqrt{\pi}\, Q_0) + 1]$ is the maximum optical depth of the aerosol, $Q_0$ is the energetic parameter of the beam, and $T_{tr}^*$ is the transmission to be achieved in the 'clearing' process.

The problem of stationary beam self-action in evaporating aerosols was investigated in [5-7] using the method of parabolic equations. In [7] the solution of the parabolic equation was achieved by using the difference method of two-cycle component splitting. The solution of this equation was obtained for the case of stationary self-action of a collimated Gaussian beam from a $CO_2$ laser propagating through the atmospheric layer below cloud level. The calculations were made using standard models of the atmosphere. It was shown in this paper that, for the parameters of the beam and medium used in the calculations ($R_0 = 50$ cm, $I_0 \leq 10^3$ Wcm$^{-2}$, $V_0 = 5$ to 10 ms$^{-1}$, $q_0 = 0.2$ gm$^{-3}$, cloud height $x_{cloud} = 1$ km), the thermal distortions of the beam limiting the penetration of the beam into a cloud take place mainly in the atmospheric layer below cloud level. The intensity of the beam in an aerosol can be calculated using the Glickler formula (4.4.12), in which the beam's intensity at the cloud boundary should be taken as the initial value. In [5, 6] the integration of the parabolic equation was carried out numerically using the finite element method. It was shown in [5] that, under certain conditions, an increase of peak intensity of the beam can occur in the cleared zone. The authors of [5] explain this observation by suggesting the process of light diffraction on a soft diaphragm distributed along the beam's axis. The calculations in [6] showed that slight beam splitting can occur in the windward side of the channel due to nonlinear refraction.

The above discussion allows one to draw the following conclusions:

(1) The main factor limiting the dissipation of the aerosol in the beam channel is the nonlinear distortion of the beam caused by refraction within the channel, since the mean profile of the channel's dielectric constant - formed as a result of heating of the gases due to heat losses from evaporating droplets and molecular absorption of light - is not uniform. The beam broadening caused by fluctuations can affect the beam's energy parameters only in the cases of focused beams with long paths and when the thermal gaseous nonlinearity is weak.

(2) In the case of nonstationary beam self-action under conditions of weak gaseous thermal nonlinearity, the aerosol dissipation process has a wave-like character. The characteristic time of the extinction coefficient's decrease depends on such beam parameters as size, focusing, and beam divergence due to partial coherence, and on the propagation conditions in a clear atmosphere. The location of the aerosol dissipation front is determined by the above factors, along with the beam defocusing taking place in the zone of droplet vaporization. The nonlinear distortions in the inter-

mediate zone are minimal for broad beams. In this case, the limitation of the aerosol dissipation level is related to the effects of the thermal nonlinearity of the surrounding gas.

In the case of stationary beam self-action, the nonlinear distortions of the beam in the intermediate zone lower the efficiency of the dissipation process. The main factor limiting the efficiency of aerosol dissipation when using broad beams is the thermal nonlinearity of the surrounding gas in the aerosol-free regions of the beam.

In conclusion, we will discuss some aspects of the problem of the propagation of sounding beams with wavelengths in the visible region through the cleared channels.

The studies of laser beam propagation in the channels burnt through dense aerosols by high-power laser beams are of paramount importance for such applications as light energy transportation, optical communication, ranging, and remote, contact-less sensing of these channels. Discussions of various aspects of this wide problem can be found in many papers, e.g., [3, 25-29].

The refraction of narrow beams of visible radiation in the zones of interaction between high-power laser beams and artificial aerosols was studied experimentally in [25] using cw $CO_2$ laser. Figure 4.4.5 presents the results of measurements of He-Ne laser beam displacements occurring in

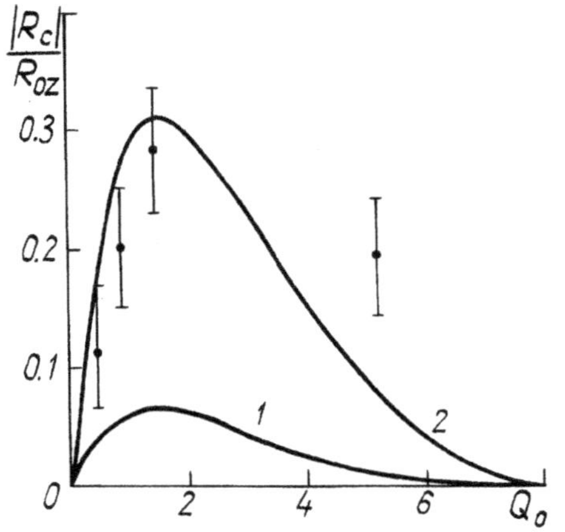


Fig. 4.4.5. The dependence of the sounding beam's displacement in the cleared channel on the parameter of thermal action with $\tau = 1.2$; $R_0/R_{0z} = 5.75$; $\xi = 0.261$ (curve 1); $\xi = 1.34$ (curve 2); the dots show experimental data from [25].

the cleared channel at different parameters of the thermal action of the beam. The experiments were carried out using a cw $CO_2$ laser beam (800 W power) with $\lambda = 10.6$ μm with radius $R_0 = 2.3$ cm, and a sounding beam with

$\lambda = 0.63$ μm with $R_0 = 0.4$ cm in an aerosol with $\alpha$ ($\lambda = 10.6$ μm) $= 0.3$ $m^{-1}$ and a path length of 4 m. The velocity of regular movement of the medium varied from 0.1 to 0.5 $ms^{-1}$. The experimental set-up allowed for a coaxial configuration of the two beams. In this same figure the results are presented of theoretical calculations of the sounding beam gravity center wanderings near the high-power beam axis, carried out using the nonaberrational approximation [27]. The basic parameters of the process are $Q_0 = \beta I_0 (R_0/V_0)$, $\xi_1 = (|\tilde{\varepsilon}_{max}/R_0^2\alpha_0^2)^{1/2}$, $R_0/R_{0z}$. Curve 1 in this figure presents the calculational results obtained using the process parameters realized in the experiment. Satisfactory agreement between theoretical and experimental results was obtained only after the parameter $\simeq\xi_1(R_0/R_{0z})^{1/2}$ was taken instead of $\xi_1$ (see curve 2). This fact is explained as being due to the influence of inhomogeneities of the dielectric constant in the interaction zone. The scale of these inhomogeneities was comparable with the dimensions of the sounding beam. These inhomogeneities could be due to the multi-mode structure of the high-power beam used in this experiment. The data presented in Figure 4.4.6 illustrate the dependence of the effective parameters of

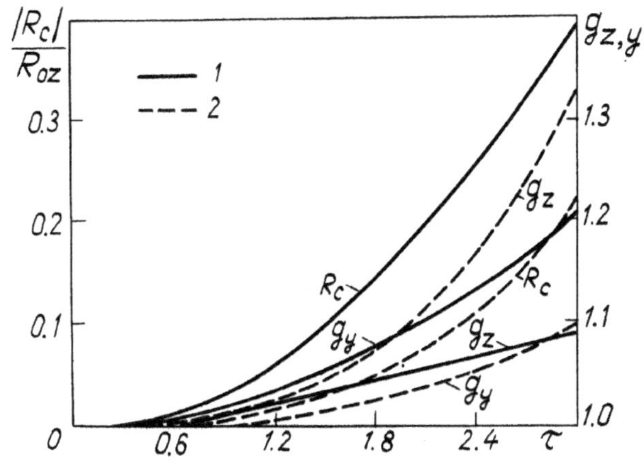


Fig. 4.4.6. Dependence of the parameters of the sounding beam in a cleared channel on the channel's optical depth for $\lambda = 10.6$ μm, with $R_0/R_{0z} = 5.75$, $\xi = 0.26$, $Q_0 = 1.5$ (curve 1) and $Q_0 = 4.5$ (curve 2).

the sounding beam, propagating coaxially with the high-power beam, on the optical depth of the fog measured at $\lambda = 10.6$ μm. The data presented in this figure were calculated for the conditions prevailing in the experiment carried out in [25].

The paper [2] presents experimental and theoretical results relating to to the fluctuations of the sounding beam intensity in the cleared channel produced in an artificial fog by a high-power $CO_2$ laser. It was found in this investigation that the intensity fluctuations observed in the zone of droplet vaporization are mainly due to fluctuations of the aerosol extinction coefficient.

### 4.4.3. The Influence of the Turbulent Motion of the Medium on the Dissipation of Water Aerosols by Laser Beams

Turbulent motion of the cloud medium makes the time during which a droplet is in the energetically active zone of the beam a random value. As a consequence, the mean level of the complex effective dielectric constant changes and its fluctuating component appears (see § 4.2); that, in turn, can cause changes in the high-power beam energetics and in the whole clearing process.

If the beam has no phase-amplitude distortions caused by diffraction and perturbations of the real part of the complex dielectric constant, then the solution for an instantaneous value of the beam intensity can be presented in the Bouguer form:

$$I(x, \vec{R}, t) = I_0(\vec{R}, t) \exp[-\tau_N(x, \vec{R}, t)], \tag{4.4.19}$$

where $\tau_N = \int_0^x \alpha(x', \vec{R}, t)\, dx'$ is the nonlinear optical depth. In the case of an exponential approximation of the aerosol extinction coefficient's dependence on the density of light energy in a localized volume of turbulent medium (the energetic variable), one obtains

$$J(x, \vec{R}, t) = \int_0^t I(x, \vec{R}(t'), t')\, dt', \tag{4.4.20}$$

where $\vec{R}(t') = \vec{R} - \int_{t'}^{t} V_\perp(x, \vec{R}(t''), t'')\, dt''$ is the trajectory of particle vaporization in the velocity field.

The equation for the nonlinear optical depth is

$$\begin{aligned} \frac{\partial \tau_N}{\partial t} + \vec{V}_\perp \nabla_{\vec{R}} \tau_N &= -\beta I_0(\vec{R}, t)(1 - \exp(\tau_N)) + \\ &+ \int_0^x dx' \, \frac{\partial \vec{V}_\perp(x', \vec{R}, t)}{\partial x'} \nabla_{\vec{R}} \tau_N . \end{aligned} \tag{4.4.21}$$

If $\vec{V}_\perp = \mathrm{const}$, then the solution of (4.4.21) is

$$\tau_N = \ln[(e^{\tau} - 1)\, e^{-\beta J_0} + 1], \tag{4.4.22}$$

where

$$J_0(\vec{R}, t) = \int_0^t I_0(\vec{R} - \vec{V}_\perp(t - t'), t')\, dt'$$

is the boundary value of the energetic variable on the plane $x = 0$. The distribution of the beam's intensity in the interaction channel, in this case, is described by the Glickler formula (4.4.12).

The investigation of the effect of wind speed fluctuations on the process of aerosol dissipation by laser beams (discussed in Chapter 3) was carried out based on the study of the behavior of the energetic variable. The representation of the energetic variable used in Chapter 3 differs a little from that described by (4.4.20). Using the representation (4.4.20), one can obtain the following equation for the evolution of the energetic variable along the x-axis:

$$\frac{\partial J}{\partial x} + \int_0^J \alpha(J')\,dJ' - \int_0^t dt' \int_{t'}^{t} dt'' \,\frac{\partial \vec{V}_\perp(x, \vec{R}(t''), t'')}{\partial x} \times \times \nabla_{\vec{R}} I(x, \vec{R}(t''), t'') = 0. \quad (4.4.23)$$

If the atmospheric situation satisfies the requirements of the hypothesis of 'frozen' turbulence, and if the scale of wind speed fluctuations is large enough for the spatial variations of the wind speed within the beam's cross-section to be considered as being negligible, then $\vec{V}_\perp(x, \vec{R}, t) = \vec{V}_\perp(x, \vec{R} - \langle\vec{V}_\perp\rangle t)$ and one can assume that $\vec{V}_\perp(x, \vec{R}(t'), t') = \vec{V}_\perp(x)$ in equation (4.4.23).

If the process is stationary, then the wind does not alternate in its direction and, if this direction coincides with the z-axis, then one can obtain from (4.4.23)

$$\frac{\partial J}{\partial x} + \int_0^J \alpha(J')\,dJ' + \frac{\partial \ln v_\perp}{\partial x} J = 0. \quad (4.4.24)$$

The influence of large scale inhomogeneities of the wind speed on the configuration of cleared zones in clouds was considered in Chapter 3 for situations where (4.4.24) is valid. If the scales of the wind speed fluctuations are comparable with the radius of the laser beam, then the analysis of the process can be better made using the equation for the nonlinear optical depth $\tau_N$. Let us restrict our discussion to the case when the speed of the regular motion of the medium is constant along the beam channel, $\vec{v}_\perp = \text{const}$. In the region of weak fluctuations of the beam intensity and energetic variable, the beam intensity averaged over the wind speed fluctuations, which are assumed to be of a Gaussian type, is described as follows:

$$\langle I(x, \vec{R}, t)\rangle = I_0(t)\,[(e^{\tau} - 1)\,\exp[-\beta\langle J_0(x, \vec{R}, t)\rangle] + 1]^{-1}, \quad (4.4.25)$$

where

$$\langle J_0(x, \vec{R}, t)\rangle = (2\pi)^{-1} \int_0^t dt' \iint_{-\infty}^{\infty} d^2R' \,\frac{I_0(\vec{R}', t - t')}{D(t')} \times \times \exp\left(-\frac{|(\vec{R} - \vec{R}' - \langle\vec{v}_\perp\rangle)t'|^2}{2D(t')}\right), \quad (4.4.26)$$

here

$$D(t) = \tfrac{1}{2}\int_0^t\int_0^t dt'\, dt''\langle \vec{v}_\perp'^{L}(t')\vec{v}_\perp'^{L}(t'')\rangle$$

is the variance of Lagrangian particle drifts in the field of the fluctuating component of the Lagrangian wind speed. $\vec{v}_\perp'^{L}(t) = \vec{v}_\perp'(\vec{R}(t), t)$. If $I_0(\vec{R}, t) = I_0(t)\exp(-R^2/R_0^2)$, then

$$\langle J_0(x, \vec{R}, t)\rangle = \int_0^t dt' \frac{I_0(t - t')}{1 + 2D(t')/R_0^2}\exp\left\{-\frac{|\vec{R} - \langle\vec{v}_\perp\rangle t'|^2}{1 + 2D(t')/R_0^2}\right\}.$$

The expression (4.4.25) generalizes the Glickler formula for the case of beam self-action in an aerosol media with a fluctuating wind speed. Let the correlator be $\langle\vec{v}_\perp'^{L}(t')\vec{v}_\perp'^{L}(t'')\rangle = \tfrac{1}{2}\langle\vec{v}_\perp'^{2}\rangle\exp(-|t' - t''|/t_0)$. In this case, $D(t) = 2\chi_t(t - t_0(1 - \exp(-t/t_0)))$. Here, as introduced in (4.2.2), $t_0$ is the time of correlation of the Lagrangian wind speed, and $\chi_t$ is the coefficient of turbulent diffusion. It is assumed in further discussions that $\vec{v}_\perp = \vec{v}_z = \vec{v}_0$, $I_0 = \text{const}$.

By introducing the dimensionless time $\bar{t} = V_0 t/R_0$ $(V_0 \neq 0)$, $\bar{t} = I_0 t$ $(V_0 = 0)$, one finds that, at a fixed optical depth of the medium, the problem is characterized by the following parameters:

$$\Gamma = \frac{4\chi_t}{R_0 V_0};\qquad \bar{t}_0 = \frac{t_0 V_0}{R_0};\qquad Q_0 = \frac{\beta I_0 R_0}{V_0},\qquad (V_0 \neq 0);$$

$$\Gamma = \frac{4\chi_t}{R_0^2\beta I_0};\qquad \bar{t}_0 = \beta I_0 t,\qquad (V_0 = 0).$$

The calculation of the beam channel transmission carried out using (4.4.25) showed that the influence of turbulence on the process of beam self-action is strongest at $V_0 = 0$ and at large values of the parameters $\Gamma \gtrsim 10$.

If $\bar{t} > \Gamma^{-1}$, then the turbulence can prevent the penetration of high-power radiation into a cloud. This fact was first discussed in [20], where the problem was solved using the diffusion approximation. If one takes into account the finiteness of the correlation time of wind speed fluctuations, then one can see that the influence of turbulence on the process of aerosol dissipation by a laser beam is weak, even at large values of $\Gamma$ (see Figure 4.4.7).

For $\Gamma > \Gamma_0$, where $\Gamma_0$ is some characteristic value, the random wind 'smearing' of the beam dominates over the regular one. In this case, the transmission of the beam channel decreases. The value of $\Gamma_0$ depends on $t_0$ and it decreases with a decrease in $t_0$. For $t_0 = 0$, $\Gamma_0 \simeq 1$ (by the diffusion approximation).

In the case of stationary self-action $(v_0 \neq 0)$ of the beam in such atmospheric conditions, and for selected beam parameters so that $\Gamma_0 > 10$, $\bar{t}_0 > 10$, the atmospheric turbulence does not affect the process of aerosol dissipation by high-power radiation. For $V_0 = 0$ and $\bar{t} \gg \bar{t}_0$, $\Gamma^{-1}$, the influence of atmospheric turbulence on the beam channel transmission is significant.

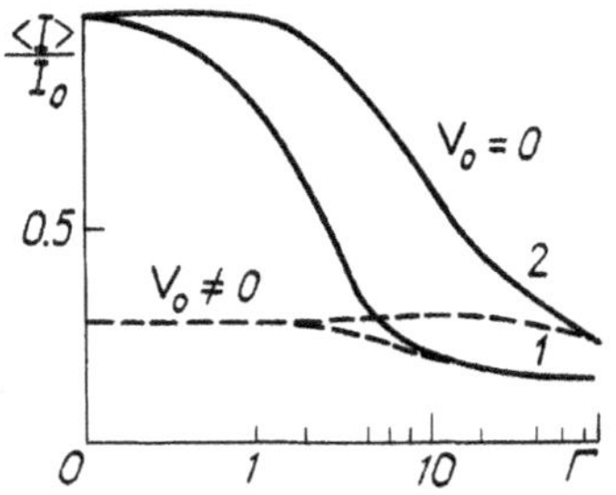


Fig. 4.4.7. The influence of random wind smearing on the process of water aerosol dissipation by a laser beam, with $\tau = 2$; $Q_0 = 1$, $\bar{t} = 20$, $\bar{t}_0 = 0.1$ (curve 1); and $\bar{t}_0 = 10$ (curve 2).

## REFERENCES: CHAPTER 4


[1] V. E. Zuev, Yu. D. Kopytin, and A. V. Kuzikovsky: Nonlinear Optical Effects in Aerosols (Nauka, Novosibirsk, 1980), in Russian.

[2] O. A. Volkovitsky, Yu. S. Sedunov, and L. P. Semenov: Propagation of High-Power Laser Radiation in Clouds (Gidrometizdat, Leningrad, 1982), in Russian.

[3] A. A. Zemlyanov and A. V. Kuzikovsky: 'Laser beam self-action in randomly inhomogeneous water aerosols', in Optical Wave Propagation through a Randomly Inhomogeneous Atmosphere (Nauka, Novosibirsk, 1979), pp. 104-112.

[4] A. A. Zemlyanov, V. V. Kolosov, and A. V. Kuzikovsky: Kvant. Elektron. **6**, 1148-1153 (1979) (Sov. J. Quantum Electron.).

[5] K. D. Egorov, V. P. Kandinov, and M. S. Prakhov: Kvant. Elektron. **6**, 2562-2566 (1979) (Sov. J. Quantum Electron.).

[6] S. A. Armand and A. P. Popov: Radiotekh. Elektron. **25**, 1793-1800 (1980) (Radio Eng. Electron.).

[7] M. P. Gordin, V. P. Sadovnikov, and G. M. Strelkov: Radiotech. Elektron. **27**, 1457-1461 (1982) (Radio Eng. Electron.).

[8] M. V. Vinogradova, O. V. Rudenko, and A. P. Sukhorukov: The Theory of Waves (Nauka, Moscow, 1979).

[9] V. I. Tatarsky: Wave Propagation in a Turbulent Atmosphere (Nauka, Moscow, 1967), in Russian.

[10] S. M. Rytov: Introduction to Statistical Radiophysics (Nauka, Moscow, 1966).

[11] S. M. Rytov, Yu. A. Kravtsov, and V. I. Tatarsky: Introduction to Statistical Radiophysics, Part. 2: Stochastic Fields (Nauka, Moscow, 1978), in Russian.

[12] V. I. Klyatskin: The Statistical Description of Dynamic Systems with Fluctuating Parameters (Nauka, Moscow, 1975), in Russian.

[13] Yu. S. Sedunov: Physics of the Formation of the Water Droplet Phase in the Atmosphere (Gidrometizdat, Leningrad, 1972), in Russian.

[14] D. C. Smith: Proc. IEEE **6**, 1679-1714 (1977).

[15] J. A. Fleck, J. R. Morris, Jr., and M. D. Feit: Appl. Phys. **10**, 129-160 (1977).

[16] V. L. Mironov: Laser Beam Propagation through a Turbulent Atmosphere (Nauka, Novosibirsk, 1981), in Russian.

[17] V. L. Mironov, V. V. Nosov, and B. N. Chen: Izv. Vyssh. Uchebn. Zaved. Fiz. **4**, 37-40 (1981) (Sov. Phys.).

[18] L. D. Landau and E. M. Lifshits: Statistical Physics, Part 1 (Nauka, Moscow, 1976), in Russian.

[19] A. S. Monin and A. M. Yaglom: Statistical Hydromechanics, Part II (Nauka, Moscow, 1967), in Russian.

[20] A. P. Sukhorukov and E. P. Shumilov: Zh. Tekh. Fiz. **48**, 1029-1041 (1973) (J. Sov. Tech. Phys.).

[21] M. Born and E. Wolf: Principles of Optics (Pergamon Press, Oxford, 1968), p. 719.

[22] H. C. van de Hulst: Light Scattering by Small Particles (Wiley, New York, 1957).

[23] G. I. Marchuk: Methods of Computational Mathematics (Nauka, Moscow, 1980), in Russian.

[24] G. Strang and J. Fix: The Theory of the Finite Element Method (Nauka, Moscow, 1980).

[25] V. A. Belts, O. A. Volkovitsky, L. F. Nerushev, and V. P. Nikolaev: in Atmospheric Optics: Proc. Inst. Experiment. Meteorol. (Gidrometizdat, Moscow, 1978), pp. 67-77, in Russian.

[26] M. S. Belen'ky and A. A. Zemlyanov: Kvant. Elektron. **6**, 853-855 (1973) (Sov. J. Quantum Electron.).

[27] A. A. Zemlyanov: 'Propagation of a Narrow Sounding Beam in the Interaction Channel between a High-Power Light Beam and its Medium', in Remote Sensing of Atmospheric Physico-Chemical Parameters Using High-Power Lasers (Inst. Atm. Opt., Tomsk, 1979), pp. 102-106, in Russian.

[28] V. V. Kolosov and A. V. Kuzikovsky: Kvant. Elektron. **8**, 490-493 (1981) (Sov. J. Quantum Electron.).

[29] V. P. Bisyarin, V. V. Efremenko, M. A. Kolosov, V. N. Pozhidajev, A. V. Sokolov, G. M. Strelkov, and L. V. Fedorova: Izv. Vyssh. Uchebn. Zaved. Fiz. **2**, 23-45 (1983) (J. Sov. Phys.).

CHAPTER 5

# LASER BEAM PROPAGATION THROUGH AN EXPLOSIVELY EVAPORATING WATER-DROPLET AEROSOL

Regular (surface) regimes of vaporization of droplets heated by laser radiation are changed, under certain conditions, by droplet explosion. As experiments have shown, the explosion processes vary, depending on the laser beam energetics, the absorption index of the droplet substance, and the droplet's diffraction parameter. The explosion of an absorbing droplet is caused by the phase transition from a liquid to a vapour in regions where the electromagnetic wave energy is dissipated in the form of heat. The above factors determine the character of the phase transition within the droplet.

The explosion of droplets heated by radiation causes an essential nonlinearity in the interaction of high-power laser radiation with aerosols. Methodically, the description of laser beam propagation through the explosively evaporating water-droplet aerosol does not differ from that used in the problems of laser beam self-action in aerosols under regular regimes of droplet vaporization.

The dynamics of the optical properties of the exploding particles, as well as the relations between these properties, the beam parameters, and the characteristics of the medium, are the key questions for the solution of this problem.

This chapter is devoted to a systematic treatment of the vast amount of material devoted to different aspects of the explosion of absorbing particles irradiated by high-power laser radiation. Great attention is paid in this chapter to the physical nature of the phenomenon and to the basic concepts used in the construction of the explosion models, and to the optical parameters of the exploding droplets. The experimental data on the propagation of laser radiation through aerosol media with exploding droplets are considered.

## 5.1. DROPLET EXPLOSION INITIATED BY HIGH-POWER LASER RADIATION

Droplet explosion initiated by high-power laser radiation was predicted theoretically in [1] and investigated experimentally in many works [10, 12, 16-30, 39]. Theoretical treatment of the process based on models which take into account the most essential features of the process can also be found in the literature [1, 4, 10-15, 32-34, 36].

The problem of a droplet explosion initiated by laser radiation is the subject of a branch of laser physics - the physics of the non-resonant interaction of high-power laser radiation and condensed matter [2].

### 5.1.1. Droplet Explosion as an Optothermodynamic Process

In terms of methodology, the optothermodynamic approach is the most important aspect in the problem of the non-resonant interaction of high-power laser radiation and condensed matter [52]. This approach is based on the study of the influence of laser radiation characteristics on thermodynamic processes within the substance, this approach allows one to qualitatively analyze the problem. That, in turn, makes it possible to perform quantitative investigations more efficiently, as well as to make a correct interpretation of the experimental results and, last but not least, the approach allows one to predict new physical effects stimulated by laser radiation in condensed matter.

The application of the optothermodynamic approach to the problem of the interaction of high-power optical radiation and aerosols has been studied and developed in various papers [3, 4, 34]. The thermodynamic transitions of a volume of droplet substance can follow different paths, depending on the amount of light energy dissipated in the form of heat and on the rate of its release. In the subcritical region the trajectory of the transition from the liquid to the vapor phase always crosses the region of unstable states, and is accompanied by jumps in the values of some of the thermodynamic parameters that are connected with the fundamental properties of any phase to have the phase boundary.

In the region corresponding to supercritical states the system is not divided into regions of coexistent phases. In this case, the fluctuation inhomogeneities appearing inside the initial phase have no surface tension [9].

The role of each type of transition is eventually revealed in the dynamic behavior of the droplet. Thus, the thermal effects taking place inside a droplet irradiated with a high-power light flux can be treated as the effects of optothermodynamic transitions of certain type.

The characteristic feature of optothermodynamic processes in liquids is their nonlinearity. When increasing the laser beam energetics, one observes, at a certain stage, a qualitative change in the process. Thus, volume vaporization is observed instead of surface vaporization, and consumption of the energy of the beam decreases. The nonlinearity of the processes inhibits the existence of different regimes of laser interaction with a liquid aerosol, which can exist if certain threshold conditions are fulfilled. The existence of different regimes of droplet explosion are also due to this nonlinearity.

Curves plotted in Figure 5.1.1 illustrate qualitatively different optothermodynamic transitions in the elementary volume of a droplet. Isobar $LL_1$ and the line of saturation $GG_1$ on the phase plane correpond to the surface diffusion regimes of evaporation. The transition $GG_1$ represents a convective regime of droplet vaporization, when the pressure of saturated vapors at the droplet surface equals the atmospheric pressure.

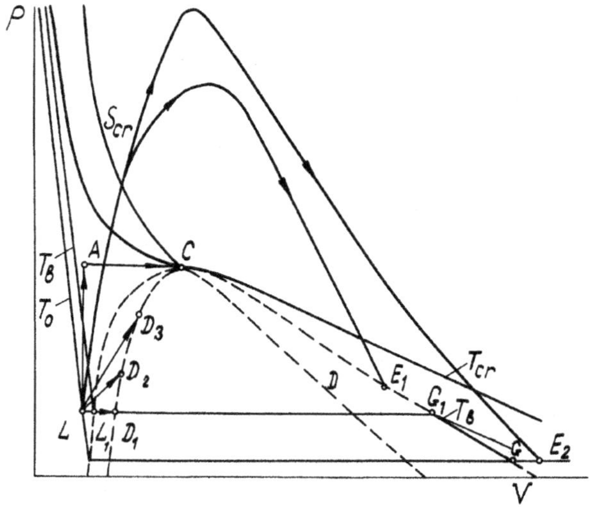


Fig. 5.1.1. Qualitative representation of the optothermodynamic transitions of an elementary volume of liquid. C - critical point; $L_1CG$ - binodal; D - spinodal; $T_0$ - initial isotherm; $T_b$ - isotherm of boiling; $T_{cr}$ - critical isotherm; $S_{cr}$ - segment of critical isentrope.

With significant superheating of the droplet, deep penetration into the region of metastable liquid states occurs (transition $LD_1$) and the phase coordinates of the liquid state increase (transitions $LD_2$, $LD_3$). The transitions $LE_1$ and $LE_2$, crossing the critical isentrope, correspond to the drift of liquid in the region of a one-phase thermodynamic state. Penetration into the region of metastable states from the gas phase results in recondensation of the expanding vapor.

Let us give some calculated results concerning the optothermodynamic trajectories for the case of uniformly-absorbing particles.

First, we will find the conditions under which the droplet reaches the boundary of absolute instability (spinodal) when it undergoes isobaric heating (see transition $LD_1$ in Figure 5.1.1). This transition is described by the following thermodynamic relationship: $dH = dQ$.

Let the effects of surface evaporation and heat conductivity be negligible during the time of heating, then one has

$$dQ = k_{ab}(T, V)VI\ dt.$$

The energy density of a light pulse can be written as

$$w_p(t) = (k_{ab}V)^{*-1}(H_{ai} - H_0), \tag{5.1.1}$$

where $H_0$ = 80 J/g; [8] $H_{ai}$ = 1500 J/g [5], the asterix denotes that this value is characteristic for the region of metastable states. Here, $H_{ai}$ is value H on the spinodal.

An intensity of about $10^5$ to $10^6$ W/cm$^2$ is quite sufficient for such heating to be realized in water droplets of less than 10 μm diameter with the use of $CO_2$-laser pulses of $10^{-5}$ - $10^{-6}$ s duration.

Let us now consider the energetics under which particles expand within a region corresponding to a single-phase state. In this case, the regime of heat release should be such that the path of the thermodynamic process passes around the critical point, being then above the curve representing the phase equilibrium. The movement in the phase plane along the curve of phase equilibrium is energetically optimal. As the estimates in [8] have shown, using tables of the termodynamic parameters of water, the amount of heat required for a substance to achieve a near-critical state from its initial state is of an order of magnitude equal to the heat of vaporization at the normal boiling point, $Q_e^b = Q_e$ ($T_b$, P = 1 bar). Meanwhile, the movement of the substance inside a single-phase region requires that an amount of heat not less than $2Q_e^b$ is absorbed by the liquid. In the case of other paths like e.g., isochoric heating to the supercritical state with further unloading along the adiabatic curve lying above the critical one, calculations show that the energy required for droplet expansion in the supercritical region is also not less than $2Q_e^b$. These calculated results are in good agreement with the qualitative estimations [7] of the energy consumption necessary for the substance (droplet) to expand in the single-phase region. Figure 5.1.2 shows the adiabatic curves for water, which we calculated using the equation of water state obtained by Kuznetsov [41].

Assessments of the laser pulse parameters required to sustain the thermodynamic transitions of the liquid in a droplet within a single-phase region are presented below.

The intensity of short laser pulses with pulse duration $t_p \leqslant r_0 c_{s0}$ (here $r_0$ is the initial radius of a droplet and $c_{s0}$ is the speed of sound in the droplet material), sufficient for evaporating a uniformly-absorbing droplet in a single-phase region, should obey the following relationship:

$$I_{s0} = 2Q_e^b / k_{ab} V_0 t_p. \tag{5.1.2}$$

The lower limit of the intensity required for a supercritical transition at the instantaneous energy release is defined as follows:

$$I_{sc}^- = 2Q_e^b \rho_0 c_{s0} / k_{ab} r_0. \tag{5.1.3}$$

Here, the subscript '0' denotes the initial values of the parameters inside a droplet.

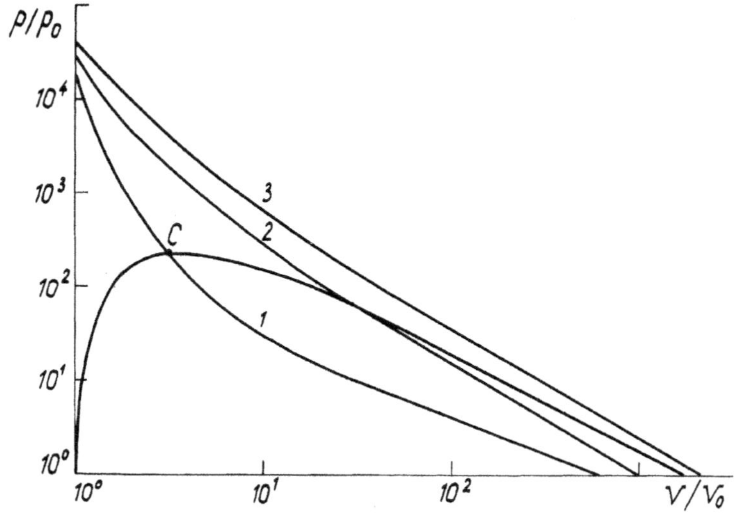


Fig. 5.1.2. Water unloading adiabat at initial pressures $P_H$ = 20 kbar (1); 30 kbar (2); 40 kbar (3).

For large droplets, $r_0 \gg k_{ab}^{-1}$ and on a plane surface this intensity is estimated as

$$I_{sc\infty}^{-} = 2Q_e^b \rho_0 C_{s0}. \tag{5.1.4}$$

For long pulses, when the nonadiabatic processes taking place during the droplet explosion become of great importance, the upper limit of the laser pulse intensity necessary for droplet expansion in the supercritical region should be estimated as follows:

$$I_{sc}^{+} = Q_e^b C_{s0} \rho_0 (k_{ab} r_0, \quad k_{ab} r_0 < 1, \quad t_p \gg r_0/C_{s0} \tag{5.1.5a}$$

or

$$I_{sc\infty}^{+} = Q_e^b C_{s0} \rho_0, \quad k_{ab} r_0 \gg 1, \quad t_p \gg (k_{ab} C_{s0})^{-1}. \tag{5.1.5b}$$

The lower limit of this intensity at $t_p \gg r_0/C_{s0}$ can be found by considering the following trajectory of the substance: $(p_0, V_0) \to (p_{cr}, V_0) \to (p_{cr}, V_{cr})$ (i.e., the transition LAC in Figure 5.1.1). The intensity of the leading portion of the pulse, which can provide isochoric heating up to the pressure $p_{cr}$, should satisfy the condition

$$I_f = C_v(V_0, T_0)(T_A - T_0)/k_{ab} V_0 t_f, \quad t_f < r_0/C_{s0}. \tag{5.1.6}$$

A substance can move along the trajectory $p = p_{cr}$ if the pulse parameters are sufficient to sustain the constant velocity of expansion, which is equal to that achieved at the moment it is heated up to $p = p_{cr}$. This velocity is determined using an acoustic approach for the plane case, as $v_c = p_{cr}/{}_0C_{s0}$. For $p = p_{cr}$, $dH = (\partial H/\partial \rho)_{p_{cr}} d\rho$, and the required intensity is

$$I = (\partial H/\partial \rho)_{p_{cr}} (d\bar{\rho}/dt)\bar{\rho} k_{ab}^{-1}(T, \bar{\rho}), \tag{5.1.7}$$

where $\bar{\rho} = \rho_0(r_0^3/r_c^3)$ is some average density of the droplet, and $r_c$ is the coordinate of a droplet (boundary) surface. Taking into account that $d\bar{\rho}/d\rho = - 3\bar{\rho}v_c/r_0$, and assuming that $k_{ab} = k_{ab}^0(\rho/\rho_0)^2$ [35], one obtains for $V = V_{cr}$:

$$I_{cr} = -(\partial H/\partial \rho)_{p_{cr}} 3p_{cr}\rho_0/r_0 k_{ab} C_{s0}. \tag{5.1.8}$$

The characteristic duration of the portion of the pulse sufficient to provide such heating, is estimated as $t_{cr} \simeq (r_{cr} - r_0)/v_c$. For $r_0 = 10$ μm, one obtains $t_f \simeq 5 \times 10^{-9}$ s; $I_f \simeq 4 \times 10^7$ Wcm$^{-2}$, $I_{cr} \simeq 10^7$ Wcm$^{-2}$, and $t_{cr} \simeq 10^{-7}$ s.

Thus, optothermodynamic analysis shows that there exist two types of transitions in a droplet, namely two-phase and single-phase transitions. The explosion of an absorbing droplet is a process involving both kinetic and dynamic factors that accompany the optothermodynamic transitions. According to contemporary concepts, these factors are the following: shock boiling of liquid [5, 6], cavitation [5 - 7], evaporation of liquid in the wave of rarefaction [5, 7], and thermal instability of the phase transitions fronts [2, 12]. These factors will be discussed below.

### 5.1.2. Experiments

Experimental studies of explosive droplet vaporization initiated by high-power laser radiation have been carried out for a wide range of droplets sizes from 5 to $10^3$ μm, covering the range of droplets sizes found in natural meteorological objects. Both cw and pulsed laser sources with wavelengths λ, λ = 10.6, 2.36, 1.06, and 0.69 μm were used in the experiments. The power density in the region of the droplets varied from $10^2$ to $10^9$ Wcm$^{-2}$.

We do not propose to discuss all of these experiments. Instead, let us consider the results of those which give us an idea of the basic physical regularities in the process of explosive droplet vaporization.

It should be noted that information on the initial distribution of heat sources inside the particles is of principal importance for the interpretation of experimental results. The local release of heat per mass unit per time unit is defined as $Q_{ab} = k_{ab}I_0VB$, where the function $B = I_a/I_0$ characterizes the nonuniformity of the light field intensity distribution. Figure 5.1.3 presents the typical behavior of the function B [12, 13] along the principal diameter of a droplet. It follows from the figure that there exist characteristic heat release scales which depend on the diffraction parameter of a droplet and the complex refraction index of its constituent substance.

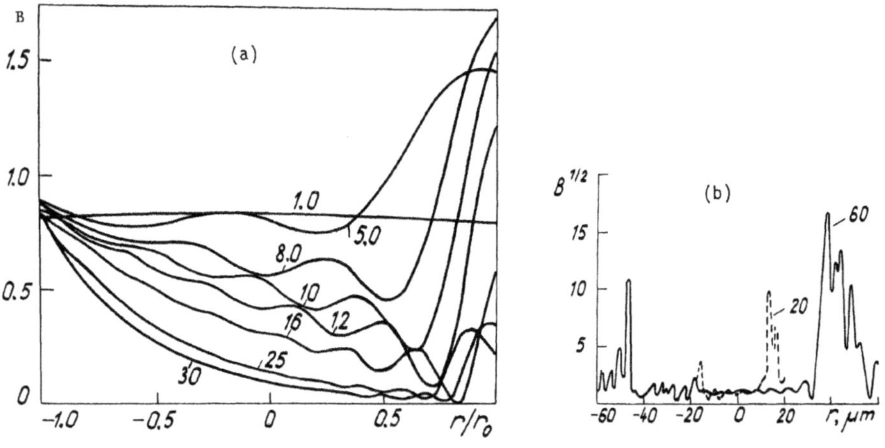


Fig. 5.1.3. Relative intensity distribution of a light field inside a droplet along the diameter coinciding with the direction of incident radiation (from left to right). The numbers near the curves are the values of $r_0$ in μm (a) λ = 10.6 μm, (b) λ = 0.69 μm.

A large number of experiments have been carried out on the action of laser radiation with λ, λ = 0.69, 10.6, 2.36 μm on large particles ($r_0 \simeq 25 - 400$ μm) of both pure and colored water, using laser sources operating both in a free generation mode (λ, λ = 0.69, 1.06 μm) and in a repetitively pulsed mode with Q-switching (λ = 2.36 μm) [10, 16, 20, 21]. It is a well-established fact that the blowoff of vapor and condensed liquid from the back and front hemispheres, where the heat centers are localized, is characteristic for the explosion of weakly-absorbing droplets. The threshold of the effect is about $10^5 - 10^6$ $Wcm^{-2}$. The diagram of the explosion products' angular spreading shows a distribution stretched along the direction of incident radiation. First, the material of the back (shadowed) hemisphere is blown off, and then the material of the front (illuminated) one. This shows that the maximum value of the light field intensity in the back hemisphere is higher than that in the front hemisphere of the droplet.

The degree of destruction of weakly-absorbing particles is not high. This is due to weak vaporization in the regions of heat release. As the calculations show [14], the volume of the hot regions in the back hemisphere of a cold droplet is about $10^{-3}$ of the droplet's volume, and about $10^{-4}$ in the front hemisphere. An increase in the liquid's absorption coefficient increases the extent of the droplet's destruction. In [20] the absorption coefficient has been varied in the range $(0.5 - 2) \times 10^{-4}$. The explosion in this case is of a multi-stage character. First, the material is being blown off locally, than strong deformation of the droplet is observed and, finally, the droplet is fragmented entirely.

Figure 5.1.4 shows the experimental data for the radiation energy density when a weakly-absorbing particle is completely destroyed [20, 39].

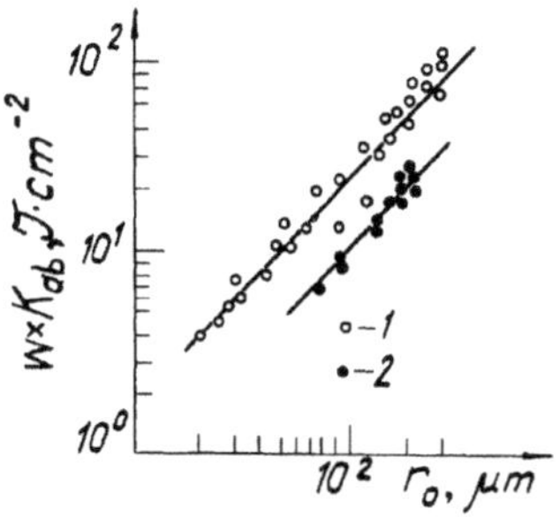


Fig. 5.1.4. Experimental dependence of the absorbed energy necessary for total destruction of droplet starting from its initial radius [20, 38].
1 – $\lambda = 0.69\ \mu m$ ($\kappa_a = 2.1\times10^{-4}-5\times10^{-5}$);
2 – $\lambda = 2.36\ \mu m$ ($\kappa_a = 8\times10^{-4}$).

The explosions of water droplets caused by pulsed $CO_2$ laser radiation were observed in [17-19, 23, 25-30, 36]. The authors of [17] took photomicrographs of the explosion of droplets irradiated with a pulse of $CO_2$ laser radiation with about 0.5 J of energy and a duration of 300 ns at half maximum of the pulse height. They studied droplets in the size range 7-50 μm radius. The average power density in the region of the water droplets was of the order of 10 $MWcm^{-2}$. For droplets with a radius of 12 μm they observed symmetrical explosions. For droplets with a radius of 15 μm the explosion begins to be asymmetric and the increase in radius gives rise to front surface blowoff. The average velocity of the expanding material after the explosion of a droplet with a radius of 20 μm was Mach 0.4 (in air), measured during a 1.3 μs time interval. A pulsed explosion type was observed for large droplets.

In [18] the investigation of the velocities of shock waves produced by the exploding droplets has been carried out using the Schlirien technique. A $CO_2$ laser used in the experiments provided a power density of about 10 to 30 $MWcm^{-2}$ in the region of focusing, and an energy density of 5 to 15 $Jcm^2$, with the pulse duration varying from 25 to 175 ns. The size range of the droplets studied was 5-70 μm. As the experiments showed, the velocity of a shock wave at the beginning of the explosion significantly exceeded the speed of sound in air (Figure 5.1.5).

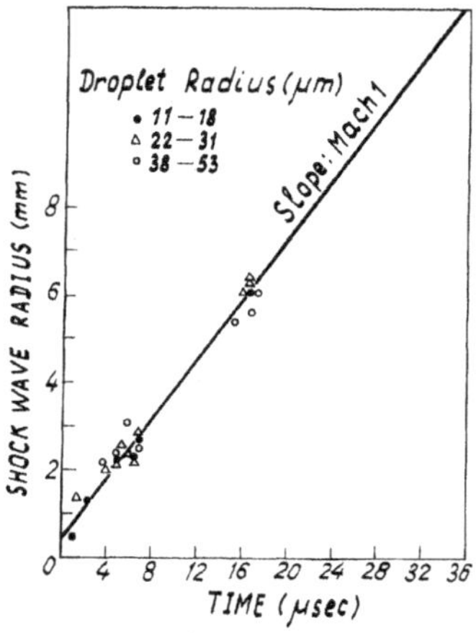


Fig. 5.1.5. Experimental time-dependence of shock wave radius during droplet explosion in a $CO_2$ laser field [18].

These experiments clearly showed that the explosion of small droplets caused by radiation with an intensity of $10^7$ $Wcm^2$ can be described as a quasi-continuous flow of the medium with a high degree of vaporization accompanied by the production of shock waves in air.

A symmetrical, angle diagram of the expansion of the material in the case of droplets with $r_0 \simeq 12$ μm indicates that mechanisms exist for smoothing the temperature field inside the droplets. As follows from the a priori estimates, the field can be uniform only when $2k_{ab}r_0 < 1$, which corresponds to $r_0 \leqslant 5$ μm. It is quite probable that the effect of a decrease in the absorption coefficient along with an increase in the thermodynamic parameters of the droplet's material can serve as such a mechanism.

Experiments carried out on the explosion of large water droplets ($r_0 \simeq 100$ μm) caused by pulsed $CO_2$ laser radiation have been documented in [36]. The laser used in the experiments delivered a 10 J pulse of 300 ns FWHM duration. The high-speed camera used for recording the explosion allowed a time resolution of 10 ns. Back-illumination of the droplet was accomplished using a laser spark generated in the focal plane of a lens. The time delay between the initiation of a laser pulse and the droplet's breakdown in the focal plane was ~1 ns. Droplets were suspended from a fiber in front of a back-illumination source. The slit of the camera was located along a line drawn perpendicular to the principal diameter of the droplet.

This system of integrated photography was also used in the following experiments. The exposure time was 3 μs. The photographs of the explosion process show that several types of droplet explosion process exist. With a radiation power density of ~ $10^7$ $Wcm^{-2}$ a slow expansion of the heated surface layer of a droplet takes place, which is then followed by its rapid spallation. During further development of the process the speed of the contact boundary layer first decreases and then increases again. Such behavior of the boundary layers is caused by the frontal (surface) character of the heat release taking place with large particles. The explosion process begins within a thin surface layer of the front (illuminated) hermisphere of a droplet. The thickness of this layer is $k_{ab}^{-1}$.

A slow, thermally induced expansion of the layer occurs due to an inflow of energy. This corresponds, from the standpoint of optothermodynamics, to a phase transition of the substance into the metastable state under almost constant pressure conditions. Then the explosion of this heated layer begins. There is a sharp threshold for this process, and this shows that the temperature inside the heated region is equal to the temperature necessary for the explosive vaporization of water corresponding to the conditions of a given experiment.

The expansion and explosion of the surface layer facilitates the penetration of incident radiation into the interior of a droplet. Consequently, the heating, thermal expansion, and explosion of the next layer takes place, which transfers more energy to the preceding layers. This process carries on until the light intensity of the laser pulse provides the necessary conditions for the stage-by-stage heating of the layers in the interior of the droplet to the temperature of explosive vaporization. The increase in radiation intensity up to $10^8$ $Wcm^{-2}$ results in the smoothing of the process of surface layer expansion. The speed of expansion becomes lower, and corresponds to the speed of the gas-dynamic flow of a heated substance.

Time-sequence photographs of the process clearly show the asymmetry of the explosion of large droplets in the initial stage. The region occupied by the explosion products in the final stage is, in practice, almost a perfect sphere. This can be explained by the symmetry of the expanding vapor during the recondensation process in the final stage of the explosion. The increase of the energy density from 30 to 50 $Jcm^{-2}$ leads to the growth of the final radius of the sphere from 4 to 8 $r_0$. There is a threshold for surface spallation in the case of large particles, and this allows one to assess the extent of a liquid transition to the metastable state under conditions of a small volume and a high rate of heating, such as, e.g., $10^{10}$ $Ks^{-1}$.

An investigation of the effects of a focused laser beam on different water targets has been carried out using a $CO_2$ laser (0.1 J pulse energy and 80 ns pulse duration). The spot occupied by the focused laser beam on the surface of the target was 0.15 mm in diameter. The water targets investigated were a free surface, thin films (0.1-2.5 μm thick), films of water streaming down a copper plate, cylindrical and stripwise flows, and suspended droplets. The characteristic size of a target's cross-section was

~ 0.3 mm. The Schlieren technique was used for recording the process, along with photographs taken with an Imacon image-converter tube. A Hg-flash lamp was used as a back-illumination source. The flash pulse duration was about 20 μs.

The interaction of laser pulse with thin films resulted in the films breaking, followed by the appearance of a barely-discernible small vapor cloud. The Schlieren set-up did not record any shock wave in the air.

The blow-off of the vapor and condensed water was observed from the front surface of thick targets being irradiated by the laser pulse. A shock wave was also produced. In the particular case of a stripwise flows, spallation of water from the back surface was observed along with blow-off of material from the front surface. The initial velocity of the shock wave was 3.8 Mach (in air). The velocity of the liquid water on both sides of the strip did not exceed the velocity of sound in air (see Figure 5.1.6).

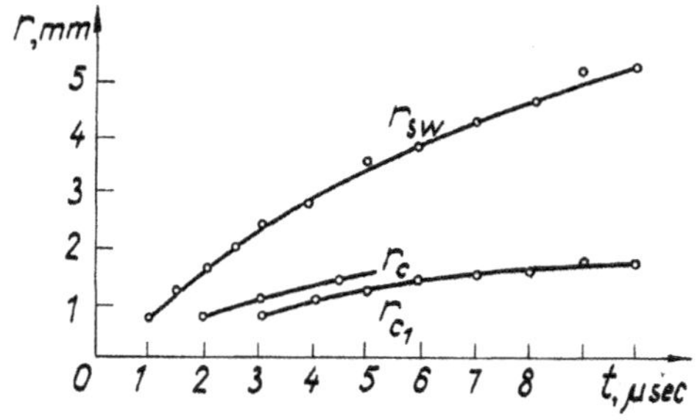


Fig. 5.1.6. Experimental time-dependence of shock wave radius $r_{sw}$, front $r_c$ and back $r_{c1}$ limits of tape current during the action of a $CO_2$ laser pulse [23].

It was revealed in [47] that a radiation pusle with an energy of 1.67 J and a duration of 75 ns concentrated in a beam of 0.8 cm diameter produced a compression wave in the liquid and a shock wave in the air of about 6.4 Mach initial velocity when directed on to a plane water surface. There was no optical break-down observed in any of the above cases.

The experimental study of the effect of cw $CO_2$ laser radiation on large water droplets has been documented in [21, 22]. Pulsating blow-off of stream-condensate from the surface layers of large particles was observed. The explosion of freely-falling droplets caused by cw $CO_2$ laser radiation was investigated in [24]. The explosion threshold was found experimentally in this work. For droplets with radii ranging from 12 to 33 μm, it lies within the range from $1.5 \times 10^4$ to $3 \times 10^4$ Wcm$^{-2}$. The time necessary for the explosion to begin varies from $10^{-4}$ s to $3.5 \times 10^{-4}$ s.

A classification of droplet explosions can be proposed, based on experimental results, the results of optothermodynamic analysis, and information concerning the structure of the internal optical field of a droplet. Classification of droplet explosions:

(a) Fragmentation - a gas-dynamic process of droplet destruction resulting in the appearance of condensate particles accompanied by local heat release. This process occurs at two-phase transitions in limited regions of a droplet.

(b) Gas-dynamic explosion - a flow of a one- or two-phase medium caused by the regimes of heating leading to high temperatures and pressures nearly everywhere through the bulk of a droplet. This process if characterized by quasi-uniform heating and the explosion is of a detonation type. In the latter case, surface heating is observed first, which results in the removal of the surface layer, thus facilitating the penetration of the electromagnetic field into the droplet. This process takes place at one- and two-face optothermodynamic transitions.

In the case of large, strongly-absorbant drops, fragmentation and the gas-dynamic explosion of surface layers (surface explosion) are observed.

## 5.2. DROPLET EXPLOSION REGIMES

### 5.2.1. Fragmentation

The effects induced by laser radiation inside a droplet can lead to the explosive vaporization of the droplet (or to fragmentation, as it is called). The effects of droplet overheating under conditions of constant pressure, the expansion of underheated liquid in the rarefaction wave occurring due to the gas-dynamic expansion of high pressure regions within a droplet, as well as other more complicated gas-dynamic and kinetic effects caused by the transition of a liquid into a metastable state under a quasi-equilibrium pressure, should be mentioned first as prime examples.

The construction of a model of the fragmentation process implies the determination of the relationships existing between the parameters of the laser radiation and the parameters of the droplet material, the droplet's size, and the dynamic characteristics of the process. As regards the latter characteristic, the time interval of the explosion, the characteristic velocities of expansion, the size spectrum of the explosion products, and the degree of vaporization of the droplet should be mentioned as being important, relative importance depending on the requirements of the problem being investigated.

The explosion-like character of liquid vaporization is due to an avalanche-type increase of the number of fluctuation ceters of the vapor phase in the superheated liquid when the temperature reaches a value close to that of the temperature of absolute instability of the substance. The frequency of spontaneous nucleation (homogeneous nucleation) can be estimated according to the kinetic theory of boiling (for details see [5, 6]) as follows:

$$J = N_1 B_K \exp(-G - t_s/t) = J_1 \exp(-t_s/t), \tag{5.2.1}$$

where $J_1$ is the velocity of the stationary process, $t_s$ is the time of relaxation to a steady state ($t_s \simeq 10^{-9}$-$10^{-8}$ c); $N_1$ is the number density of the molecules of the metastable liquid; $B_K$ is the kinetic factor ($B_K \simeq 10^{10}$ $s^{-1}$, for more detail see [5]); $G = \Delta\phi^*/K_B T$ is the Gibbs number; $K_B$ is the Boltzmann constant; and $\Delta\phi^*$ is the energy required for the formation of a critical nucleus of the vapor phase.

For explosive boiling of the superheated liquid to take place, a large number of fluctuation centers of boiling (vapor bubbles) should appear in the system, consistent with the inequality

$$\int_0^t dt' \int_V d\vec{r}\; J(t';\; T(\vec{r},\; t');\; p(\vec{r},\; t')) \gg 1. \tag{5.2.2}$$

When equality takes place, this expression defines the parameters for the achievable superheating of the liquid in a droplet under the given regime of heat release, i.e., the temperature $T_{sh}$, and the lifetime of the superheated liquid $t_m$.

The maximum temperature of superheating is the corresponding value on the curve of absolute instability of the liquid $T_{ai}$ and, at a pressure of 1 bar, is equal to 593 K (according to [5]). With the stationary process and a constant pressure, the achievable temperature of superheating is determined by the inversion $J_1$ $(T_{sh}) = J_{max}$, where $J_{max}$ is the value of $J_2$ in the region of a droplet having the maximum temperature. The mean lifetime of a metastable state can be assessed as follows: $t_m = (J_{max}V_{sh})^{-1}$, where $V_{sh}$ is the volume of the superheated region. For water under a pressure of 1 bar with $T_{sh}$ = 304.9, 310.1, and 318.7 °C, $J_1 = 10^4$, $10^{14}$, $10^{24}$ $cm^{-3}s^{-1}$ [5], respectively. A value of $V_{sh} = 10^{-9}$-$10^{-11}$ $cm^3$ corresponds to $t_m = 10^5$-$10^7$, $10^{-5}$-$10^{-3}$, $10^{-15}$-$10^{-13}$ s, respectively. A small value of $t_m$ in the last case shows that high rates of nucleation take place and, as a consequence, a large number of critical boiling centers appear in the superheated region of a droplet.

The theoretical calculations carried out so far aim chiefly at the elucidation of the energetics required for reaching the temperature of achievable superheating in the inner regions of a droplet under normal pressure, i.e., at its center, see [1] (following the approach of isotropic heat release) and in the regions of temperature field maxima for particles with a large diffraction parameter [14]. According to [1], the threshold intensity of continuous radiation for the case of isotropic absorption is

$$K_{ab}I = 8\chi_\ell r_0^{-1} c_p \rho_0 (T_{sh} - T_b), \quad t \gg r_0^2/4\chi_\ell; \tag{5.2.3a}$$

$$K_{ab}It = 4/3 r_0 c_p \rho_0 (T_{sh} - T_0), \quad t \ll r_0^2/4\chi_\ell. \tag{5.2.3b}$$

In [14], the time required to reach the temperature of explosive boiling in a certain region of a droplet, as well as the dependence of the amount of energy absorbed on the droplet's size and the intensity of incident radiation, are determined by analyzing calculated data concerning the temperature field inside the droplet, also taking into account the inhomogeneity of the heat release and also comparing the results with the experimental data derived from the explosion of freely-falling droplets caused by cw $CO_2$ laser radiation [24].

$$t_{expl} = 4.82 I_0^{-1.11} r_0^{0.094}; \qquad W_{ab} = (7.45 \times 10^{-8}) I_0^{-0.112} r_0^{2.35}$$

$$(I_0[\mathrm{Wcm^{-2}}],\ r_0[\mu\mathrm{m}],\ t_{expl}[\mathrm{s}],\ W_{ab}[\mathrm{J}]) \tag{5.2.4}$$

For strongly-absorbant large particles ($K_{ab} r_0 \gg 1$), the conditions necessary for the explosion to occur are met in a thin surface layer of thickness $K_{ab}^{-1}$. The values of the threshold intensity of the radiation are estimated using the following relationships:

$$K_{ab} I = 4\chi_\ell k_{ab} C_p \rho_0 (T_{sh} - T_0), \quad t \gg (4\chi_\ell k_{ab}^2)^{-1}; \tag{5.2.5a}$$

$$K_{ab} I t = k_{ab}^{-1} C_p \rho_0 (T_{sh} - T_0), \quad t \ll (4\chi_\ell k_{ab}^2)^{-1}. \tag{5.2.5b}$$

Important aspects in the modelling of the fragmentation explosion are the search for the mechanism of droplet destruction and the construction of expressions for estimating the characteristic times and rates of this destruction.

A model of droplet destruction by means of a vapor bubble growing outwards from the region of maximum superheating is suggested in [10]. The escape of an individual bubble from the droplet's surface corresponds to local destruction. The speed of bubble growth can be used for estimating the rate of spallation of the surface layers of the liquid.

The simplest solution to the problem is the Rayleigh solution for the stage of the process, corresponding to the bubble's movement in a nonviscous liquid, which allows the estimation of an upper limit of the bubble's speed. In this case, its size $r_b \gg r_b(t = 0)$. Here, the difference between the vapor pressure inside the bubble and in the liquid is assumed to be constant. It is also assumed that T is constant. It can be shown, for this case, that $V_b = [2/3(\Delta p/\rho_0)]^{1/2}$ at $\Delta p \leqslant 10^2$ bar and $V_b \leqslant 10^4$ cms$^{-1}$.

As follows from the estimations and from the calculations [15] based on the model, the time interval necessary for a bubble to grow to the size of a droplet is too short compared with that required to heat the droplet, or any part of it, to the temperature necessary for explosive boiling, $T_{sh}$.

This circumstance, together with the fact that the lifetime of a metastable state is short, leads one to consider that the time interval required for heating a droplet to the temperature of explosive boiling is an important parameter in the explosive fragmentation process. This time correlates with the time of the explosion itself.

Since the process of explosion if a multistage process, then it is expedient to consider the time interval during which it takes place. The data on this interval can be obtained experimentally. In particular, if a volume filled with a quasi-monodispersed calibrated aerosol, whose particles are uniform with respect to the absorption of high-power radiation, is illuminated with visible light, then this time interval is measured as the time between the beginning of the increase in volume turbidity and the moment when it reaches its maximum value. The turbidity is characteristic of this fragmentation explosion regime.

The suggested treatment of droplet explosion is based on a consideration of the process of thermal instability of phase transition boundaries [12] (free surface, bubbles' surfaces). The physical explanation of this effect is the following. The vaporization front always moves in the direction of the temperature gradient. Evaporation, as is known, leads to cooling of the liquid layers adjacent to the phase transition boundary, therefore the temperature gradient runs into the liquid. Thus, any random shear of the boundary in the direction of the liquid results in an increase of the temperature gradient in the region of the shear, and hence to an increase of the local evaporation rate that, in turn, makes the initial shear larger. The development of the process of thermal instability can cause the destruction of the liquid layers adjacent to the phase transition boundary. For a plane boundary surface, the threshold intensity of $CO_2$ laser radiation for this effect to take place is $10^2$-$10^3$ $Wcm^2$ [2]. It is possible that just this effect causes the destruction of the surface layers of large particles ($r_0 \simeq 0.5$-$1$ mm) irradiated with laser radiation with an intensity of about $10^3$ $Wcm^{-2}$ [22], this is lower than the necessary threshold for the fragmentation regime of droplet explosion (i.e., $\sim 10^4$ $Wcm^{-2}$).

The mechanism of thermal instability leading to droplet destruction due to boiling is quite a realistic mechanism, since it provides an explanation for the destruction of thin layers which can occur between growing bubbles, or between bubbles and the free surface of a droplet.

The fragmentation of a droplet can occur due to the explosive boiling of hot or cold (cavitation) liquid in the rarefaction wave. This effect is analogous to shock heating; the temperature of explosive boiling is achieved here not due to rapid heating, but as a consequence of sharp depressurization in the 'underheated' liquid. The destruction of a condensed substance irradiated with laser radiation by means of this mechanism has been assumed in [31]. This effect, is basically similar to the unloading effect taking place in the substance under shock pressure when the shock wave arrives at the free surface [7]. This mechanism of destruction can operate by means of the expansion of localized hot regions near the surface of a droplet.

The explosive boiling effects described above, which occur in liquid droplets irradiated with high-power laser radiation, are observed at standard atmosphere pressure. As is known, see [6], an increase in atmospheric pressure leads, correspondingly, to an increase in the temperature of the explosive boiling, and hence to a decrease in superheating $(T_{ai} - T_s)$; $T_s$ is the temperature on the saturation curve. The role of hydrodynamic perturbations caused by vapor bubbles also becomes less important. This is connected with a decrease in the pressure difference between the layers of liquid near the bubble's boundary and those far removed from it.

Weakening of the explosion under higher external pressures has been noted in [6] when analyzing experiments on the boiling of droplets floating in acid. In the experiments on the explosion of droplets caused by $CO_2$ laser radiation this effect was revealed indirectly by a decrease in the boundary surface velocity with an increase in the laser radiation power from $10^7$ Wcm$^{-2}$ to $10^8$ Wcm$^{-2}$ [36].

Nonisobaricity means that, during the time required for depressurizing a heated region of a scale $\ell$, the pressure increase due to heating exceeds the equilibrium pressure. Since $\Delta p = \Gamma_0 k_{ab} I \ell C_{s0}^{-1}$, then $\Delta p > p_0$ if

$$I > I_{gas} = p_0 C_{s0}/\Gamma_0 k_{ab} \ell \tag{5.2.6}$$

Here, $\Gamma_0$ is the Grüneisen coefficient of the liquid. In water under standard conditions $\Gamma_0 \simeq 0.1$. The sphericity of the expansion process results in the appearance of the factor 3 in $I_{gas}$. For $\lambda = 10.6$ µm, $r_0 = 10$ µm, $I_{gas} = 5 \times 10^5$ Wcm$^{-2}$.

Under high pressures, the regime of droplet explosion moves from fragmentation to the regime of two-phase liquid flow. In this case, the kinetics of the process follow the speed of expansion of a heated region. The velocity of a bubble is determined only by the evaporation rate within it. The explosion of a droplet in this case can be described using the gas-dynamic equations.

### 5.2.2. Gas-Dynamic Explosion

We will use the mechanism of a continuous flow of the medium as the basic model of gas-dynamic explosion and, therefore, we will describe all stages of the explosion using gas-dynamic equations. This approach is most applicable when applied to the description of the explosion of uniformly-absorbant droplets that have undergone significant superheating, in which case the phase trajectory of the substance is in the supercritical region or in the near vicinity of the critical point (see Figure 5.1.1).

The system of one-dimensional Lagrangian equations of gas dynamics has the form

$$V = (r/\Lambda)^2 \partial r/\partial \Lambda; \qquad \partial v/\partial t = -(r/\Lambda)^2 \partial p/\partial \Lambda; \qquad \partial r/\partial t = v;$$

$$\frac{dU}{dt} = -p\,\frac{dV}{dt} + q_{ab}(U,\ V,\ t), \quad p = p(V,\ U), \tag{5.2.7}$$

where $r$ is the Euler coordinate; $\Lambda = \left(3 \int_0^{r \leqslant r_1} \rho(r',\ t = 0) r'^2\, dr'\right)^{1/3}$ is the

Lagrangian coordinate; $r_1$ is the boundary of the region under consideration;

$$q_{ab} = \begin{cases} k_{ab}(U, V)IV, & \Lambda \leqslant \Lambda_0 \\ 0 & \Lambda > \Lambda_0 \end{cases}$$

is the specific power of a thermal source; and $\Lambda_0 = \rho_0^{1/3} r_0$ is the Lagrangian coordinate of the boundary of the explosion products.

It is assumed that $p = p_e + p_h$, and $U = U_e + U_h$, where $p_h$, $U_h = \int_0^T C_v \, dT$ are the thermal components of corresponding values, while $p_e = -dU_e/dV$, $U_e$ are the elastic ones. The initial conditions for this system of equations are: $p = p_0$; $V = V_0$ at $\Lambda \leqslant \Lambda_0$; and $p = p_1$, $V = V_1$ at $\Lambda > \Lambda_0$. The boundary condition is $\Lambda = 0$, $v = 0$. Subscript 1 denotes the parameters of the undisturbed surrounding medium. Since this problem deals with the destruction of an arbitrary discontinuity in the process of heat release inside the sphere, its solution will be determined by the generalized solution of the gas-dynamic equations, and can be obtained only using numerical techniques.

Below, we shall discuss the results of numeric simulations of the explosion process of an isotropically absorbant droplet. At this stage certain assumptions must be made concerning the equation describing the water state. Van der Waals equation was taken as the model in these numerical experiments. We used this equation because of the lack of an equation of state to adequately describe the state of water for a wide range of thermodynamic parameters. Meanwhile, the Van der Waals equation is qualitatively correct for describing isotropic phases and is used as a model of the two-phase state.

The air conditions are described by the equation of state of an ideal gas with the adiabatic exponent $\gamma_1 = 1.4$. The radiation pulse shape is described by

$$I(t) = I_{max} t/t_0, \quad t \leqslant t_0; \qquad I(t) = I_{max} \exp\{-(t - t_0)/nt_0\}, \quad t > t_0,$$

where $I_{max}$, $t_0$, and n are parameters. This expression is quite adequate for $CO_2$ laser pulses. Numeric integration of the gas-dynamic equations was performed using the explicit difference scheme, according to the Neumann-Richtmayer method of artificial viscosity [40, 41]. This technique allows the calculations to be carried out without localization of singularities. The value of $C_v$ was assumed to be constant in these calculations. In this case the Van der Waals equation can be reduced to the following form: $p = (\gamma - 1)(U + (a/b))/(v-b) - (a/V^2)$, where a and b are constants, $\gamma - 1 = R_\mu/\mu_n C_v$. The value of $\gamma$ used in these calculations is 4/3. The initial conditions are $p_0 = p_1 = 1$ bar and $T_0 = T_1 = 293$ K. It was assumed that $k_{ab} = k_{ab}^0 (V_0/V)^2$ [35], $k_{ab}^0 = 800$ cm$^{-1}$.

Figure 5.2.1 shows the pressure in the explosion products and in the air as a function of a dimensionless spatial coordinate. The spherical contact surface of a droplet works like a piston compressing the air, since the movement of the surface is accelerated due to the inflow of heat energy converted from the light energy consumed by a droplet. A shock wave appears in the air, and an increase in the peak pressure is observed in the early stages of the explosion. This process terminates at some given moment due to the slowing down of this'piston'effect, which is caused by a decrease in the absorbtivity of the explosion products as they expand, as well as because of the spherical expansion pattern and the transfer of heat energy to the shock wave.

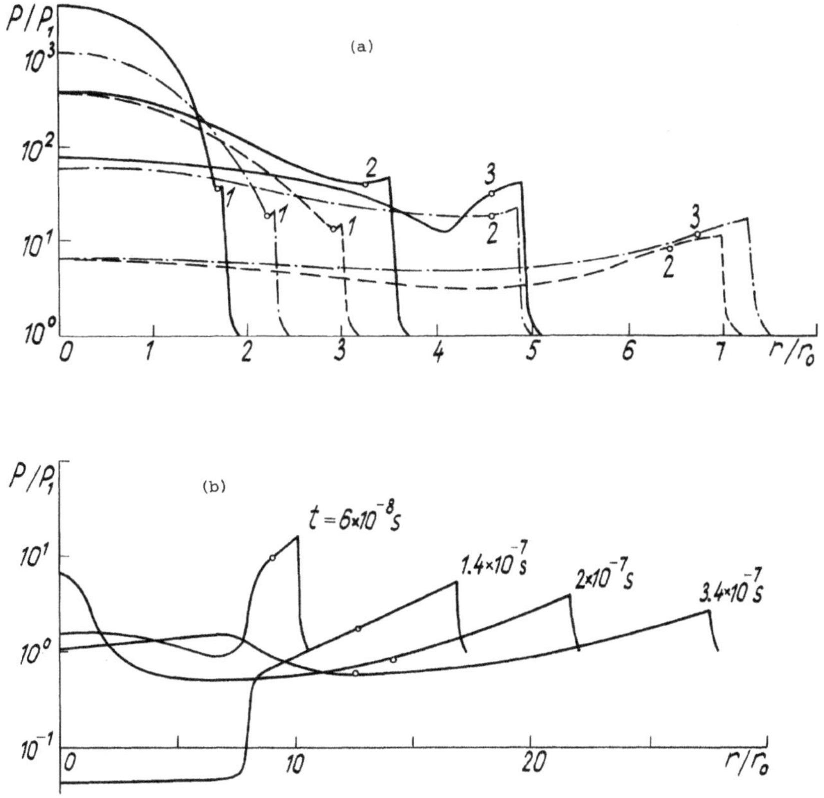


Fig. 5.2.1. Spatial pressure distribution during the explosion of a droplet with (a) $r_0$ = 2.5 (dashed line), 5 (dot-dashline), 10 μm (solid lines - circles denote the positions of a contact surface); the other parameters are $I_{max}$ = $10^9$ $Wcm^2$, $t_0$ = 10 ns, n = 2, t = 10 (1), 20 (2) and 30 ns (3). t is varied in 5.2.1(b).

For weaker thermal sources the explosion process is qualitatively the same as in the case of an 'instantaneous' spherical explosion [41, 32]. As follows from the calculations, the maximum achievable value of the peak pressure in the shock wave in the air increases with an increase in $r_0$. Within the framework of the model describing a uniform distribution of the electromagnetic field inside the sphere, such behavior takes place only for small droplets where $r_0 < k_{ab}^{0-1}$. This dependence of peak pressure on droplet size is the result of the process of nonstationarity, due to which the larger particles spend a longer time in the stage of light power absorption. For $r_0 \gg k_{ab}^{0-1}$, this feature disappears, since under these conditions the rate of absorbtion increase decreases with an increasing $r_0$.

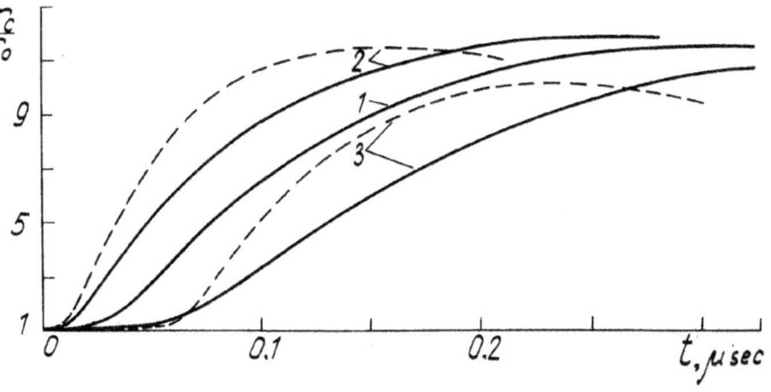


Fig. 5.2.2. Time-dependence of contact surface radius for $r_0 = 5$ (dashed) and 10 μm (solid lines), with $n = 2$, $I_{max} = 5 \times 10^8$ (1), $2.5 \times 10^8$ (2), $3 \times 10^7$ Wcm$^2$ (3), and $t_0 = 200$(1,3) and 10 ns (2).

Figure 5.2.2 illustrates the dynamics of the boundary surface coordinate $r_c(t)$. The curve obtained from calculations agrees qualitatively with the results of the experimental study of the explosion of large droplets ($r_0 \simeq 100$ μm) caused by $CO_2$ laser radiation [36]. As the calculations showed, the process of the explosion products stopping near the hydrodynamic equilibrium state is oscillatory. The limiting radius of a sphere is $r_\infty = (\rho_0/\rho_\infty)^{1/3} r_0$, where $\rho_\infty$ is the vapor density at the end of the process. The value of $r_\infty$ increases with an increase in the amount of energy absorbed. If, by the end of the process, the vapor is saturated, i.e., $\rho_\infty = \rho_\infty^s$, then $r_\infty = 11.92\ r_0$. For droplets with $r_0 = 5$-$10$ μm, the time when oscillations of the sphere achieve their first maximum varies from 0.1 to 0.3 μs, depending on the parameters of the radiation pulse. The time taken for the explosion products to stop is of the order of 1 μs.

Figure 5.2.3. shows the thermodynamic trajectories of the droplet's Lagrangian coordinates. Owing to the irregular conditions of heat release, different regions of the droplet have their own trajectories on the p-V plane. As calculations have shown, the energy of the leading part of a pulse train chiefly determines the nature of the expansion process. The main bulk of a droplet will expand within a single-phase region if only the following condition is fulfilled:

$$I_{max}[\mathrm{Wcm}^{-2}] \geqslant 5\times 10^{9}\ t_0[\mathrm{s}]/r_0^2[\mathrm{cm}^2]. \qquad (5.2.8)$$

The peripheral regions of the droplet are practically always in a two-phase state.

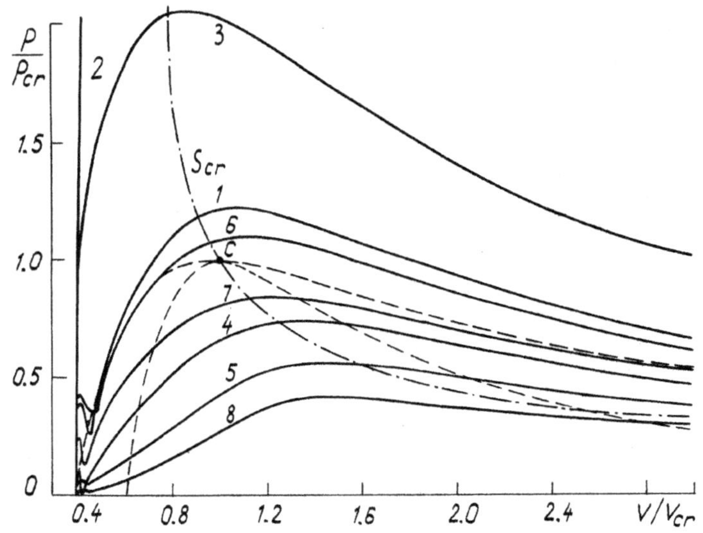


Fig. 5.2.3. p-V diagram of droplet explosion, with $r_0$ = 10 m, n = 2, $I_{max}$ = 1.25 × $10^8$ (1,5), 5 × $10^8$(2,3,6,7), 3 × $10^7$ $\mathrm{Wcm}^2$ (4.8), $t_0$ = 200 (1,3,4,5,7,8) or 100 ns (2,6) for the center (1-4) and the segment of the droplet with 75% of its mass (5-8); the dashed lines are the binodal line (outside curve) and spinodal (inside curve); the dot-dash line is the line of the critical isentrope.

Qualitative considerations show that two-phase regions undergo certain changes during expansion. For small specific volumes, the two-phase region is the part of the liquid full of vapor cavities, and after significant expansion it is composed of droplets inside vapor. The dynamics of such a medium can hardly be described. So, the only possible way is to use the effective thermodynamic parameters. The van der Waals model is the simplest solution of the problem which can take into account the energy losses due to droplet evaporation occurring in the wave of rarefaction, i.e., the energy required for doing the work against the cohesive factors $\int_{V_0}^{V} |p_e|\,dV$

(here $p_e = -a/V^2$).

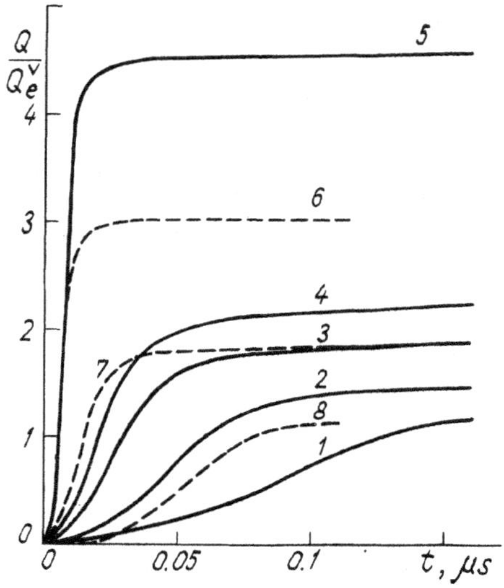


Fig. 5.2.4. Time-dependence of normalized absorbed energy upon the droplet's explosion, with $r_0$ = 5 (dashed line), 10 μm (solid lines); n = 2 (1-7) and 9,5 (8); $I_{max}$ = 3 × $10^7$ (1,8), 1.25 × $10^8$ (2), 5 × $10^8$ (3, 4, 7), $10^9$ Wcm$^2$ (5, 6); $t_0$ = 200 (1-3), 100 (4), 50 (7, 8) and 10 ns (5, 6).

Now consider the explosion energetics. Figure 5.2.4 represents the temporal behavior of the specific absorbed light energy,

$$Q = 4\pi \int_0^t dt \int_0^{r_c} \rho(r, t)\, q_{ab}(r, t) r^2\, dr/M_0,$$

normalized relative to the van der Waals heat of vaporization $Q_e^v$. Here, $M_0$ is the mass of a droplet. The value of the heat of vaporization $Q_e^v = a(V_L^{-1} - V_g^{-1}) + p(V_g - V_L)$ was calculated at the boiling point $V_L$ and $V_g$ being the specific volumes of the condensed and gaseous phases, respectively, on the saturation line. From the calculations, an amount of heat $Q = 2Q_e^v(T_b)$ is released during the expansion of a droplet irradiated with a light pulse obeying the criterion (5.2.8). In the case of spherical droplets expanding mainly in the two-phase region, the absorbed light energy is almost entirely consumed by the process of droplet vaporization in the wave of rarefaction. Figure 5.2.5 shows the relative value of the excess pressure at the front of a shock wave $\bar{p}_{sw} = (p_{sw} - p_1)/p_1$ as a function of the dimensionless coordinate $r(p_1/W_{sw})^{1/3}$, where

$W_{sw} = 4\pi \int_{r_c}^{r_1} \rho(r,\ t)(u(r,\ t) - U_1 + v^2(r,\ t)/2)r^2\ dr_{t\to\infty}$ is the full energy transferred to the shock wave at the time of explosion of a droplet (the explosion energy). The curves presented in the figure reveal the similarity in the behavior of shock waves in the late stages.

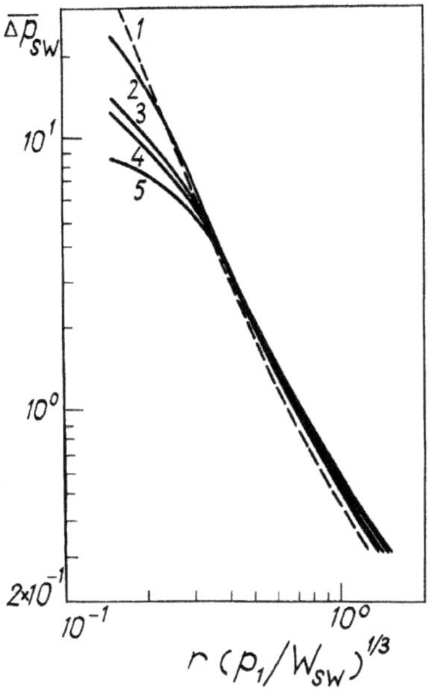


Fig. 5.2.5. Dependence of the relative overpressure at a shock-wave front on the dimensionless radius, with $n = 2$, $r_0 = 10$ μm for a point explosion (1); $I_{max} = 10^9$ (2), $5 \times 10^8$ (3, 5), $2 \times 10^8$ $Wcm^2$ (4); $t_0 = 10$ (2, 4), 100 (3) and 200 ns (5).

The asymptotic arrival of a shock wave at the limiting stage of the process is described by $r_{ws} = r^0 + c_{s1}t$, where $c_{s1}$ is the speed of sound in air. It follows from the property of similarity that the following equality is valid for two processes having explosion energies $W_{sw.1}$ and $W_{sw.2}$:

$$(r_1^0/r_2^0) = (W_{sw.1}/W_{sw.2})^{1/3}.$$

A comparison of calculated and experimental results confirms the validity of this equation.

The reliability of the calculated results is supported by the following facts: the integral laws of the conservation of mass and energy are obeyed within the accuracy limits of 1%, and the calculated results remain the same when the spatial increment of the network is diminished.

Some aspects of the problem of the effects of lasers on liquid dispersed media have been discussed above. A satisfactorily complete picture of the theoretical and experimental studies on the explosive vaporization of water droplet aerosols can be extracted from the summary of main results available in the literature.

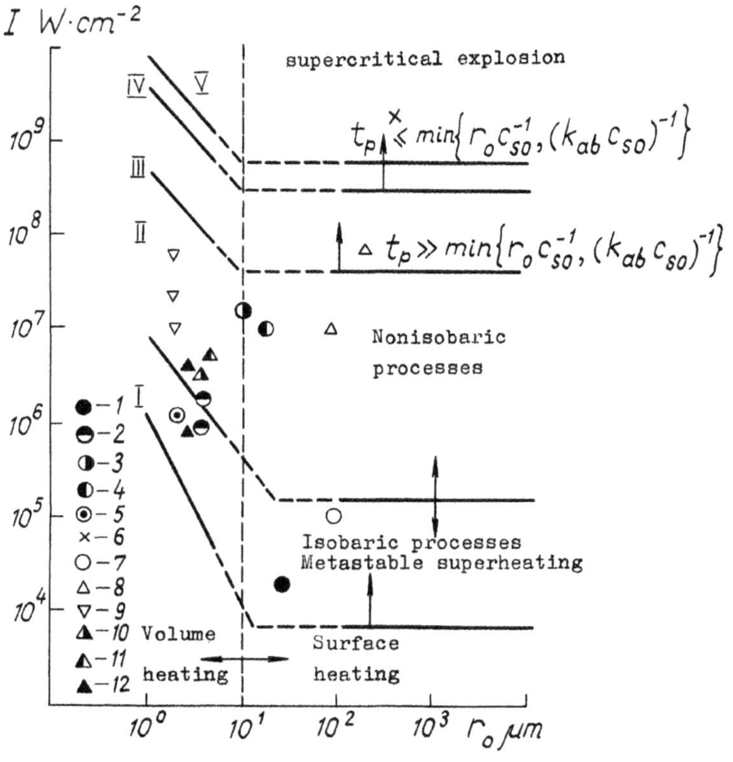


Fig. 5.2.6. Experimental data on a droplet's explosion in a radiation field, with λ = 10.6 µm. 1 - [24], 2 - [27], 3 - [18], 4 - [17], 5 - [30], 6 - [23], 7 - [19], 8 - [36], 9 - [29], 10 - [25], 11 - [28], 12 - [26].

Figure 5.2.6 presents experimental data describing water droplet explosions caused by $CO_2$ laser radiation. The dependence of the characteristic intensity of radiation (either the mean or the peak power, according to the situation) on the particle's radius is presented in this figure and is characteristic for a given experiment.

The threshold intensities calculated theoretically for various explosion regimes are also presented in this figure. These curves have the following meanings:
Curve I corresponds to a stationary regime of particle heating to its temperature of explosion vaporization (e.g., (5.2.3(a)), (5.2.5(a)). The region enclosed by curves II and III is the region corresponding to nonstationary processes of heating the liquid to a metastable state. The level II is defined according to (5.2.6). The region above this line corresponds to the unsteady movement of heated liquid.

In the region between curves II and III the transition from the fragmentation regime of droplet explosion to the gas-dynamic regime takes place (i.e., from droplet vaporization due to boiling to liquid vaporization in the rarefaction wave). Line III is defined by (5.2.6) for $t_f = r_0/C_{s0}$ in the case of a sphere, or by a corresponding relationship for a plane surface ($k_{ab}r_0 = 1$ in (5.2.6)). The level IV characterizes the upper limit for the supercritical explosion of the super heated region when $t_p \gg \min r_0/C_{s0}, k_{ab}^{-1}/C_{s0}$ (see (5.1.5)), while level V represents the corresponding lower limit when $t_p \ll \min r_0/C_{s0}, k_{ab}^{-1}/C_{s0}$, see (5.1.3) and (5.1.4).

Taking into account the difficulties involved in making a quantitative description of the explosion process, one can arrive at the conclusion that there is a good correlation between theoretical assessments and experimental data.

## 5.3. ATTENUATION OF LIGHT BY AN EXPLODING DROPLET

The particular regime of droplet explosion determines the temporal behaviour of the coefficient of light extinction in an aerosol medium. A supercritical regime of droplet explosion results in the decrease of the droplets' optical density.

A two-phase explosion regime is characterized by the destruction of the droplet followed by the creation of a two-phase medium composed of droplets and vapor. This leads to an increase in the geometrical effects of the light-scattering by droplets and to the expansion of the spherical region occupied by vapor and droplets.

The explosion process is accompanied by recondensation of the vapor, initiation of shock waves in the air, and by both heat and mass transfer in air.

The study of the optical 'consequences' of such an interaction for the case of a single particle can form the basis for the construction of a theoretical model of nonlinear laser propagation through a dispersed medium, thus allowing the interpretation of the experimental data.

## 5.3.1. Extinction Coefficient of a Droplet Exploding in the Supercritical Regime

For a known density distribution $\rho$, one can easily find the spatial distribution of the components of the complex refractive index of a droplet, at any moment, using the Lorentz-Lorenz equation for the real part of the refractive index $n'_a = n_a \sqrt{\varepsilon_0}$ [42],

$$[(n'^2_a - 1)/(n'^2_a + 2)] = C_n \rho, \tag{5.3.1}$$

and a model representation [35],

$$\kappa_a = \kappa_{a0}(\rho/\rho_0)^2, \tag{5.3.2}$$

for the imaginary one, and assuming the constant $C_n$ to be independent of temperature. As the calculations show, one can neglect the influence of a shock wave on the cross-section of extinction of light by an exploding droplet till the moment when the explosion products stop.

The extinction cross-section of an exploding droplet is calculated using the approach for lage 'soft' scattering centers [43]:

$$\sigma(t, \lambda) = 2 \text{ Re} \int d\vec{R}(1 - \exp[(-ik/2)\int_{-\infty}^{\infty} dx'(m_a^2(x', \vec{R}, t) - 1)]). \tag{5.3.3}$$

Figure 5.3.1 shows the extinction cross-section and extinction efficiency factor of the expanding sphere as functions of the dimensionless surface, as calculated using (5.3.3). It was assumed that $\lambda = 10.6$ μm and $m_{a0} = 1.144 - i0.067$. The spatio-temporal distribution of density used in the calculations was obtained by numerically solving the problem for a supercritical explosion with $Q_{ab} = 2Q_e^b$. As seen in this figure, the $\sigma$ and K values at $\lambda = 10.6$ μm decrease with an increase of the radius of the contact surface. At the very beginning of the process the values $\sigma$ and K change mainly because of the decrease of $\kappa_a$, then extinction is determined only by scattering, and $\sigma \simeq r_c^{-2}$.

Figure 5.3.2 represents the calculated results for the extinction cross-section when $\lambda = 0.63$ μm and $m_a = 1.33$. The calculations show that, in the initial stage of the process, any increase in $\sigma$ is observed to be due to the growth of the sphere, then $\sigma$ reaches a maximum and after that the value of $\sigma$ decreases and finally reaches a value which is determined by the state of the vapor in the sphere at the moment when its expansion stops. Since in the final stage of the process the spherical surface oscillates around its equilibrium position, then the temporal behavior of the extinction coefficient is also oscillatory. The value of the scattering coefficient for the visible range at the moment when the sphere stops strongly depends on the final stage of the vapor inside the sphere which, in turn, is determined by the explosion parameters in the initial stage and by the phase thermodynamic phase trajectory of the expanding substance.

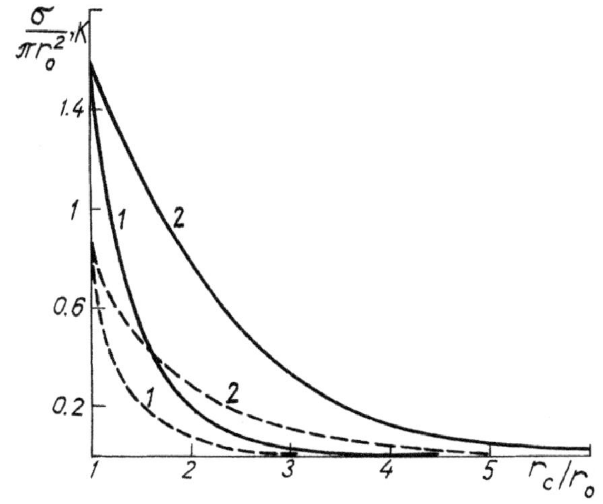


Fig. 5.3.1. Dependence of the extinction coefficient (1) and cross-section (2) (with $\lambda = 10.6$ μm) on the dimensionless radius of the contact surface during homogeneous droplet explosion, with $r_0 = 5$ (dashed lines) and 10 μm (solid lines).

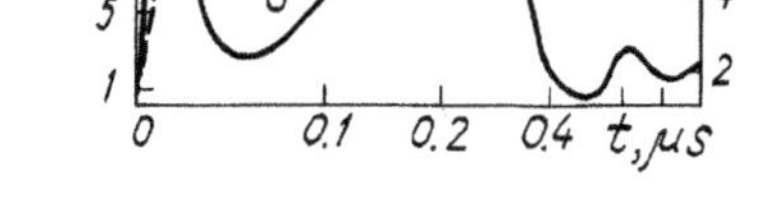


Fig. 5.3.2. Time-dependence of the extinction cross-section for $\lambda = 0.63$ μm during supercritical droplet explosion, with $r_0 = 10$ μm. The dashed line shows the time-dependence of the radius of the droplet's contact surface.

Let the vapor be saturated at the end of the process, i.e., for $\rho = 1$ bar, $V_\infty = 1.69 \times 10^3$ cm$^3$g$^{-1}$. Such a limiting volume corresponds to a sphere of radius $r_\infty = 11.92\ r_0$, and the scattering cross-section in this case is

$\sigma = 2\pi k^2 r_\infty^4 (n-1)^2$, where $n_\infty = \varepsilon_0^{-112}(1 + 1.5C_n(V_0/V_\infty))$. Thus, for $r_0 = 10$ μm, one obtains $\sigma/2\pi\ r_0^2 = 2.7$. The extinction cross-section is equal to zero only when $V_\infty = 1.024 \times 10^3$ $cm^3g^{-1}$ $(r_\infty/r_0 = 10.08)$.

The light extinction cross-section of the vapor-air inhomogeneity is written as follows:

$$\sigma(t, \lambda) = 4\pi^4 k^2 \iint_{-\infty}^{\infty} d^2q|F_{\varepsilon'}(q, t)|^2 \tag{5.3.4}$$

where $F_{\varepsilon'}$ is the three-dimensional spectrum of the dielectric constant at a point near the centre of the explosion, and

$$\varepsilon'(r, t) = \varepsilon_0^{-1}\{C_n(\rho_n - \rho_{n\infty}) + C_1(\rho_1 - \rho_{1\infty})\},$$

where subscript '1' refers to the air. The disturbance $\rho_n - \rho_{n\infty}, \rho_1 - \rho_{1\infty}$ is determined for $t > t_\infty$, where $t_\infty$ is the time when the generation of explosion products stops, this is obtained by solving the Cauchy problem for the diffusion equation. Applying the Fourier transform to $\varepsilon'$, one finds that changes in the spectrum over time are described by the expression

$$F_{\varepsilon'}(q, t') = F_{\varepsilon'}(q, 0)\exp\{-q^2Dt'\}, \quad t' = t - t_\infty. \tag{5.3.5}$$

As a consequence, the scattering cross-section can be written as

$$\sigma(t') = \sigma(0)\psi(t')/\psi(0) \tag{5.3.6}$$

where $\psi(t') = \int_0^\infty d\zeta(\sin\zeta - \zeta\cos\zeta)^2\zeta^{-5}\exp\{-2\zeta^2 Dt'/r_\infty^2\}$.

It follows from (5.3.6) that, for $t' \gg r_\infty^2/D$,

$$\sigma(t') \sim \sigma(0)\ r_\infty^2/Dt'. \tag{5.3.7}$$

We will now evaluate the effects of the recondensation of the expanding vapor on the extinction coefficient.

The adiabatic equation for a two-phase system is as follows:

$$\overline{C}_V\, dT + p(T, V_n)\, dV = \{Q_e(T) - p(V_\pi - V_L)\}dX, \tag{5.3.8}$$

where $\overline{C}_V = [(1 - X)C_V^n + XC_V^L]$, the subscripts '$\pi$' and 'L' indicate the gas and the liquid (condensed) phases, respectively, and X is the degree of condensation, which is found from the kinetic equation of droplet growth [44].

An estimate of the maximum effect of this recondensation can be obtained by taking into account the fact that, in the equilibrium stage of the process, temperature changes more slowly than density. It can be assumed, in this case, that $dT = 0$ in (5.3.8), and then $X = 1 - (V^*/V)^\delta$, where $\delta = R_nT^*/(Q_e(T^*) - R_nT^*)$, here the asteriks denotes values corresponding to the junction point of the unloading adiabatic curve and the line of vapor saturation. $R_\pi$ is the gas constant for the vapor.

As follows from Figure 5.1.2, the relief adiabat 2 intersects the equilibrium curve between the liquid and vapor phases at $p^* = 50$ bar and $V^* = 39.41\ cm^3g^{-1}$. As a result, $\delta = 0.178$. It is also seen from this figure that the final value of the specific volume of vapour expanding along the adiabat of a two-phase system is between $10^3$ and $1.69 \times 10^3\ cm^3g^{-1}$. Using the value $V = 10^3\ cm^3g^{-1}$, one finds that an estimate of $X_\infty = 0.44$ is valid. For determining the extinction coefficient for light with a wavelength $\lambda = 10.6$ μm passing through an aerosol of condensed vapor droplets, we shall use an approach which is based on the use of the water content parameter:

$$\sigma = k_{ab}M_0X_\infty V_0. \tag{5.3.9}$$

It follows from (5.3.9) that $\sigma = 1.47 \times 10^{-6}\ cm^2$ when $\lambda = 10.6$ μm, which is 3.4 times less than the initial extinction coefficient of the droplet. When $r_0 = 5$ μm, $\sigma = 1.84 \times 10^{-7}\ cm^2$, i.e., 3.7 times less than the initial one. It is assumed here that $X_\infty = 0.44$.

The determination of the extinction coefficient of the condensate when $\lambda = 0.63$ μm requires the knowledge of the size distribution function of the ensemble of droplets. This is because the extinction coefficient of the ensemble, even if it is a monodispersed ensemble of Rayleigh scattering centers, is determined by the number of condensed droplets, $N_d$, and by the second power of the droplets' volume, $V_d$.

$$\sigma = \frac{3k^4}{2\pi}\left(\frac{n_a^2 - 1}{n_a^2 + 2}\right)^2 N_dV_d^2. \tag{5.3.10}$$

At a fixed degree of recondensation, $V_d = M_0X_\infty V_0/N_d$.

Thus, the transparency of a monodispersed aerosol at the moment of a supercritical explosion and at the moment of time $t \gg r_\infty^2/D$ is entirely determined by the degree of recondensation of the vapor at the end of the process. In the IR range, e.g., when $\lambda = 10.6$ μm, recondensation can reduce the transparency by a maximum of 25%. In the visible range, clearing can be observed only if the condensed particles are Rayleigh scattering centers and their number for every exploding droplet is

$$N_d > \frac{3k^4}{2\pi}\left(\frac{n_a^2 - 1}{n_a^2 + 2}\right)^2 \frac{M_0^2X_\infty^2}{\rho_0^2\ (t = 0;\ \lambda)}\ .$$

### 5.3.2. The Extinction Coefficient in the Case of a Two-Phase Explosion

It is difficult at present to calculate the extinction coefficient for the explosion of a droplet in the subcritical region. This is caused by the difficulties in describing the size spectrum of the particles forming the two-phase system in this process. However, certain qualitative assessments can be made in this case. As mentioned above, in the beginning of the process the droplet expands until it 'decays' into a two-phase mixture. Since the combined surface area of the many droplets produced in this process of droplet fragmentation is greater than the surface of the droplet before the explosion, due to small amount of the vapor phase, then one can expect an increase in the scattering cross-section for visible light. The decrease in the optical depth of the aerosol medium for visible light can be expected, in this regime, due to the vaporization of parts of droplets in the two-phase medium which has appeared as a result of the explosion.

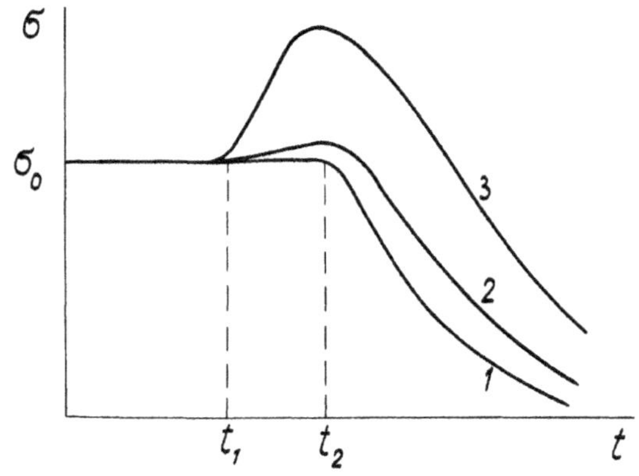


Fig. 5.3.3. Qualitative time-dependence of the extinction coefficient of a droplet exploding in a two-phase region: 1, 2 - $\lambda$ = 10.6 μm; 3 - $\lambda$ = 0.63 μm.

Figure 5.3.3 qualitatively presents the dependence of the extinction coefficient of an exploding droplet on time at different wavelengths of the incident radiation. A turbidity interval is shown on the time axis. For $\lambda$ = 0.63 μm, $\Delta t = t_1 - t_2$, where $t_1$ is the time necessary for a droplet to heat up to its temperature of explosive destruction, and $t_2$ is the time of cessation of the explosion.

It is assumed that the duration of exposure to the laser beam is sufficient to evaporate the irradiated condensed water. Therefore, for $t > t_2$, surface vaporization of the droplet's fragments will occur and, as a consequence, the dissipation (i.e., a decrease in the extinction cross-section) of the medium will also occur.

The extinction of laser radiation with $\lambda$ = 10.6 μm (and in the case of small particles with $r_0$ < 10 μm) by the explosion products will be determined by their total mass (the 'water content' regime). Any change in the extinction coefficient with time under conditions of a low rate of explosive vaporization is practically negligible up to the moment $t_2$. After this moment a decrease in the extinction coefficient is observed due to the vaporization of the condensate (curve 1 in Figure 5.3.3).

In the case of large particles ($r_0 > 10$ μm), an increase in the extinction cross-section for radiation with $\lambda = 10.6$ μm can occur during the initial stage of the interaction when $t < t_2$ (curve 2 in Figure 5.3.3).

The use of model calculations [45] for assessing the extinction coefficient when $\lambda = 0.63$ μm, and for a set size-distribution function of the droplet fragments, does not make any improvements in the quantitative description of the process. These approaches need additional experimental information on the microstructure of the explosion products.

As was shown in §5.1 and §5.2, the explosion of droplets caused by pulsed laser radiation is followed by the generation of shock waves. The shock waves create additional optical inhomogeneities in the medium which cause the scattering of light. The energy of these shock waves reaches maximum values during the supercritical explosion and, hence, their contribution to the optics of the process is at its maximum during this period. It is shown in [46], based on the acoustic model of the explosion, that shock waves essentially attenuate radiation during the time interval after the explosion, and become optically inactive when the sound waves come out of the interaction zone.

Summarizing the assessments above, as well as the material in the preceding sections of this chapter, one can arrive at a conclusion that provides a sufficiently complete picture of the optical effects resulting from droplet explosion caused by high-power laser radiation.

(1) Rapid clearing (comparable with the times necessary for gas-dynamic processes in the droplet) of an aerosol irradiated by radiation with $\lambda = 10.6$ μm can occur as a result of gas-dynamic droplet explosions in the one-phase region. For laser pulses with the temporal profile $t_p \gg r_0/c_{s0}$, the time necessary for dissipation of an aerosol composed of droplets 5 to 10 μm in radius does not exceed $10^{-7}$ s. Since the energetic efficiency of the gas-dynamic explosion is too low, it is expedient to use short, high-power laser pulses with $\lambda = 10.6$ μm ($W \simeq 10^8$ Wcm$^{-2}$, $t_p \simeq 10^{-8}$ s) to dissipate the aerosols.

The dissipation of the aerosol droplets using radiation in the visible range by means of one-phase gas-dynamic explosions can be achieved only during the after-pulse effect ($t > t_p$).

(2) The subcritical explosions (two-phase gas-dynamic explosions and fragmentation explosions) do not cause aerosol dissipation during the explosions themselves. An increase in the medium's optical transmission, in this case, can be expected only after (or as a result of) the evaporation of the liquid fraction appearing due to the explosion of the droplet. There is a significant increase in the medium's turbidity with respect to visible light at the moment of explosion.

Pulsed irradiation of aerosols, aimed at dissipating them, necessitates the use of pulses with 'energetic tails' sufficient to evaporate the droplets' fragments. As regards the visual range, the use of more powerful laser pulses for this same purpose has no beneficial effects.

## 5.4. EXPERIMENTAL INVESTIGATIONS OF LASER BEAM PROPAGATION THROUGH EXPLOSIVELY EVAPORATING AEROSOLS

The experimental investigations previously discussed concerned laser beam propagation through an ensemble of exploding aerosol particles, and were aimed at the study of the dynamics of the medium's optical characteristics, describing both the integral optical state (transmission) and local behavior (scattering cross-section of small volumes) of the medium [25-30]. The majority of the experiments were carried out using the high-power radiation generated by $CO_2$ lasers, while the information concerning optical processes was obtained using radiation of a wavelength at $\lambda = 0.63$ μm generated by He-Ne lasers. As was shown in the above discussion, a theoretical analysis can provide the physical information necessary for forecasting the optics of the explosion process, i.e., to find the key parameters of the process. Based on such an analysis, one can assume that the volume extinction coefficient is defined by the following function: $\sigma = \sigma(t, w, I, \vec{a})$, where $w = \int_0^t I(t')\,dt'$, and $\vec{a}$ is the microphysical parameter of the aerosol ensemble. The experimental determination of the form of this function for particular cases is the final goal of the problem of constructing a semiempirical model of the nonlinear extinction coefficient of the medium, which is one of the most important components of the radiation transfer equation.

Experimental studies of the changes in optical transmission of dense water droplet folgs ($\tau \sim 1$) irradiated with pulsed $CO_2$ laser radiation have been carried out in [25-28, 30]. It was shown here that an increase in the medium's turbidity with respect to visible radiation is observed during the irradiation with a high-power laser beam. High-energy laser pulses can cause the clearing of an aerosol, after the vaporization of the explosion products, by means of a relatively low-energy tail of the pulse [26, 30].

Investigations carried out in [29] concerned the study of the dynamics of scattered visible radiation in the volume occupied by optically-thin water aerosols ($\tau \leqslant 0.1$) irradiated by TEA $CO_2$ laser pulses with an intensity of up to 50 $Jcm^{-2}$. The small optical depth of the aerosol allowed one to obtain unequivocal information on the local optical characteristics of the exploding fog droplets. As was pointed out earlier, the knowlegde of these characteristics forms the basis for constructing models of the nonlinear propagation of laser radiation through aerosols. The fog investigated in this experiment was an aerosol stream 2.5 mm in diameter. The speed of the stream was 8 $ms^{-1}$. The modal radius of the droplets in this stream was 2.3 μm, while the maximum radius was 5 μm. A gas-discharge $CO_2$ laser, emitting pulses of about 10 J energy and 300 ns duration (at half-maximum level) was used as the source of high-power radiation. The sounding beam was directed into the interaction zone at an angle of 45° with respect to the direction of high-power beam propagation. The light flux scattered at an angle of 15° was recorded using a PMT.

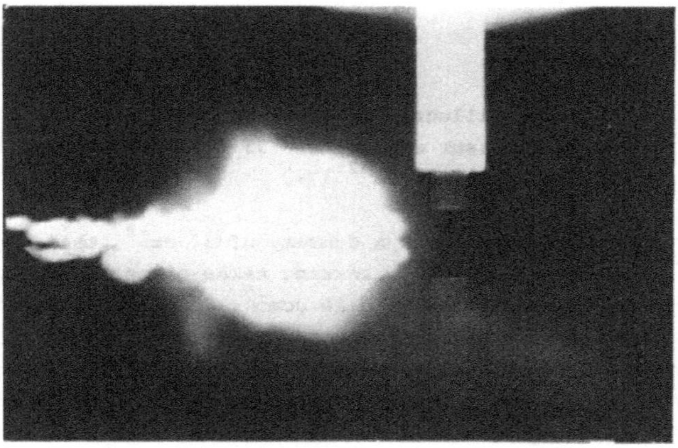

Fig. 5.4.1. Integrated photography of the process of laser action. A connecting pipe belonging to a mist generator is at the top right. The region of mist current discontinuity is the cleared zone. Radiation is travelling from right to left. The luminous region to the left of the mist current is the region of optical breakdown of the air.

Figure 5.4.1 presents an integrated photograph of the fog stream irradiated with a $CO_2$ laser radiation pulse of more than 25 $Jcm^{-2}$ energy. The exposure time was 5 μs. The absence of scattered radiation in the zone where the fog stream and the high-power laser beam intersect proves that clearing does take place during irradiation with the laser beam.

Figure 5.4.2 shows the temporal behavior of the radiation intensity (when $\lambda$ = 0.63 μm) scattered in the irradiated zone. The shape of the oscillograms strongly depends on the energy density distribution of the incident radiation.

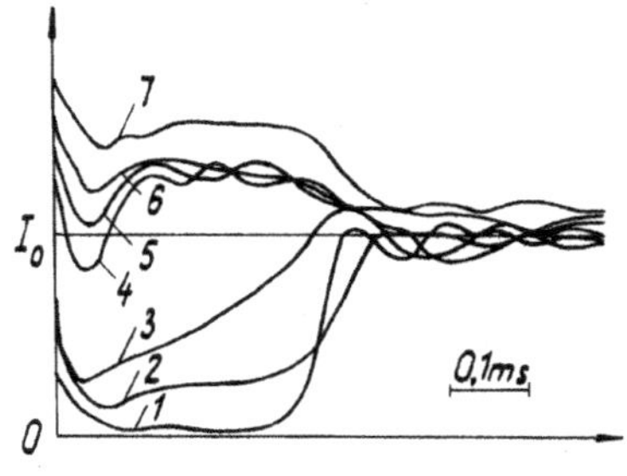


Fig. 5.4.2. Intensity oscillograms of the sounding radiation scattered in the zone of laser action: w = 33 (1), 21 (2), 13 (3), 9 (4), 8 (5), 7 (6) and 6 $Jcm^{-2}$ (7).

There is no observed clearing at a density of 8 $Jcm^{-2}$, this agrees well with the results of [27, 28]. Partial clearing takes place in the region with energy densities ranging from 10 to 20 $Jcm^{-2}$. Complete clearing is observed only in regions with densities in excess of 20 $Jcm^{-2}$. The oscillograms display several stages of the process. In every case the action of the high-power laser pulse gives rise to the appearance of a minimal (relative to the succeeding stages) level of light scattering. The duration of this stage is ~ 50 μs, this significantly exceeds the duration of the laser beam action (1.5 μs) and of the explosion. If the minimum of the scattered radiation is not equal to zero (curves 2 to 7 in Figure 5.4.2) then, in the succeeding stage, an increase in turbidity takes place in the interaction zone. The rate of this increase depends upon the energy of the incident radiation. The lifetime of the consequent perturbations of the optical transmission in the observation region was determined by the velocity of the stream, and for this experiment it was ~ 0.3 ms. If the duration of the turbidity stage is within these limits (curves 4-7), then we enter a stage of constant optical transmissivity. For the opposite case (curves 2, 3) the increase in turbidity is terminated by the flow of the stream, which mimics a wind drift effect. If complete clearing is achieved, this state of the interaction zone remains unchanged during an interval determined by the wind drift.

As can be seen from this figure, complete clearing of the turbid aerosol medium takes place at a threshold intensity of the order of 30 $Jcm^{-2}$. This means that, under conditions of a constant absorption coefficient of condensed water (800 $cm^{-1}$), a unit volume would consume 24 kJ of light energy, which is 10 times greater than the normal heat of vaporization. Theoretical estimates show (see §5.2) that the complete vaporization of a small droplet ($K_{ab}r_0 < 1$) in a gas-dynamic explosion takes place only if the energy density of incident radiation exceeds 25 $Jcm^{-2}$. The amount of energy consumed by the medium in this case is twice as much as the normal heat of vaporization. The experiments, as well as the theoretical analysis, reveal a decrease in the efficiency of the absorption process during the interaction.

All the oscillograms presented in Figure 5.4.2 show the influence of after-effects, since the pulse duration is shorter than the experimental timing error. It should be noted, however, that in all cases the increase in turbidity is observed in the first stage of the process, which corresponds to the moment of incidence of the laser pulse. According to the theoretical forecast (see §5.3), this increase in turbidity for radiation in the visible range is mainly due to an increase in the geometrical effects of an exploding droplet, this influence is most prominent in the initial stage.

The question of the lifetime of the cleared zone, as well as the physical factors involved in the destruction of this state, are of importance. As seen from the oscillograms, the increase in turbidity is observed only if the fog is not completely dissipated. The estimates made in [29] show that the level of probable supersaturation is insufficient for homogeneous recondensation to occur, while the rate of heterogeneous condensation on the condensation nuclei does not explain the observed level of turbidity.

From the standpoint of the optothermodynamics of the process, curves 2-7 in Figure 5.4.2 correspond to thermodynamic transitions in the region of a two-phase state, while curve 1 represents transitions in a single-phase region.

The above results show that the possibility of the complete clearing of a small-droplet fog by TEA $CO_2$ laser radiation pulses with a duration of microseconds exists. The completely-cleared zone is maintained during a time interval which is determined by the extent of wind blurring of this zone. The energy threshold for complete clearing of about 30 $Jcm^{-2}$ reveals a low local efficiency of the transformations involving consumed energy. The energy threshold for complete clearing is lower than the threshold for optical breakdown. Under conditions of incomplete clearing, the after-effect results in a turbidity incerase due to heterogeneous recondensation on the droplets that appeared during the explosion and confined in the finite sphere. Supersaturation in the interaction zone appears due to the diffusion and thermal relaxation of the finite spheres [29].

## REFERENCES: CHAPTER 5

[1] A.V. Kuzikovskii: 'Dynamics of a spherical particles in a high-power optical field', Izv. Vyssh. Uchebn. Zaved. Fiz. 5, 89-94 (1970), in Russian.

[2] F.V. Bunkin, M.I. Tribelskii: 'Nonresonance interaction of high-power

optical radiation with liquid', Usp. Fiz. Nauk 130, 193-239 (1980), in Russian.

[3] V.E. Zuev and A.A. Zemlyanov: 'Water droplet explosions under the effect of intense laser radiation', Izv. Vyssh. Uchebn. Zaved. Fiz. 2, 53-65 (1983), in Russian.

[4] A.A. Zemlyanov, A.V. Kuzikovskii, V.A. Pogodaev, and L.K. Chistyakova: 'Macroparticle in Intense Optical Field', in Problems of Atmospheric Optics (Nauka, Novosibirsk, 1983), in Russian.

[5] V.P. Skripov, E.N. Sinitsin, P.A. Pavlov, et al.: Thermal Physical Characteristics of Liquids in the Metastable State; Handbook (Atomizdat, Moscow, 1980), in Russian.

[6] V.P. Skripov: Metastable Liquid (Nauka, Moscow, 1972), in Russian.

[7] Ya.B. Zeldovich and Yu.P. Raizer: Physics of Shock Waves and High-Temperature Hydrodynamic Phenomena (Nauka, Moscow, 1966), in Russian.

[8] S.L. Rivkin and A.A. Aleksandrov: Thermodynamical Characteristics of Water and Water Vapor (Energiya, Moscow, 1973), in Russian.

[9] V.K. Semenchenko: Selected Chapters of Theoretical Physics (Prosveshcheniye, Moscow, 1966), in Russian.

[10] V.V. Barinov and S.A. Sorokin: 'Water droplet explosions under the effect of optical radiation', Kvant. Elektron 2 (14), 5-11 (1973), in Russian.

[11] A.V. Korotin, L.P. Semenov and P.N. Svirkunov: 'Liquid Droplet Explosion due to Strong Superheating, in Atmospheric Optics. Trans. Inst. Exp. Meteorology 11 (54), 24-33 (Gidrometeoizdat, Moscow, 1975), in Russian.

[12] A.M. Iskoldskii, Yu.E. Nestherikhin, Z.A. Patashinskii, V.K. Pinus and Ya.G. Appelbaum: 'On the instability of gradient explosion', Dokl. Akad. Nauk SSSR 236, N6, 1346-1349 (1977), in Russian.

[13] N.V. Buksdorf, V.A. Pogodaev, and L.K. Chistyakova: 'On the connection of inhomogeneities of the internal optical field of a droplet with its explosion, Kvant. Elektron. 2, N5 (1973), in Russian.

[14] A.P. Prishivalko: Optical and Thermal Fields within Light Scattering Particles (Nauka i Tekhnika, Minsk, 1983), in Russian.

[15] V.S. Loskutov and G.M. Strelkov: 'Explosive Vaporization of Weakly-Absorbant Droplets under the Effect of Laser Pulses', Preprint N12 (295) (Inst. Radioengineering and Electronics, U.S.S.R. Acad. Sci., Moscow, 1980), in Russian.

[16] V.A. Pogodaev, V.I. Bukaty, S.S. Khmelevtsov and L.K. Chistyakova: 'Dynamics of the explosive vaporization of water droplets in an optical radiation field', Kvant. Elektron. 4, 128-130 (1971), in Russian.

[17] P. Kafalas and A.P. Ferdinand: 'Fog droplet vaporization and fragmentation by a 10.6 μm laser pulse', Appl. Opt. 12, N1, 29-33 (1973).

[18] P. Kafalas and J. Hermann: Dynamics and energetics of the explosive

vaporization of fog droplets by a 10.6 μm laser pulse', Appl. Opt. 12, N4, 772-775 (1973).

[19] J. Reilly, P. Singh, and S. Glickler: 'Laser Interaction Phenomenology for a water aerosol at $CO_2$ laser wavelengths', AIAA Paper, N659, 1-7, (1977).

[20] V.A. Pogodaev, A.E. Rozhdestvensky, S.S. Khmelevtsov, and L.K. Chistyakova: 'Thermal explosion of water droplets under the effect of high-power laser radiation', Kvant. Elektron. 4, N1, 157-159 (1977), in Russian.

[21] V.A. Pogodaev, V.V. Kostin, S.S. Khmelevtsov, and L.K. Chistyakova: 'Some problems of the explosion regime of water droplet vaporization', Izv. Vyssh. Uchebn. Zaved. Fiz. 3, 56-60 (1974), in Russian.

[22] M.A. Kolosov, V.K. Rudash, A.V. Šokolov, and G.M. Strelkov: 'Experimental study of the effect of intense IR radiation on large water droplets', Radiotekh. Elektron. 1, 45-50 (1974), in Russian.

[23] D.C. Emmony, and M.A. Engelberts: 'High-speed study of laser-liquid interaction', J. Photographic Science 25, N1, 41-44 (1977).

[24] V.Ya. Korovin and E.V. Ivanov: 'Experimental studies of the effect of $CO_2$ laser radiation on water droplets', in Abstracts: 3rd All-Union Symposium on Laser Radiation Propagation in the Atmosphere (Tomsk, U.S.S.R., 1975), 00. 93-94, in Russian.

[25] V.A. Belts, A.P. Dobrovolskii, V.P. Nikolaev, and S.S. Khmelevtsov: 'Variation of water aerosol transmittance under the effect of a $CO_2$ laser radiation pulse', in Abstracts: 6th All-Union Symposium on Laser Radiation Propagation in the Atmosphere (Nonlinear Effects of Laser Radiation Propagation in the Atmosphere) (Tomsk, U.S.S.R., 1977), pp. 36-40, in Russian.

[26] V.I. Bukaty and M.F. Nebolsin: 'Study of the transmittance of artificial fog under the effect of $CO_2$ laser pusle radiation', ibid, pp. 22-26, in Russian.

[27] V.P. Bisyarin, I.P. Bisyarina, and A.I. Fatievskii: 'Variation of the optical depth of water aerosol as a result of irradiation by 10.6 μm pulsed radiation propagation', ibid, pp. 41-45, in Russian.

[28] V.A. Belts, A.F. Dobrovolskii, and V.P. Nikolaev: 'Pulsed radiation propagation with $\lambda$ = 10.6 μm through artificial droplet fog', in Abstracts: 3rd All-Union Symposium on Laser Radiation Propagation in the Atmosphere (Tomsk, U.S.S.R., 1975) pp. 102-103, in Russian.

[29] A.V. Kuzikovskii, V.I. Kokhanov, and L.K. Chistyakova: 'Clearing of an artificial water aerosol by $CO_2$ laser radiation pulses', Kvant. Elektron. 8, N10, 2090-2096 (1981), in Russian.

[30] J.E. Lowder, H. Kleiman, and R.W. O'Neil: 'High-energy $CO_2$ laser pulse transmission through fog', J. Appl. Phys. 45, N1, 221-223 (1974).

[31] A.A. Kolmykov, V.N. Kondratiev, and M.V. Nemchinov: On the separation of simultaneously-heated substances and the determination of the equation of State by the values of pressure and pulse parameters, Applied Mech. and Techn. Phys. 5, 3-16 (1966), in Russian

[32] N.V. Bukzdorf, A.A. Zemlyanov, A.A. Kuzikovskii, and S.S. Khmelevtsov: 'Spherical droplet explosion under the effect of high-power laser radiation', Izv. Vyssh. Uchebn. Zaved. Fiz. 5, 36-40 (1974), in Russian.

[33] A.A. Zemlyanov and A.V. Kuzikovskii: 'Limiting characteristics of processes in the gas-dynamic explosion of droplets in a high-power light field', Abstracts: 2nd Conf. on Atmospheric Optics (Tomsk, U.S.S.R., 1980) pp. 186-189, in Russian.

[34] A.A. Zemlyanov and A.V. Kuzikovksii: 'Model description of the gas-dynamic explosion of water droplets in a high-power pulsed light field field', Kvant. Elektron. 7, N7, 1523-1530 (1980), in Russian.

[35] F.D. Feiock and L.H. Goodwin: 'A calculation on the laser-induced stress in water', Appl. Phys. 43, N12, 5061-5064 (1972).

[36] A.A. Zemlyanov, A.V. Kuzikovskii, and L.K. Chistyakova: 'Water droplet explosion in a $CO_2$ laser radiation field', in Study of Complex Heat Exchange (Inst. Thermal Physics, Siberian Branch, U.S.S.R. Acad. Sci., Novosibirsk, 1978), pp. 106-111, in Russian.

[37] Kh.S. Kestenboim, G.S. Roslyakov, and L.A. Chudov: Point Explosion. Tables (Nauka, Moscow, 1974), in Russian.

[38] V.A. Pogodaev and L.K. Chistyakova: 'Experimental Study of Explosion Regimes of Water Aerosol Vaporization', Proc. 13th International Symposium on the Dynamics of Rarefied Gases (Novosibirsk, 1982) pp. 542-543, in Russian.

[39] P. Richtmaier and K. Morton: Difference Methods for Solving Boundary Value Problems (Mir, Moscow, 1972), in Russian.

[40] G. Broud: Calculations of Explosions Using a Computer. Gas Dynamics of Explosions (Mir, Moscow, 1976), in Russian.

[41] N.M. Kuznetsov: 'Equation of state and specific heat of water, within a wide range of thermodynamic parameters', J. Appl. Mech. and Tech. Phys., N1, 112-120 (1961), in Russian.

[42] M. Born and E. Wolf: Foundations of Optics (Nauka, Moscow, 1970), in Russian.

[43] G. van de Hulst: Light Scattering by Small Particles (Inostran. Literatura, Moscow, 1961), in Russian.

[44] Yu.P. Raizer: 'On condensation in a cloud of evaporated substance being expanded into vacuum', Zh. Eksp. Teor. Fiz. 37, (12), 1741-1750 (1959), in Russian.

[45] Yu.N. Grachev and G.M. Strelkov: 'Water aerosol transmittance variation under the effect of $CO_2$ laser radiation pulse', Kvant. Elektron. 3, N3, 621-625 (1976), in Russian.

[46] A.A. Zemlyanov, V.V. Kolosov, and A.V. Kuzikovskii: 'Light propagation during aerosol explosion due to laser beams', J. Tech. Phys. 51, N4, 776-781 (1981), in Russian.

[47] C.E. Bell and B.S. Maccabee: 'Shock wave generation in air and in water using a $CO_2$ TEA laser', Appl. Opt. 13, N3, 605-609 (1974).

CHAPTER 6

# PROPAGATION OF HIGH-POWER LASER RADIATION THROUGH HAZES

## 6.1. NONLINEAR OPTICAL EFFECTS IN HAZES: CLASSIFICATION AND FEATURES

Atmospheric hazes are the most frequently observed type of 'optical weather'. Hazes are characterized by a relatively large meteorological visual range (from one to tens of kilometers) and this is just the right situation for the operation of the different types of laser and other opto-electronic devices.

The characteristics of the self-action of high-power laser radiation under these conditions are caused by a great variety of physico-chemical properties of the haze as well as the various number densities, size spectra and particle shapes in different types of hazes. Natural hazes can be divided into three basic groups according to the mechanisms of various optical nonlinearities: (1) dry, dusty hazes; (2) chemically reactive hazes; and (3) humid hazes.

The main peculiarities of high-power laser self-action in hazes of the first and second types are connected with the spatially-localized character of the energy runoff into the surrounding medium through fast-melting absorbing centers. This fact causes significant thermal, acousto-hydrodynamic, and thermochemical disturbances of the medium's refractive index at scales corresponding to both the gaps between particles and the size of the whole laser beam. Sections 6.1.1, 6.1.2, 6.2-6.4, and 6.5.2 of this chapter are devoted to a review of investigations carried out into this question. Section 6.5.3 presents the results of investigations into the non-linear distortions of high-power laser beams in the humid hazes occupying the atmospheric ground layer.

The nonlinear optical effects characteristic of the finely-dispersed water aerosols of humid hazes can be compared, in certain cases, with the thermal effects occurring in gases. This determines the specific nature of the joint influence of these effects on high-power laser beams propagating along long atmospheric paths.

### 6.1.1. Characteristic Relaxation Times in Hazes Irradiated with High-Power Lasers

An initial classification of nonlinear thermal effects accompanying the propagation of a laser beam through a haze can be made by comparing the characteristic times of thermal and acoustic disturbance transfer in the space between absorbing centers and over the beam's cross-section $R_0$ [1]. Figure 6.1.1 presents the dependences of characteristic times $t_c$ on the effective radius $R_0$ of the laser beam. The curves in Figure 6.1.1 are constructed for typical values of the atmospheric haze parameters: $a = 10^{-4}$ cm; $N_0 = 10^3$ cm$^{-3}$; $v = 10^2$ cms$^{-1}$; $\langle v_1^2 \rangle^{1/2} \simeq 0.1$ v; and $X_T = 10^2$ cm$^2$s$^{-1}$, where a is the effective particle radius, $N_0$ is the concentration of particles in the haze; v, $\langle v_1^2 \rangle$ are the mean velocity and the variance of the wind speed fluctuations, respectively; and $X_T$ is the coefficient of the thermal conductivity of the air. Dotted lines 1 and 3 in this figure correspond to the times $t_1$ and $t_3$ of the development of the quasistationary regime of heat transfer into the medium through the particle surface, occurring by means of molecular thermal conductivity and the heating of the particle to the steady temperature. The characteristic times are $t_1 \simeq a^2/4X_T$ and $t_3 \simeq (a^2/3X_T)(C_a\rho_a/C_p\rho)$; here $C_a\rho_a$ and $C_p\rho$ are the volume specific heats of the particulate matter and the air, respectively.

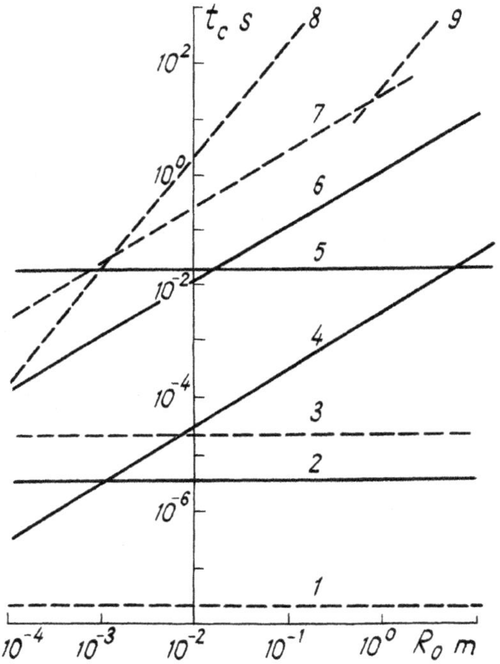


Fig. 6.1.1. Characteristic times of thermohydrodynamic processes in a laser beam channel containing light-absorbing particles.
1 - $t_1 \approx a^2/4X_T$; 2 - $t_2 \approx (N^{1/3}C_s)^{-1}$; 3 - $t_3 \approx (a^2/3X_T)\times$ $\times(C_a\rho_a/C_p\rho)$; 4 - $t_4 \approx R_0/C_s$; 5 - $t_5 \approx (4N^{2/3}X_T)^{-1}$; 6 - $t_6 \approx R_0/V_\perp$; 7 - $t_7 \approx R_0^2/4X_{eff}(\beta \ll 1)$; 8 - $t_8 \approx R^2/4X_T$; 9 - $t_9 \approx R_0^2/4X_{eff}(\beta \gg 1)$.

The lines 2 and 5 represent averaged time intervals during which light-induced thermal and acoustic disturbances (aureoles) overlap in the space between the absorbing centers. The characteristic times in this case are $t_2 \simeq (N^{1/3}C_s)^{-1}$ and $t_5 \simeq (4N^{2/3}X_T)^{-1}$, where N is the particles' number density and $C_s$ is the speed of sound in air.

Lines 4 and 6 represent the time of transfer of the acoustic disturbances and the wind shear across the beam $v_\perp$, respectively ($t_4 \simeq R_0/C_s$; $t_6 \simeq R_0/v_\perp$).

The broken lines 7-9 represent the dependences of the temperature relaxation time on $R_0$ due to molecular (8) and turbulent (7 and 9) thermal conductivity, respectively. The effective coefficient of thermal conductivity $X_{eff}$ in the case of turbulent heat transfer in the beam can be estimated [3] as follows:

$$X_{eff} \simeq \begin{cases} X_t, & \beta \gg 1; \\ \frac{\sqrt{2}}{\pi} R_0 \langle v_1^2 \rangle^{1/2}, & \beta \ll 1, \end{cases} \tag{6.1.1}$$

where $X_t = t_L\langle v_1^2\rangle$ is the coefficient of turbulent thermal conductivity: $\beta = (R_0/t_L)(8\langle v_1^2\rangle)^{-1/2}$, $t_L$ is the Lagrange correlation time of the wind speed fluctuations. Taking into account (6.1.1), one can write for the characteristic times $t_{7-9}$ the following expressions: $t_7 \simeq R_0^2/4X_{eff}$ ($\beta \ll 1$); $t_9 \simeq R_0^2/4X_{eff}$ ($\beta \gg 1$); $t_8 \simeq R_0^2/X_T$.

For some applications it is more convenient to use the following time-dependent representation of $X_{eff}$:

$$X_{eff}(t) \simeq t_L\langle v_1^2\rangle[1 - \exp(-t/2t_L)],$$

which approximates Taylor's formula [3] in the region where $L_0 \gg R_0 \gg \ell_0$, where $L_0$ and $\ell_0$ are the outer and the inner scales of atmospheric turbulence, respectively. In the region where $\Delta x \ll \ell_0$ the heat transfer process is governed by molecular thermal conductivity, with a characteristic time of $t_8$.

It follows from Figure 6.1.1 that different effects of laser self-action can dominate the processes in the laser beam channel depending on the relationship between the laser pulse duration $t_p$ and the characteristic times of the various thermohydrodynamic processes. The following situations are of the greatest interest [1-2]:

$$t_4 \ll t_p \ll t_5, \tag{6.1.2}$$

$$t_4,\ t_5 \lesssim t_p, \tag{6.1.3}$$

$$t_p \lesssim t_4 \ll t_2,\ t_5, \tag{6.1.4}$$

$$t_2 \lesssim t_p \lesssim t_4 \ll t_5. \tag{6.1.5}$$

In the case of cw laser radiation acting during the time period $t > R_0/v_\perp$, one should understand by $t_p$ the characteristic time $R_0/v_\perp$ that corresponds either to the time of irradiation of the medium in set-ups involving beam scanning, or to the time of transportation of the medium across the beam. It should be noted, however, that processes involving molecular heat transfer in the laser beam can be neglected in all atmospheric situations when $t_p \ll t_8$. As analysis shows, nonlinear light scattering by spatially-localized inhomogeneities can occur only in situations like those represented by (6.1.2) and (6.1.4). It is characteristic for situations (6.1.3) and (6.1.5) that the cooperative effects of laser beam self-action become of practical importance. These effects are due to laser-induced defocusing (6.1.3) and focusing (6.1.5) gaseous lenses. The stochasticity of the medium caused by the overlapping of the distortions induced by the beam on randomly-located absorbing centers, as well as by fluctuations of their number density and sizes, also results in the above-mentioned cooperative effects. In general, these fluctuations are not Gaussian. When $t \gtrsim R_0/V$ wind deflection of the beam's axis will take place due to the axial asymmetry of the laser-induced gaseous lenses.

### 6.1.2. Propagation Equations for High-Power Radiation in Media Composed of Randomly-Distributed Centers

Chapter 4 gave an approach for solving the problem of high-power radiation propagation through a nonlinear, randomly heterogeneous medium. In general, the light-induced fluctuations of the medium's dielectric constant are not Gaussian [1, 4, 21].

In the case of a medium with 'soft' scatteres ($|\varepsilon - \varepsilon_0| \ll 1$), one can use the stochastic parabolic equation for determining the complex amplitude of the electric field of the light wave:

$$2ik\,\frac{\partial E}{\partial x} + \Delta_{\vec{\ell}}E + k^2E(\vec{r}, t)[\varepsilon(r, t, |E|^2) - 1] =$$

$$= 0(\partial^2E/\partial x^2 + |\nabla(\vec{E}\nabla\varepsilon)|), \qquad (6.1.6)$$

where $\vec{r} = (x, \vec{\ell})$; $\Delta_{\vec{\ell}}$ is the transverse Laplacian; $\varepsilon$ is the complex dielectric constant of the atmosphere;

$$\varepsilon(\vec{r}, t, |E|^2) = 1 + \varepsilon_1(\vec{r}) + \sum_{k=1}^{\infty} \varepsilon_p(\vec{r} - \vec{r}_K, a_K, t, |E(\vec{r}_k)|^2); \qquad (6.1.7)$$

where $\varepsilon_1$ is the fractional random deviation of the dielectric constant caused by atmospheric turbulence: $\langle\varepsilon_1\rangle = 0$; $\varepsilon_p$ is the profile of the dielectric constant in the vicinity of a solid particle,

$$\varepsilon_p = \begin{cases} \varepsilon_a = \varepsilon_a' - i\varepsilon_a'', & |\vec{r} - \vec{r}_K| \leqslant a, \\ (d\varepsilon/d\rho)(\rho - \rho_0), & |\vec{r} - \vec{r}_K| > a; \end{cases} \tag{6.1.7}$$

and $\rho = \rho(\vec{r} - \vec{r}_K, a_K, t, |E(\vec{r}_K)|^2)$ is the profile of distortions in the density of the k-th optical inhomogeneity, written assuming the absence of interference between the center of heating. Equations which take into account the interference mentioned above can be found in [1]. The usual limitations of the scalar parabolic equation (of the type (6.1.6)) are written as follows: $|\varepsilon - 1|$, $\lambda/a$, $\lambda/r_N$, $R_0/x \ll 1$, where $r_N$ is the characteristic radius of the light-induced inhomogeneities ($\varepsilon_N = (d\varepsilon/d\rho)\times\times(\rho - \rho_0)$). In the following, the fluctuations $\varepsilon_1$ are considered to be Gaussian.

The probabilistic description of the $\Sigma\varepsilon_p$ field uses the concept of a characteristic functional:

$$\psi^x_{\varepsilon_p}[v; v^*] = \left\langle \exp\left\{i \int_0^x dx' \int_{-\infty}^{\infty} d^2\ell' \left[v(x',\ell')\Sigma\varepsilon_p(x' - x_K, \vec{\ell} - \vec{\ell}_K, a_K) + \right.\right.\right.$$
$$\left.\left.\left. + v^*(x', \vec{\ell}')\Sigma\varepsilon_p^*(x' - x_K, \vec{\ell}' - \vec{\ell}_K, a_K)\right]\right\}\right\rangle. \tag{6.1.8}$$

Functional series expansion of this functional or its logarithm $\ln \psi^x \varepsilon_p[v; v^*]$ gives all of the momenta and the cumulant functions of $\varepsilon_p$. The Dirac brackets < > in (6.1.8) mean averaging over all the possible values of $\vec{r}_K$ and $a_K$. The particles are considered to be statistically independent and uniformly distributed over the space $\vec{r}_K$. The normalized size-distribution function of the particles is $f(a_K)$. If the mean number of aureoles within the limits of the Fraunhofer diffraction zone $kr_N^2$ is much greater than one, then [2] fluctuations of the random field $\Sigma\varepsilon_p$ are Poissonian in the general sense, and their characteristic function is

$$\psi^*_{\varepsilon_p}[v; v^*] = \exp\left\{-N_0 \int_0^x dx' \int_{-\infty}^{\infty} d^2\ell'[1 - \psi_a[\int_{x'}^x dx''(v(x'', \vec{\ell}') \times\right.$$
$$\times \varepsilon_p(x'' - x', \vec{\ell}'' - \vec{\ell}', t, |E(x', \vec{\ell}')|^2 + v^*(x'', \vec{\ell}') \times$$
$$\left.\times \varepsilon_p^*(x'' - x', \vec{\ell}'' - \vec{\ell}', t, |E(x', \vec{\ell}')|^2)\right\}, \tag{6.1.9}$$

where $\psi_a[\phi] = \int_0^\infty da_K f(a_K) \exp(ia_K\phi)$ is the characteristic function of the random value $a_K$, and $N_0$ is the number density of the particles.

The problem formulated in the form of stochastic equation (6.1.6) can be reduced to differential equations for different moments of the field E by using the mathematical techniques developed for arbitrary, non-Gaussian random processes delta-correlated with respect to x [1, 4]. The equation for the moment

$$\langle U_{n,m}\rangle = \langle E(x, \vec{\ell}_1) \ldots E(x, \vec{\ell}_n)E^*(x, \vec{\ell}'_1) \ldots E^*(X, \vec{\ell}'_m)\rangle$$

has the form

$$2ik \frac{\partial}{\partial x} \langle U_{n,m}\rangle + [\sum_{k=1}^{n} \Delta_{\vec{\ell}_K} - \sum_{j=1}^{m} \Delta_{\vec{\ell}_j}]\langle U_{n,m}\rangle + \frac{ik^3}{4} \langle U_{n,m}\rangle \times$$

$$\times Q_{n,m} + 2ik \langle U_{n,m}\rangle \frac{\partial}{\partial x} \ln \psi^{x}_{\varepsilon_p} [\frac{k}{2} \sum_{k=1}^{n} \delta(\vec{\ell} - \vec{\ell}_K); \tag{6.1.10}$$

$$- \frac{k}{2} \sum_{j=1}^{m} \delta(\vec{\ell} - \vec{\ell}'_j)] = 0,$$

where

$$Q_{n,m} = \sum_{k=1}^{n} \sum_{j=1}^{m} [A_{\varepsilon_1}(x, \vec{\ell}_K - \vec{\ell}_j) - 2A_{\varepsilon_1}(x, \vec{\ell}_K - \vec{\ell}'_j) + A_{\varepsilon_1}(x, \vec{\ell}'_K - \vec{\ell}'_j)];$$

$$A_{\varepsilon_1}(x, \vec{\rho}) = (2\pi) \int_{-\infty}^{\infty} d^2\kappa \exp(i\vec{\kappa}\vec{\rho})\phi_{\varepsilon_1}(x, \vec{\kappa}).$$

Here, $\phi_{\varepsilon_1}(x, \vec{\kappa})$ is the spectral density of the fluctuations of the atmospheric dielectric constant $\varepsilon_1$.

The assumption that the typical sizes $r_N$ and $r_{\varepsilon_1}$ of the inhomogeneities of $\varepsilon_N$ and $\varepsilon_1$, respectively, are much less than X is of principal importance for the derivation of (6.1.10). This condition controls the property of delta-correlation of the fluctuations of the dielectric constant along the direction of propagation. Fluctuations of $\varepsilon_1$ and $\Sigma\varepsilon_p$ are assumed to be statistically independent. In addition, it is assumed that $r_N \ll R_0$ and the function $\varepsilon_p(x, \vec{\ell}, t, |E|^2)$ depends on the mean intensity $\langle |E|^2\rangle$.

The quantitative analysis of the dynamics of laser beam self-action is based on the equation for the second moment of the field

$$\Gamma(x, \vec{\ell}, \vec{\rho}) = \langle U_{1,1}\rangle = \langle [E(x, \vec{\ell} + \vec{\rho}/2)E^*(x, \vec{\ell} - \vec{\rho}/2)]\rangle.$$

If, for simplicity, one considers a monodispersed aerosol for which $a/r_N \ll 1$, then one can obtain from (6.1.9) and (6.1.10) that

$$\frac{\partial\Gamma}{\partial x} - \frac{i}{K} \nabla_{\vec{\ell}}\nabla_{\vec{\rho}} + \frac{k^2}{4} \Gamma D_{\varepsilon_1}(x, \vec{\rho}) + \Gamma D_a(x, \vec{\rho})\exp[i\delta\phi_N(x, \vec{\ell}, 0, \vec{\rho})] +$$

$$+ \Gamma D_N(x, \vec{\ell}, \vec{\rho}) = 0, \tag{6.1.11}$$

where

$$D_{\varepsilon_1}(x, \vec{\rho}) = A_{\varepsilon_1}(x, 0) - A_{\varepsilon_1}(x, \vec{\rho}); \tag{6.1.12}$$

$$\delta\phi_N(x, \vec{\ell}, \vec{\ell}', \vec{\rho}) = \frac{k}{2}\int_0^\infty dx'\Big[\varepsilon_N(x, \vec{\ell} - \vec{\ell}', \Gamma(x, \vec{\ell}, 0)) - $$
$$- \varepsilon_N^*(x, \vec{\ell} - \vec{\ell}' - \vec{\rho}, \Gamma(x, \vec{\ell} - \vec{\rho}, 0))\Big]; \qquad (6.1.13)$$

$$D_N(x, \vec{\ell}, \vec{\rho}) = N_0\int_{-\infty}^{\infty} d^2\ell'\{1 - \exp[i\delta\phi_N(x, \vec{\ell}, \vec{\ell}', \vec{\rho})]\}. \qquad (6.1.14)$$

The function $D_a(x, \vec{\ell}, \vec{\rho})$ refers to an undisturbed particle, and is defined by an equation analogous to (6.1.14), except for the following fact. The function $\delta\phi_N$ entering the equation, in this case, is determined by (6.1.13), in which $\varepsilon_N$ is replaced by $\varepsilon_a$.

If we expand the exponential term of (6.1.14) into a series and truncate it at the second-order terms, then this allows us to make a limiting transformation of (6.1.11). As a result of this transformation, we obtain an expression which is valid for the particular case of Gaussian spatial fluctuations of $\Sigma\varepsilon_p$, while in its initial form (6.1.11) is valid for the more general case of spatial fluctuations of the dielectric constant $\Sigma\varepsilon_p$ as described by generalized Poisson statistics. Physically, this corresponds to the situation of a small phase change of the plane wave on the thermohydrodynamic aureole of a particle.

$$k\int_0^\infty dx'|\varepsilon_N(x', \vec{\ell}, \Gamma)| \ll 1. \qquad (6.1.15)$$

In this case, the function $D_N$ is written as

$$D_N(x, \vec{\ell}, \vec{\rho}) = 4\pi^3 kN_0\{F_N(0, \Gamma(x, \vec{\ell}, 0)) - F_N(0, \Gamma(x, \vec{\ell} - \vec{\rho}, 0))\} +$$
$$+ \frac{k^2}{4} N_0\{A_N(0, \Gamma(x, \vec{\ell}, 0)) - A_N(\vec{\rho}, \Gamma(x, \vec{\ell}, 0))\}, \qquad (6.1.16)$$

where $F_N(\vec{q}, \Gamma)$ is the Fourier transform of $\varepsilon_N(\vec{r} - \vec{r}_k, \Gamma(x, \vec{\ell}, 0))$ with respect to the vector difference $(\vec{r} - \vec{r}_k)$:

$$F_N(q, \Gamma) = (2\pi)^{-3}\int_{-\infty}^{\infty} d^3r \exp(-i\vec{q}\vec{r})\cdot\varepsilon_N(\vec{r}, \Gamma(x, \vec{\ell}, 0)). \qquad (6.1.17)$$

$A_N(\vec{\rho}, \Gamma)$ is the two-dimensional correlation function of the induced fluctuations $\Sigma\varepsilon_N$, and

$$A_N(\vec{\rho}, \Gamma) = (2\pi)\int_{-\infty}^{\infty} d^2\kappa \exp(i\vec{\kappa}\vec{\rho})F_N^2(\vec{\kappa}, \Gamma), \qquad (6.1.18)$$

where $\vec{\kappa} = (\kappa_2, \kappa_3)$.

The analysis of (6.1.16) shows that its first term accounts for the nonlinear effect of beam defocusing on the statistically-averaged profile of the dielectric constant $\bar{\varepsilon}_N = \langle\Sigma\varepsilon_N\rangle$ originating due to the non-uniform heating of aerosol particles located at different points within the beam's cross-section. The second term of (6.1.16) describes the effect of the nonlinear scattering of light on the beam-induced perturbations of the dielectric constant $\Sigma\varepsilon_N - \bar{\varepsilon}_N$.

The situation when the macroscopic number density of the particles $N_0(\vec{r})$ is also a random value having the characteristic spatial scale $r_n \gg N_0^{-1/3}$ has been discussed in [6].

The small-angle approximation of the radiation transfer equation for the case of a medium scattering nonlinearly with regular refraction can be obtained from (6.1.11) and (6.1.16) by using the ray intensity $I_n(x, \vec{\ell}, \vec{\omega})$ instead of the coherence function $\Gamma(x, \vec{\ell}, \vec{\rho})$. The acceptability of such a substitution is based on the relationship between these functions through the Fourier transform:

$$I_n(x, \vec{\ell}, \vec{\omega}) = (2\pi)^{-2} \int_{-\infty}^{\infty} d^2\rho \, \exp(-ik\vec{\omega}\vec{\rho})\Gamma(x, \vec{\ell}, \vec{\rho}). \qquad (6.1.19)$$

Using the approximation of single scattering by nondistrubed aerosol particles, and assuming $\varepsilon_1 = 0$, the function $I_n(x, \vec{\ell}, \vec{\omega})$ has the following form:

$$\left[ \frac{\partial}{\partial x} + \vec{\omega}\nabla_{\vec{\ell}} + \frac{1}{2}\nabla_{\vec{\ell}} \; \bar{\varepsilon}_N \nabla_{\vec{\omega}} + \alpha + \alpha_N \right] I_n(x, \vec{\ell}, \vec{\omega}) =$$

$$= \int_{-\infty}^{\infty} d^2\omega' G_N(\vec{\omega} - \vec{\omega}', \Gamma) I_n(x, \vec{\ell}, \vec{\omega}'), \qquad (6.1.20)$$

where $\alpha$ and $\alpha_N$ are, respectively, the volume extinction coefficient of the undisturbed aerosol and the volume scattering coefficient of the thermohydrodynamic aureoles localized in the near vicinity of the absorbing centers:

$$\alpha = \frac{k^2}{4} N_0 A_a(x, 0) + \pi a^2 N_0 K_{ab}; \qquad \alpha_N = \frac{k^2}{4} N_0 A_n(0, \Gamma(x, \vec{\ell}, 0)). \quad (6.1.21)$$

$G_N(\vec{\omega}, \Gamma)$ is the normalized scattering phase function:

$$G_N(\vec{\omega}, \Gamma(x, \vec{\ell}, 0)) = \frac{2\pi}{\alpha_N} F_N^2(\vec{\omega}k, \Gamma(x, \vec{\ell}, 0)). \qquad (6.1.22)$$

The mean profile of the perturbations of the dielectric constant entering (6.1.20) can be defined as $\bar{\varepsilon}_N = 4\pi^3 N_0 F_N(0, \Gamma(x, \vec{\ell}, 0))$ if the conditions (6.1.2) and (6.1.4) are fulfilled. In the regions limited by (6.1.3) and (6.1.5), as well as for $t \gtrsim t_6$, the function $\bar{\varepsilon}_N$ can be found by solving the thermohydrodynamic equations for the medium in the beam channel; relevant calculated results were published in [2] and will be discussed in §6.3-§6.5.

Finally, (6.2.10) can be reduced, providing the approximation of a single scattering by the aureoles holds, and with the absence of any refraction of the beam, to the differential form of the Bouguer law.

$$(\partial/\partial x)I_n(x, 0, 0) = -(\alpha + \alpha_N)I_n(x, 0, 0) \quad (6.1.23)$$

## 6.2. NONLINEAR SCATTERING OF LIGHT BY THERMAL AUREOLES AROUND LIGHT-ABSORBING PARTICLES

### 6.2.1. Introduction

Particles capable of absorbing radiation are the sources of local perturbations of the medium's refractive index due to both heat and mass exchange between the particles and the medium, as well as molecular thermal conductivity, diffusion, and other processes. Already, initial investigations have shown [7-13], that the effect of the nonlinear scattering of light by the thermal aureoles of aerosol particles is the dominating mechanism in high-power laser beam self-action, even if the intensity of the beam is not sufficiently high to cause phase transitions or to change the chemical composition of the particulate matter. The effect is at its strongest when the material of the solid particles of the haze irradiated by the laser radiation has a high boiling point.

The problem of the scattering of light by localized thermal aureoles, considered in this section, has physical meaning only if the aureoles of individual particles do not overlap in the gaps between the particles. The spatial separation between particles is, on average, proportional to $N_0^{1/3}$, where $N_0$ is the number density of the particles. Characteristic scales of the regions displaying refractive index perturbations around the light-absorbing centers are estimated as $r \simeq (X_T t)^{1/2}$; $(D_n t)^{1/2}$ for thermal and vapor aureoles, and $r_s \simeq c_s t$ for the acoustic ones; here t is the time interval during which the laser beam interacts with a particle [2].

Calculations of the nonlinear extinction of radiation by auereoles made using the radiation transfer equation require that the condition of discrete, separate aureoles is replaced by a more stringent restriction: that scatterers are located, on average, in the Fraunhofer diffraction zone relative to each other, i.e., $N_0$ must be less than or equal to $(r_{T(s)}^4 k)^{-1}$. This condition is fulfilled for short laser pulses. Thus, for example, if $N_0 = 10^{-6}$ $m^{-3}$ and $\lambda = 1060$ nm, then the laser pulse duration should not exceed $10^{-3}$ s.

Below, we shall derive the relationships which serve as the basis for calculations of the optical characteristics of the localized inhomogeneity comprising the particle itself and its thermal aureole, within the framework of the approach for 'soft' scatteres (i.e. $|\varepsilon_p - \varepsilon_0| \ll 1$). The sizes of the particles are assumed to be much greater than the radiation wavelength. The expression for the complex amplitude of the scattered field in the zone of Fraunhofer diffraction $(R \gg kr_{T(s)}^2)$ is written as

$$\vec{E}(\vec{R}, t) = \vec{E}_0 \exp(-i\vec{k}\vec{R}) + [\vec{n}, [\vec{E}_0, \vec{n}]]\frac{\exp(ikR)}{R} S(\vec{\omega}, t), \qquad (6.2.1)$$

where $\vec{E}_0$ is the amplitude of the incident radiation; $\vec{R}$ is the radius-vector centered at the particles center, i.e., $\vec{R} = \vec{r} - \vec{r}_K$; $\vec{n} = \vec{R}/R$; $\vec{\omega} = \vec{k}/k - \vec{n}$; $\vec{k}$ is the wave vector; $S(\vec{\omega}, t)$ is the amplitude scattering function of the inhomogeneity,

$$S(\vec{\omega}, t) = -\frac{ik}{2}\int_{-\infty}^{\infty} d^2\ell \exp(-ik\vec{\ell}\vec{\omega}_\perp)\times[1 - \exp(-\frac{ik}{2}\int_{-\infty}^{\infty} dx' \times$$
$$\times (\varepsilon_p(x', \vec{\ell}, t) - \varepsilon_0))], \qquad (6.2.2)$$

where $\vec{\ell} = (R_y, R_z)$; $x = R_x$ is the axial coordinate; and $\vec{\omega}_\perp = (\vec{\omega}_y, \vec{\omega}_z)$. If the phase change of a plane wave in the region occupied by a thermal aureole $(ka|\varepsilon_N| \ll 1)$ is small, then (6.2.2) can be reduced to a simpler form:

$$S(\vec{\omega}, t) = S_a(\vec{\omega}) \exp(- i\phi_N(0)) + S_N(\vec{\omega}, t). \qquad (6.2.3)$$

Here, a is the radius of a particle, $\varepsilon_N$ is the deviation of the dielectric constant from its equilibrium value $\varepsilon_0$ in the region of the thermal aureole; and $S_a(\vec{\omega})$ is the amplitude scattering function of a nondisturbed particle.

The function

$$\phi_N(\vec{\ell}) = \frac{k}{2}\int_{-\infty}^{\infty} dx\ \varepsilon_N(x, \vec{\ell}, t) \qquad (6.2.4)$$

is the phase change of a plane wave taking place in the thermal aureole along the laser beam; $\vec{\ell}$ is the radial coordinate; and $S_N(\vec{\omega}, t)$ is the amplitude scattering function of a thermal aureole described by an expression analogous to (6.2.2), in which $\varepsilon_p$ is replaced by $\varepsilon_N$ over the entire region of integration. Further simplifications of the expression for $S_N(\vec{\omega}, t)$ can be made if the phase change within the termal aureoles is much less than unity, i.e.,

$$|\phi_N(\vec{\ell})| \ll 1. \qquad (6.2.5)$$

In this case, by substituting the first non-vanishing term of the series expansion of the exponent entering (6.2.2), (in brackets), one obtains for $S_N(\vec{\omega}, t)$ an equation which corresponds to the approximation of the Rayleigh-Gans scattering law:

$$S_N(\vec{\omega}, t) = 2\pi^2k^2F_N(k\vec{\omega})[1 - O(a^2/4X_Tt)], \qquad (6.2.6)$$

where $F_N(\vec{\kappa})$ is the Fourier transform of the function $\varepsilon_N(\vec{R}, t)$ with respect to the vector $\vec{R}$:

$$F_N(\vec{\kappa}) = (2\pi)^{-3} \int_{-\infty}^{\infty} d^3R \exp(-i\vec{\kappa}\vec{R})\varepsilon_N(\vec{R}, t).$$

Thus, the approximation of the Rayleigh-Gans scattering law needs only the knowledge of the Fourier transforms of the dielectric constant for determining the optical characteristics of thermal aureoles. The latter problem is mathematically more simple in the majority of practical applications.

Let us determine the total efficiency factor of light scattering by a particle and its beam-induced thermal aureole as the ratio of scattered flux to light flux incident only on the geometrical cross-section of the particle:

$$K_s(t) = a^{-2} \int_0^{\pi} d\theta \sin\theta(1 + \cos^2\theta)|S(2 \sin\frac{\theta}{2}, t)|^2 \qquad (6.2.7)$$

For simplicity, the incident radiation is considered to be unpolarized. The scattering phase function is written as

$$G(\theta, t) = \frac{2}{a^2 K_s}(1 + \cos^2\theta)|S(2 \sin\frac{\theta}{2}, t)|^2.$$

According to the optical theorem, the extinction efficiency factor is determined by the imaginary part of the complex amplitude of the forward-scattered field.

Using optical theorem and (6.2.3), one can write the expressions describing the extinction efficiency factor of an optical inhomogeneity in two limiting cases of small ($ka \ll 1$) and large ($ka \gg 1$) absorbing particles as follows [16]:

$$K(t) = K_s^N(t) + \begin{cases} 4(ka) \, \mathrm{Im}\left(\frac{1 - \varepsilon_a}{2 + \varepsilon_a} \exp(i\phi_N(0))\right); & ka \gg 1, \\ 2 \cos(\mathrm{Re}\, \phi_N(0)); & ka \gg 1, \end{cases} \qquad (6.2.8)$$

where $K_s^N$ is the scattering efficiency factor of the thermal aureole only.

### 6.2.2. An Analysis of Thermohydrodynamic Perturbations of the Medium due to the Absorption of Radiation by Solid Aerosol Particles

Consider one practically important case of a solid particle, a motionless with respect to the medium, with a high melting point and a radius a. Assume also that the radiation-induced heating does not lead to the particle melting or evaporating. In this case, the heat flow into the medium through the boundary of a particle is due only to molecular conductivity. At the moment $t \gg a^2/4X_T = t_1$ this heat flow becomes quasi-stationary and is proportional to the difference between the temperature at the particle's surface $T_a$ and the temperature of the surrounding air $T_0$. Thus, for

$a = 1\ \mu m$, $t_1 \simeq 10^{-8}$ s.

Homogeneous heating of a particle, in this quasi-stationary approach, can be described by the following equation [1]:

$$ac_a\rho_a \frac{4}{3}\frac{\partial T_a}{\partial t} = IK_{ab} + 4\left(\int_{T_0}^{T_a} \lambda_T\, dT\right)/a + 4\sigma_B\varepsilon_B \cdot (T_0^4 - T_a^4), \tag{6.2.9}$$

where the second and the third terms of the right-hand side of (6.2.9) account for the energetic losses due to molecular thermal conductivity and re-emission, respectively; $c_a\rho_a$ and $\varepsilon_B$ are the volume specific heat and the volume coefficient of grayness of the particle, respectively, $\lambda_T$ is the thermal conductivity coefficient of air, and $\sigma_B$ is the Stefan-Boltzmann constant.

The system of thermohydrodynamic equations describing the behavior of the medium surrounding the particle is written as follows [1]:

$$C_p\rho \frac{\partial T}{\partial t} = \nabla(\lambda_T \nabla T) + \frac{\partial p}{\partial t} + O(C_p\rho\vec{v}(\nabla_T + \vec{v}\nabla_p)); \tag{6.2.10}$$

$$(\partial\vec{v}/\partial t) + (\vec{v}\nabla)\vec{v} = -\rho^{-1}\nabla p + O(\alpha_{ac}\nabla^2\vec{v}); \tag{6.2.11}$$

$$\frac{\partial\rho}{\partial t} + \operatorname{div}(\rho\vec{v}) = 0; \qquad p = R_b\rho T; \qquad T(r = a) = T_a. \tag{6.2.12}$$

The values $\rho$, $T$, $p$, $\vec{v}$, $C_p$, $r_s$, $R_b$ are the density, temperature, pressure, velocity of the hydrodynamic flow, isobaric specific heat, absorption coefficient for sound waves, and the specific gas constant.

The system of equations (6.2.9)-(6.2.12) can be reduced [1] to one linearized equation that follows the thermo-acoustic approach:

$$((\partial^3/\partial t^3) - \gamma X_T \nabla^2(\partial^2/\partial t^2) - c_s^2 \nabla^2(\partial/\partial t) + X_T c_s^2 \nabla^4)\delta\rho =$$

$$= (\gamma - 1)\nabla^2(q_n\delta(R))[1 + O(a^2/4X_T t)], \tag{6.2.13}$$

where $\delta\rho = \rho - \rho_0$; $\gamma$ is the adiabatic exponent; $\delta(\vec{R})$ is the delta function of the three-dimensional argument; and $q_n$ is the source function, which for the incident laser radiation of constant intensity $I_0$ is

$$q_n = \pi a^2 K_{ab} I_0 \beta_0 [1 - \exp(-t/t_3)], \tag{6.2.14}$$

where $t_3$ is the characteristic time of particle heating $t_3 \simeq (a^2/3X_T)\times$ $\times(C_a\rho_a/C_p\rho)$, and $\beta_0 = (1 + \sigma_B\varepsilon_B T_0^3/\lambda_T)^{-1}$. The fact that, under conditions of quasi-stationary heat transfer from a particle to the medium, the particle can be represented by an energetically-equivalent point source allows one to transfer the results calculated for sphere to particles with arbitrary shapes, if by $\sigma_{ab} = \pi a^2 K_{ab}$ one means the absorption cross-section of non-spherical particles.

A solution of (6.2.13) is generally sought using the method of integral transformations. In a particular case involving incident laser radiation of a constant intensity $I_0$, a solution for the Fourier transform $F_\rho(\vec{\kappa}, t)$ of the function $\delta\rho(\vec{R}, t)$ has the form

$$F_\rho(\vec{\kappa}, t) = \frac{\sigma_{ab} I_0 \beta_0 (\gamma - 1)}{(2\pi)^3 C_s^2} \left\{ \frac{\cos(\kappa a) - \exp(-\kappa^2 X_T t)}{\kappa^2 X_T} - \right.$$

$$- \frac{t_3}{(\kappa^2 X_T t_3 - 1)} \left[ \exp(- \frac{t}{t_3}) - \exp(-\kappa^2 X_T t) - \right. \tag{6.2.15}$$

$$\left. \left. - \frac{\kappa^2 X_T t_3 (1 + 3\gamma) + 2}{2\kappa C_s (\kappa^2 C_s^2 t_3^2 + 1)} \exp(-0.2\ \kappa^2 X_T t) \right] \sin(\kappa C_s t) + O(X_T / C_s^2) \right\}.$$

Thus, the perburbations in the density of the surrounding medium in the vicinity of a particle are described, within the framework of the linear acoustic approach, as an additive contribution of the thermal aureole (the first and the second terms in braces) and the acoustic perturbation (the third term in braces in (6.2.15)).

In the case of small relative perturbations in the density of the medium, i.e., when $\delta\rho/\rho_0 \ll 1$, the changes of dielectric constant $\varepsilon_N$ and $\delta\rho$ are related as follows: $\varepsilon_N = (d\varepsilon/d\rho)\delta\rho$, where $(d\varepsilon/d\rho) \simeq 0.233\ cm^3/g$ under normal atmospheric conditions.

Note that the values $(d\varepsilon/d\rho)$ and $(d\varepsilon/dT)_{p=const}$, widely used in the literature treating thermal self-action of laser radiation, are related to each other as follows:

$$(d\varepsilon/d\rho) = (C_p \rho (\gamma - 1))^{-1} C_s^2 (d\varepsilon/dT)_{p=const}.$$

The amplitude function of light scattering by a thermal aureole is described, in the Rayleigh-Gans approximation based on the use of (6.2.5), by the following relationship:

$$S_N(\vec{\omega}, t) = 2\pi^2 k^2 (d\varepsilon/d\rho) F_\rho(k\vec{\omega}, t). \tag{6.2.16}$$

As estimates have shown, for $t_3 \gg X_T/C_s^2$ the relative contribution of the acoustic perturbations to the total intensity of light scattering is small compared with that from the thermal aureoles.

It follows from (6.2.15) and (6.2.16) that the characteristic scattering angles due to perturbations of the medium arising from nonlinear interactions are $\theta_N \simeq (k\sqrt{X_T t})^{-1}$. So, e.g., for $\lambda = 1.06\ \mu m$ and $t = 10^{-4}$ s,

$\theta_N \simeq 10^{-3}$ rad, that means thah light scattering by the particles' aureoles only occurs in a very narrow angle around the forward direction. By substituting (6.2.15) and (6.2.3) into (6.2.6) and integrating, one obtains approximately [2] that

$$K_s(t) = K_s^N(t) + K_s^{(0)}[1 + O(\phi_N(0))], \qquad (6.2.17)$$

where

$$K_s^N = I_0^2 t_3 b_0 \; 16[\exp(-2t/t_3)(1 + t/t_3)(E_i^*(2t/t_3) - E_i^*(t/t_3)) + \exp(-t/t_3)(1 + C + \ln(t/t_3) - E_i^*(t/t_3) - \exp(-t/t_3)/2) - \ln 2(1 - t/t_3) - 1].$$

$K_s^{(0)}$ is the scattering efficiency factor of an undisturbed particle; C is the Euler constant (C = 0.772); and

$$b_0 = (ak\ \beta_0 K_{ab}(\gamma - 1)(d\varepsilon/d\rho)/C_s^2)^2/4X_T.$$

If $t/t_3 \gg 1$, then the expression for $K_s^N$ takes the most simple form:

$$K_s^N(t) = I_0^2 t b_0 \ln \sqrt{2}. \qquad (6.2.18)$$

For an arbitrary function I(t), and when $t/t_3 \gg 1$, the Rayleigh-Gans approach gives the following expression for $K_s^N$:

$$K_s^N(t) = \frac{b_0}{2} \int_0^t \int_0^t \frac{I(t_1)\ I(t_2)\ dt_1\ dt_2}{2t - t_1 - t_2}\ . \qquad (6.2.19)$$

Figure 6.2.1 presents the results of calculations of the extinction efficiency factor of an optical inhomogeneity (including the particle itself and its thermal aureole). In accordance with the optical theorem, the calculations were made using

$$K = 4(\pi a^2)^{-1}\ \mathrm{Im}(S(0,\ t)).$$

As can be seen from the figure (see also 6.2.8)), the total extinction cross-section can be less than the initial one due to the opposite signs of the phase changes on the particle and its aureole [1-2].

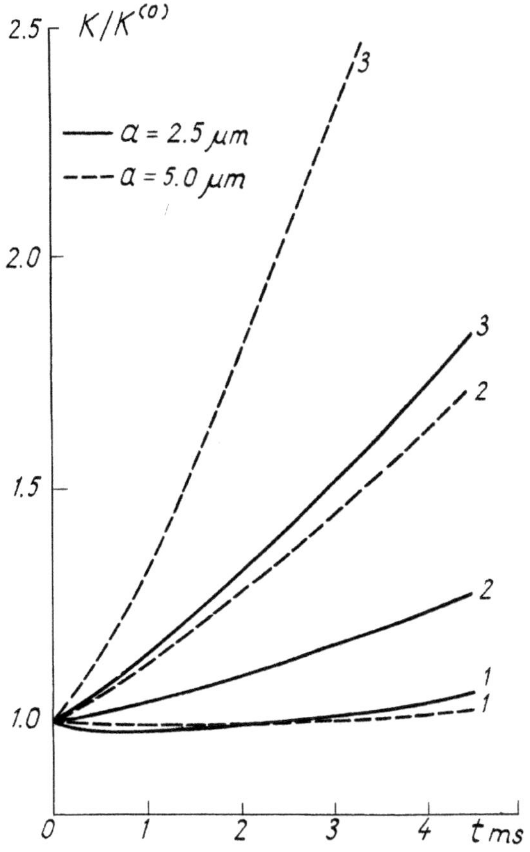


Fig. 6.2.1. Dynamics of the relative extinction efficiency factor of an optical inhomogeneity composed of an absorbing particle and its thermal aureole for high-power radiation. Solid lines represent the data for a = 2.5 μm, dashed lines for a = 5 μm. 1 - $I_a K_{ab}$ = $10^3$ W/cm$^2$; 2 - 5 × $10^3$ W/cm$^2$; 3 - 9 × $10^3$ W/cm$^2$.

## 6.2.3. The Influence of Turbulent Heat Transfer and Particle Motion Relative to the Medium on the Optical Characteristics of Thermal Aureoles.

During a long period of optical action on an absorbing particle, the process of forming the thermal aureole is determined not only by the molecular thermal conductivity, but also by turbulent diffusion, as well as by the displacements of the particle relative to the medium caused by a convective floating up of the thermal aureoles, incomplete entrainment of the particle in the medium's microdisplacements, and the acceleration of particles by incident radiation. An approximate solution for the temperature distribution in the vicinity of an absorbing particle was obtained in [16], based on the following model of the mean-square size of the thermal perturbation region:

$$x^2 \simeq 2X_T + 2d_t t^3.$$

The second term in this formula relates to the empirical Richardson law for turbulent diffusion ($d_t$ is the diffusion coefficient). The temperature distribution in this case is determined by:

$$T(\vec{R},\ t) = T_0 + \int_0^t dt' q_t(t-t')[4\pi(X_T t' + d_t t'^3)]^{-3/2} \times$$
$$\times \exp[-R^2(4(X_T t' + d_t t'^3))^{-1}], \tag{6.2.20}$$

where $q_t$ is the power of a source releasing heat into the medium through the particle's surface, and

$$q_t(t) = \pi a^2 \beta_0 I(t) K_{ab}/C_p\rho.$$

In the limiting case when $t \to \infty$ and $I(t)$ = const, one can obtain from (6.2.20) that

$$T(R) \simeq \begin{cases} q_t/(4\pi X_T R), & R \lesssim \ell_t; \\ q_t t^*(3\pi^{3/2}(4X_T t^*)^{1/3} R^{7/3})^{-1}, & R \gg \ell_t, \end{cases} \tag{6.2.21}$$

where $t^* = \sqrt{X_T/d_t}$, and $\ell_t = \sqrt{4X_T t^*}$ is a parameter very close in meaning to the inner scale of turbulence. Thus, at large distances, $R \gg \ell_t$, turbulent diffusion generates a more rapid fall in temperature than that due to molecular thermal conductivity alone. The influence of turbulence becomes significant if the interaction between radiation and absorbing particles takes place over a time period longer than t . For typical values of $\ell_t \simeq 10^{-1}$ cm and $X_T = 0.18$ cm$^2$s$^{-1}$, one finds that $t^* \simeq 10^{-2}$ s. When $t \to \infty$ and $q_t$ = const, the scattering cross-section is a finite value and, in contrast to the case where turbulence is not taken into account, it is defined as

$$K_s^N(t \to \infty) \simeq 4.8\ b_0 I_0^2 t^*.$$

The above results concerning the scattering of light by a beam-induced optical inhomogeneity are valid only if the displacement of the particle taking place during the interaction time t is smaller than the region of thermal perturbation. The convective floating up of a thermal aureole limits the interaction time t during which the problem can be considered to be spherically symmetrical. According to [1], the limiting condition is

$$t \ll 2X_T(ag(T_a - T_0)/T_0)^{-1},$$

where g is the acceleration due to gravity. Estimates show that, for $a = 10^{-3}$ cm and $(T_a - T_0)/T_0 = 2$, the interaction time $t \ll 0.2$ s.

If the particle moves along the direction of light propagation, then some corrections should be made in the calculations of the scattering efficiency factors of thermal aureoles. Such corrections were derived in [17] based on the Rayleigh-Gans scattering approximation. Movement of a particle in the laser beam can occur, e.g., due to pressure exerted by the light or through the action of radiometric forces [1]. The expression describing the efficiency factor of the scattering of light by the thermal aureole of a moving particle, in the case of unpolarized light with a constant intensity $I_0$, is written as follows:

$$K_s^N = K_s^N(v = 0)[f_v(\delta)], \tag{6.2.22}$$

where $K_s^N(v = 0)$ is the scattering efficiency factor of a stationary particle (see (6.2.18)); v is the velocity of a particle along the x-axis; and $f_v(\delta)$ is the wind factor,

$$f_v(\delta) \simeq (1 + \delta^2)^{-1}\left(1 + \frac{\ln(1 + \delta^2)}{2 \ln 2} - \frac{\delta}{\ln 2} \arctan \delta\right),$$

where $\delta = v(k X_T)$. Thus, when $\delta > 0$, the cross-section of light scattering by the thermal aureole of a moving particle is less than that of a stationary particle, other conditions being equal.

## 6.3. THERMAL SELF-ACTION OF A HIGH-POWER LASER PULSE PROPAGATING THROUGH DUSTY HAZES

The thermal perturbation appearing in the laser beam around the light-absorbing particles of dusty hazes changes the features of light propagation due to the effects of the scattering of light by thermal aureoles and self-defocusing, leading to an increased degree of extinction and beam self-broadening. From the point of view of the effeciency of the thermal beam self-action, cases of high-energy contributions from the beam to the medium (see (6.1.2), (6.1.3)) are of particular interest. Such situations occur when the pulse duration t greatly exceeds the time necessary for the relaxation of the pressure to a constant value in the heated medium due to its isobaric expansion, i.e., $t \gtrsim t_4 = R_0/C_s$. Thus, for $R_0 = 1$ cm, one finds that $t_4 \simeq 3\ 10^{-5}$ s.

Theoretically, the methods for describing laser beam self-broadening in the haze when there are thermal aureoles around the particles have been developed in [1-2, 6, 11-16, 21] based on the nonlinear equation for the function of mutual coherence, and on its modification known as small-angle approximation of the radiation transfer equation.

In [12] a method of statistical modelling is developed using the stochastic equations derived from (6.1.6) and (6.1.7) for the dimensionless beam width and weighted mean radius of the laser beam phase front curvature, and assuming that the statistics of the aerosol medium are Poissonian.

The experimental results of the study of the self-broadening effect obtained in chambers containing artificial aerosols can be found in [2, 14].

Some of the above questions are the subject of the discussion presented in this section.

### 6.3.1. A Theoretical Analysis of the Effects of Light Scattering by Thermal Aureoles and the Defocusing of the Laser Pulse in the Light-Absorbing Hazes

An approximate solution of the problem concerning light beam self-action, in the general case where the statistics of the thermal perturbations in aerosol media are Poissonian, can be obtained by moving from (6.1.11) to the Fourier transform [2]:

$$H(x,\ \vec{q},\ \rho) = (2\pi)^{-2}\int_{-\infty}^{\infty} d^2\ell\ \exp(-i\vec{q}\vec{\ell})\Gamma(x,\ \vec{\ell},\ \vec{\rho}), \tag{6.3.1}$$

which has the form

$$\left(\frac{\partial}{\partial x} + \frac{\vec{q}}{k}\nabla_{\vec{\rho}} + \frac{1}{4}k^2 D_{\varepsilon_1}(x,\ \vec{\rho}) + D_a(x,\ \vec{\rho})\left[1 + O\left(\frac{a}{2(X_T t)^{1/2}}\right)\right]\right) \times$$

$$\times\ H(x,\ \vec{q},\ \vec{\rho}) = \int_{-\infty}^{\infty} d^2q\ G_N(x,\ \vec{q} - \vec{q}',\ \vec{\rho})H(x,\ \vec{q}',\ \vec{\rho}), \tag{6.3.2}$$

where

$$G_N(x,\ \vec{q},\ \vec{\rho}) = N_0(2\pi)^{-2}\int_{-\infty}^{\infty} d^2\ell\ \exp(-i\vec{q}\vec{\ell})D_N(x,\ \vec{\ell},\ \vec{\rho}).$$

Using the method of characteristics, one can pass from (6.3.2) to the equivalent integral equation

$$H(x,\ \vec{q},\ \vec{\rho}) = H(0,\ \vec{q}(x),\ \vec{\rho}(x))\ \exp\left[-\int_0^x d\xi D^{(0)}(\xi,\ \vec{\rho}(\xi))\right] +$$

$$+ \int_0^x\int_{-\infty}^{\infty} d\xi\ d^2q\ \exp\left[-\int_0^{x-\xi} d\xi'\ D^{(0)}(\xi',\ \vec{\rho}(\xi'))\right]\times \tag{6.3.3}$$

$$\times\ H(\xi,\ \vec{q}',\ \vec{\rho}(\xi))G(\xi,\ \vec{q}(\xi) - \vec{q}',\ \vec{\rho}(\xi));$$

where $H(0, \vec{q}, \vec{\rho})$ is the boundary condition; $\vec{q}(\xi) = \vec{q}$; $x(\xi) = x$; $\vec{\rho}(\xi) = \vec{\rho} - \vec{q}(x - \xi)/k$ is the equation of corresponding characteristics; and

$$D^{(0)}(x, \vec{\rho}) = \frac{k^2}{4} D_{\varepsilon_1}(x, \vec{\rho}) + D_a(x, \vec{\rho}). \tag{6.3.4}$$

The first term of (6.3.3) is the exact solution of (6.3.2) for the case of a linear medium. A solution which takes into account the nonlinearity of the medium can be written in terms of quadratures by using the method of iteration. In the first step, the function $H(x, \vec{q}, \vec{\rho})$ of the subintegral expression in (6.3.3) (on the right-hand side) is substituted for by its representation for a linear medium $H^{(0)}(x, \vec{q}, \vec{\rho})$. If the laser beam has a Gaussian intensity distribution at the point that it enters the medium, then one can easily prove the convergence of the iteration series. The terms of the iteration series characterize the multiple scattering of light by laser-induced inhomogeneities. The number of a term corresponds to the order of multiplicity.

In a particular example of an energetically uniform cross-section of the laser beam, i.e., when $\Gamma(0, \vec{\ell}, 0) = \text{const}$, the optical characteristics of the beam-induced inhomogeneities of the medium, when calculated using the approximation of a fixed field, do not depend on the radial coordinate and, hence, (6.3.2) has an exact solution which takes into account the contribution from multiple scattering [1]:

$$H(x, \vec{q}, \vec{\rho}) = H(0, \vec{q}, \vec{\rho} - \frac{\vec{q}x}{k}) \exp[-\int_0^x d\xi\, D^{(0)}(x - \xi, \vec{\rho} - \frac{\vec{q}\xi}{K}) - \\ - \int_0^x d\xi D_N(x - \xi, 0, \vec{\rho} - \frac{\vec{q}\xi}{k})]. \tag{6.3.5}$$

If one has the expression for $H(x, \vec{q}, \vec{\rho})$, then it is possible to calculate some parameters of the laser beam, such as the effective cross-sectional area $\pi\langle R_e^2\rangle$ and the weight mean angular divergence of the beam,

$$\langle R_e^2\rangle = - \frac{[\Delta_{\vec{q}} H(x, \vec{q}, 0)]_{q=0}}{H(x, 0, 0)}; \qquad \langle\theta^2\rangle = \frac{-[\Delta_{\vec{\rho}} H(x, 0, \vec{\rho})]_{\rho=0}}{H(x, 0, 0)}. \tag{6.3.6}$$

The use of the effective laser beam parameters (6.3.6) is valid only if the angles at which light scattering by the thermal aureoles takes place ($Q_N \simeq (k\sqrt{X_T t})^{-1}$ are comparable with the initial beam divergence $\theta_0$, or the optical depth partly controlled by this scattering is close to, or greater than, unity:

$$\tau_N = \int_0^x d\xi D_N(x - \xi, 0, 0) \sim 1.$$

In [1, 2, 11] one can find an analytical solution of the problem of the nonlinear propagation of a laser beam through a haze obtained for a specific case in which there were small phase changes taking place within the region of one thermal aureole (6.2.5); this is of some practical importance. In this particular case, the spatial fluctuations of the nonlinear term for the dielectric constant of the medium are Gaussian, and that corresponds to the asymptotic representation of $D_N$ in (6.1.16). This expression involves terms which describe the nonlinear refraction of the beam and the scattering of light by thermal aureoles. The analysis is based on the use of the small-angle approximation of the radiation transfer equation for a medium exhibiting regular refraction (6.1.21), or its analog written for the function $H(x, \vec{q}, \vec{\rho}) = \iint_{-\infty}^{\infty} d^2\ell d^2\omega \exp(ik\vec{\omega}\vec{\rho} - i\vec{q}\vec{\ell}) I_n(x, \vec{\ell}, \vec{\omega})$.

With reference to [1], consider that, at the point of entrance into the medium, we have a coherent, single-mode beam

$$E(x, \vec{\ell}, t) = E_0(t) \exp[-\frac{\ell^2}{2R_0^2} + \frac{ik\ell^2}{2F_0}] , \tag{6.3.7}$$

where $E_0(t)$ is the field amplitude at the beam's axis; and $R_0$, $F_0$ are the effective radii of the amplitude and phase profiles over the beam cross-section, respectively. In order to obtain the solution (6.1.21) following the single-scattering approximation, the values of the functions $\bar{\varepsilon}_N$ and $D_N$ are sought following the approach of a set of fields in a linear medium, and the gradient of the dielectric constant $\nabla_{\vec{\ell}}\bar{\varepsilon}_N$ is approximated by the first non-vanishing term (the second-order term) of the series over $\vec{\ell}$ in the vicinity of the beam axis. The latter is known in the literature [24-33] as the paraxial, or nonaberrational, approach.

Taking into account the axial symmetry of the function $\Gamma(x, \vec{\ell}, 0)$, one can write

$$\frac{1}{2}\nabla_{\vec{\ell}}\, \bar{\varepsilon}_N(x, \vec{\ell}, t) \simeq -\vec{\rho}\eta_R^2 = -\vec{\rho}\, \frac{\pi k^2}{W_R}[\nabla_{\vec{\ell}}^2\, w(x, t)]_{\ell=0}, \tag{6.3.8}$$

where $w(x, t) = \int_0^t dt'\Gamma(x, \vec{\ell}, 0, t')$ is the density of the laser pulse energy which passes through the plane at point x and at the moment of time t; and $W_R$ is the threshold energy of beam defocusing. In the case of a Gaussian beam,

$$W_R = \frac{4\pi C_p \rho_0 \varepsilon_0^{1/2}}{k^2 \alpha_{ab} |d\varepsilon/dT|}[1 + O(\frac{t_3}{t})], \tag{6.3.9}$$

where $\alpha_{ab}$ is the volume absorption coefficient of aerosols; and $t_3$ is the characteristic delay time of heat transfer into the medium via absorbing particles (see Figure 6.1.1). In the particular case where $w(t) = w_0$ = const, the threshold energy (taking into account the correction necessary for heat transfer delay) is written as

$$W_R = W_R(t \to \infty)[1 - (t_3/t) + (t_3/t)\exp(-t(t_3)]^{-1} \tag{6.3.10}$$

where $W_R(t \to \infty)$ is calculated according to (6.3.9).

Using the method of characteristics, and taking into account (6.3.7) and (6.3.8), one can solve (6.2.21) in terms of quadratures for the Fourier transform $H(x, \vec{q}, \vec{\ell})$. The solution is analogous to (6.3.3) with the characteristics $\vec{q}(\xi)$, $\vec{\rho}(\xi)$, $x(\xi)$ being defined by

$$\frac{d\vec{q}(\xi)}{d\xi} = -\vec{\rho} n_R^2 (H^{(0)}(\xi, \vec{q}(\xi), 0)); \qquad (6.3.11)$$

$$\frac{d\vec{\rho}(\xi)}{d\xi} = \vec{q}(\xi); \qquad x(\xi) = \xi. \qquad (6.3.12)$$

Figure 6.3.1 gives an example of the calculated [6] dependences of the nonlinear corrections to the mathematical expectation of the beam intensity along the beam axis through the dusty haze, $\Gamma(x, 0, 0) = I_L + I_R + I_R$, on the intensity of incident radiation $I_0$ also measured along the beam axis, but in a linear medium; $I_R$ and $I_S$ are the corrections for beam defocusing and the scattering of light by thermal aureoles, respectively. The calculations were carried out for a Gaussian beam with $R_0 = 2$ cm, $F_0 = \infty$, and $\lambda = 1.06$ µm propagating through a haze having a volume extinction coefficient $\alpha = 1.2 \times 10^{-5}$ cm$^{-1}$ ($a = 10^{-4}$ cm; $K_{ab} = 0.75$; $K = 2$). The length of the propagation path was 200 m, and the turbulence intensity parameter was taken as $C_n^2 = 10^{-15}$ cm$^{-3}$.

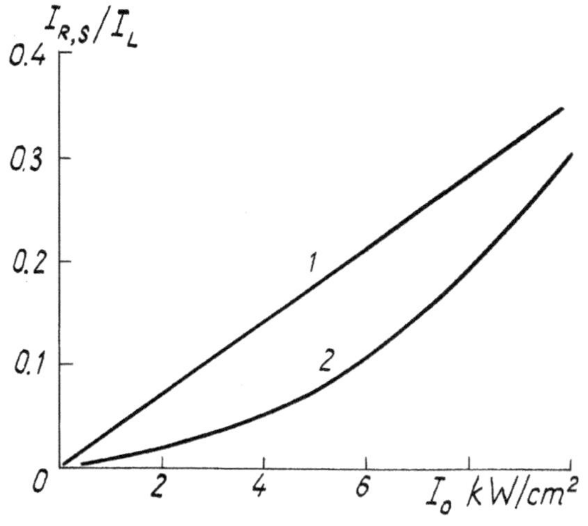


Fig. 6.3.1. The fractional contribution of self-defocusing $I_R/I_L$ (1) and light scattering by thermal aureoles (2) to the total intensity $\Gamma(x, 0, 0) = I_L + I_R + I_S$ along the axis of a laser beam propagating through a dusty haze as functions of the incident radiation intensity $I_0$. $\lambda = 1.06$ µm; $t_p = 10^{-3}$ s, $R_0 = 2$ cm; $F_0 = \infty$; $x = 2 \times 10^4$ cm; $\alpha = 1.2 \times 10^{-5}$ cm.

In [1, 11] the authors obtained expressions for the weighted mean parameters of the beam, based on an approximate solution of the radiation transfer equation (6.1.21). Thus, e.g., for the effective beam cross-sectional area $\pi\langle R_e^2\rangle$, one has

$$\langle R_e^2\rangle = R_L^2 + R_R^2 + R_S^2; \tag{6.3.13}$$

$$R_L^2 = R_0^2 + x^2\left(\frac{1}{k^2R_0^2} + \frac{R_0^2}{F_0^2}\right) - \frac{x^3}{3}\,[\Delta_{\vec\rho} D_{\varepsilon_1}(\vec\rho)]_{\rho=0}; \tag{6.3.14}$$

$$R_R^2 = -\eta_R^2(0)\,2R_0\alpha^{-2}[\alpha_0\, x - 1 + \exp(-\alpha x)]; \tag{6.3.15}$$

$$R_S^2 = B_0\sigma_t(t)\,\frac{\gamma(3,\, 2\alpha x)}{k^2(2\alpha)^3}; \tag{6.3.16}$$

where $R_L$ is the parameter describing the beam's behavior in a linear medium after its initial beam divergence, diffractional blurring, and scattering by a nonperturbed aerosol; $R_R$ and $R_S$ are the corrections for nonlinearities arising because of the effects of self-defocusing on the averaged profile of the medium's heating and light scattering by thermal aureoles around the absorbing centers, respectively; $\alpha$ is the volume extinction coefficient of an undisturbed aerosol; $\gamma(a,\, b)$ is the incomplete gamma function;

$$\eta_R^2(0) = \frac{-\pi}{2}\,\frac{(k/R_0)^2}{W_R}\left(\frac{C\varepsilon_0^{1/2}}{8\pi}\right)\int_0^t |E_0(t_1)|^2 dt_1; \tag{6.3.17}$$

$B_0 = \pi a^2 N_0 b_0/2X_T$; $b_0$ is the coefficient from (6.2.19); and $\sigma_t(t)$ is the time-profiling factor,

$$\sigma_t(t) = \left(\frac{c\varepsilon_0^{1/2}}{8\pi}\right)^2 \iint_0^t \frac{E_0^2(t_1)E_0^2(t_2)\; dt_1\; dt_2}{(2t - t_1 - t_2)^2}\,. \tag{6.3.18}$$

Thus, for example, in the case of an envelope typical for the free-generation regime of solid-body lasers,

$$I(t) = I_0(t/t_u)\,\exp(-t/t_u),$$

where $I_0$, $t_u$ are parameters. The expression for $\sigma_t(t)$ is written as follows [2]:

$$\sigma_t(t) = I_0^2(t/6t_u)\Big[(8t^2/t_u - 12t)(E_i^*(t/t_u) - E_i^*(2t/t_u)) \times$$
$$\times \exp(-2t/t_u) + (6t_u + t_u^2/t)\exp(-t/t_u) -$$

$$- (10t_u + (2t_u^2/t) - 8t - (3t^2/t_u)) \exp(-t/t_u) +$$

$$+ (4t_u + (t_u^2/t) - 4t)\Big], \tag{6.3.19}$$

where $E_i^*$ is the exponential integral function. The integral entering (6.3.17) is

$$\frac{c\varepsilon_0^{1/2}}{8\pi}\int_0^t |E_0(t_1)|^2\, dt_1 = I_0 t_u\, \gamma(2,(t/t_u)). \tag{6.3.20}$$

The solutions (6.3.13)-(6.3.16) are rigorously valid for weakly-nonlinear media and beams of a small angular width, i.e., when $R_R^2$, $R_S^2$, $(R_L - R_0)^2 \ll \ll R_0^2$.

As the analysis of the solutions obtained show, the contribution of the effect of light scattering by thermal aureoles to beam broadening is proportional to $I_0^2$, while that contribution from nonlinear refraction is proportional only to $I_0$. Moreover, the sign of the contribution due to the scattering of light by aureoles (6.3.16) is always positive, and does not depend on the intensity profile at the boundary with the medium, while the sign of the refraction term (6.3.15) is opposite (see (6.3.8)) to the sign of the intensity gradient across the beam. This means that a gaseous lens will defocus $(\eta_R^2 < 0)$ for a Gaussian beam, while it will focus $(\eta_R^2 > 0)$ for a beam with its intensity valley at the beam's axis.

In the case of short laser pulses, which are characterized by nonlinear scattering angles $\theta_N \simeq (k\sqrt{X_T t})^{-1}$ that greatly exceed the initial beam divergence $\theta_0$, the expression for estimating the intensity along the beam's axis takes the form

$$I(x, t) \simeq I_0 \frac{\exp\left[-\tau_N \dfrac{1 - \exp(-2\tau_0)}{2\tau_0} - \tau_0\right]}{(R_L^2 + R_R^2)/R_0^2}, \tag{6.3.21}$$

where $I(x, t) = \Gamma(x, 0, 0, t)$; $R_L$ and $R_R$ are determined by (6.3.14) and (6.3.15); $\tau_0 = \alpha_0 x$ and $\tau_N = \pi a^2 N_0 x K_s^N(I_0)$; and $K_s^N$ is the efficiency factor of the scattering of light by thermal aureoles (see (6.2.18) or (6.2.19).

If the beam has a uniform intensity distribution over the cross-section $I_0$, then nonlinear refraction does not take place $(R_R = 0)$ and the extinction of such a beam can be calculated according to the Bouguer law from (6.1.23). If $I_0(t)$ = const, then a self-consistent solution of (6.1.23) takes the form [13]

$$I(x, t) = I_0[g_N(1 - e^{\tau_0}) + e^{-\tau_0}]^{-1}, \tag{6.3.22}$$

where

$$g_N = K_s^N(I_0)/K^{(0)}. \tag{6.3.23}$$

### 6.3.2. Calculation of Laser Beam Self-Broadening in a Light-Absorbing Aerosol by the Method of Statistical Modelling

In the case of slightly-divergent laser beams propagating through a dusty haze of initial optical depth $\tau_0 = \alpha x$, this being less than n (where $n \lesssim 5$), the reference stochastic equation (6.1.6), which takes into account the scattering of light by the beam-induced thermal perturbations, can be written as follows [1, 12]:

$$E(x, \vec{\ell}, t) = U(x, \vec{\ell}, t) \exp\{-\tau_0[1 + 0(\frac{a}{2(X_T t)^{1/2}})]\}; \quad (6.3.24)$$

$$\hat{B}u = 2ik \frac{\partial u}{\partial x} + \Delta_{\vec{\ell}} u + k^2 u(d\varepsilon/dT) \sum_{k=1}^{\infty} \exp(-\alpha_0 X_K) \times$$

$$\times \int_0^t dt_1 G(\vec{r} - \vec{r}_K, a_K, t - t_1)|u(r_K, t_1)|^2 = 0, \quad (6.3.25)$$

where $\hat{B}$ is the operator of (6.3.25); $\vec{r}_K = (x_K, \vec{\ell}_K)$ are the coordinates of the k-th particle center; $a_k$ is the radius of the k-th particle; and G is the Green's function of the temperature profile around the particle,

$$G(\vec{r}_K, a_K, t) = \frac{a_k^2 K_{ab}(a_K, \varepsilon_{ak}, \lambda)}{64 C_p \rho (\pi X_T t)^{3/2}} \exp(-\frac{r^2}{4X_T t}). \quad (6.3.26)$$

The representation (6.3.24) corresponds to the approach which neglects the variance of the field in a linear medium.

The solution of (6.3.25) is reduced to solving the approximate variational problem [1, 12] of seeking for the minimum of the energetic functional:

$$(\hat{B}u, u) = \int_{-\infty}^{\infty} d^3 r u^* \hat{B}u. \quad (6.3.27)$$

The solution of this variational problem is sought in the class of spherical waves having a variable radius of the phase front curvature.

$$u(x, \vec{\ell}, t) = \frac{u_0}{f(x, t)} \exp[-\frac{\ell^2}{2R_0^2 f^2(x, t)} + \frac{ik\ell^2}{2F(x, t)} + i\phi(x, t)], \quad (6.3.28)$$

where f, F, $\phi$ are the effective parameters of the beam. The initial conditions are

$$f(0, t) = 1; \quad F(0, t) = F_0, \quad \phi(0, t) = 0; \quad f(x, 0) = (1 + x/F_0)^2 +$$

$$+ (X/kR_0^2)^2.$$

These conditions correspond to a Gaussian beam of the type (6.3.7).

Simple stochastic equations for the parameters f(x, t) and F(x, t) can be easily obtained from the extremum condition for the energetic potential [1]:

$$f^3 f'' = (kR_0^2)^{-2}(1 - k^2 \sum_{K=1}^{\infty} D_K[f]); \tag{6.3.29}$$

$$f = f/f', \tag{6.3.30}$$

where

$$D_k[f] = (d\varepsilon/dT)\exp(-\alpha X_k)\,\frac{C\varepsilon_0^{1/2}|U_0|^2\, a_K^2 K_{ab}(a_K)}{16C_p\rho(\pi X_T)^{1/2}} \times$$

$$\times\; 1 - \frac{\ell_K^2}{R_0^2 f^2(x,\,t)}\int_0^t dt_1\,\frac{(t - t)^{1/2}}{f^2(x,\,t_1)}\exp\left\{-\frac{(x - x_K)^2}{4X_T(t - t_1)} - \right.$$

$$\left. - \frac{\ell_K^2}{R_0^2}\,(f^{-2}(x,\,t) - f^{-2}(x,\,t_1))\right\}. \tag{6.3.31}$$

Equations (6.3.29) and (6.3.30) were solved by the Monte-Carlo method. The assumed value in this case was the mathematical expectation, which, for example, for $<f^2>$, was taken as the arithmetic mean

$$<f^2> \simeq \frac{1}{M}\sum_{j=1}^{M} f_j, \tag{6.3.32}$$

where M is the number of independent events; $f_j$ is the numerical solution of (6.3.29) obtained by the Runge-Kutta method for a j-th random realization of the spatial distribution of particle centers $\{\vec{r}_K\}$ and sizes over the region occupied by the laser beam.

A three-dimensional array of coordinates $\{\vec{r}_K\}$ was modeled using the pseudorandom number generator in order to provide the Poisson statistics of the medium. The modeling error was estimated using the Chebyshev inequality, and was proportional to $M^{-1/2}$. The size-distribution function used in the calculations was the four-parameter gamma-distribution function

$$N_a(a) = \nu_1 a^{\nu_2}\exp(-\nu_3 a^{\nu_4}), \tag{6.3.33}$$

where $\nu_1$ is the normalized constant, $\nu_1 = N_0 2(\nu_2/\nu_3\nu_4)^{-1/\nu_4}$, and $N_0$ is the number density of the particles. The modal radius of particles can be expressed in terms of the size distribution function parameters as follows: $a_m = (2/3\ \nu_3)^{1/3}$.

Figure 6.3.2 presents the dependence of the dimensionless effective beam width $<f^2(x,\,t)>^{1/2}$ normalized w.r.t. that in a linear medium on the path length, calculated for the following aerosol parameters:

$m_a = 1.43 - i0.7$ ; $N_0 = 200\ cm^{-3}$; $\nu_2 = 2$, $\nu_4 = 3$; and $a_m = 2 \times 10^{-4}$ cm.

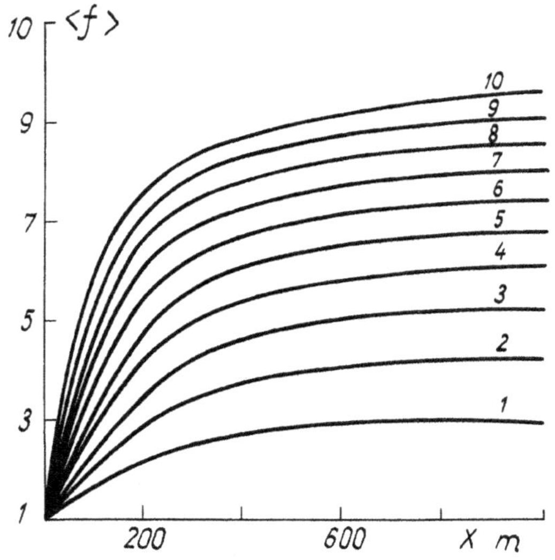


Fig. 6.3.2. Self-broadening of a high-power laser beam in a dusty haze as a function of the path length, calculated by the method of statistical modeling. Radiation parameters: $\lambda = 1.06\ \mu m$, $I_0 = 6.4 kW/cm^2$ (rectangular pulse), pulse durations $t_p = 10^{-4},\ldots,10^{-3}$ s with $10^{-4}$ s steps (curves 1 to 10 in sequence); $R_0 = 10$ cm; $\theta_0 = 0.75 \times 10^{-3}$ rad. Aerosol parameters: $N_0 = 200\ cm^{-3}$; $a_m = 2 \times 10^{-4}$ cm; $m_a = 1.43-i0.7$.

As the analysis of these results has shown, the standard deviation of the beam broadening fluctuations do not exceed 0.1% if the total number of aerosol particles within the laser beam is approximately $10^4$ to $10^5$. This allows the use of only one realization of the $\{\vec{r}_K\}$ array in the calculations.

### 6.3.3. Experimental Investigations of Pulsed Laser Self-Broadening due to Scattering by Thermal Aureoles

The self-action of laser pulses whose duration satisfies the conditions in (6.1.4) is caused by the effects of the scattering of light by thermal aureoles, as well as by nonlinear refraction, on the mean profile of the dielectric constant within the beam.

The first experimental results on the scattering of high-power radiation by the beam-induced thermal aureole of an individual particle and of an ensemble of dust particles in air were published in [18]and [1, 2, 10,

14], respectively. The data of further experiments were summarized in [2]. Similar experimental results can also be found in [19-20]. Below, we shall discuss the main results of experimental studies on the nonlinear scattering of light.

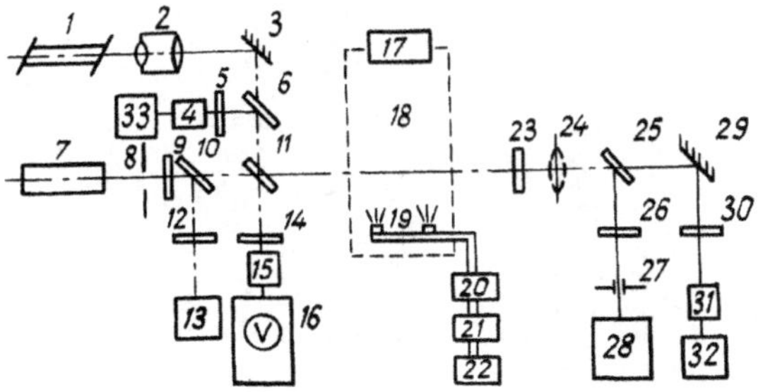


Fig. 6.3.3. Block diagram of the experimental set-up: 1 - cw laser; 2 - beam expander; 3.29 - plane mirrors; 4, 15, 31 - PMTs; 5, 9, 12, 14, 23, 26, 30 - optical filters; 6, 10, 11, 25 - beam splitters; 7 - pulsed laser; 8, 24 - variable condensing objectives; 13 - power meter; 16 - oscillograph; 17 - solid aerosol pulverizer; 18 - aerosol chamber; 19-22 - liquid aerosol pulverizer; 27 - slit diaphragm; 28 - high-speed camera; 32, 33 - milliammeters.

A typical block diagram of the experimental set-up for studying nonlinear distortions of ruby and Nd-glass laser puslse ($\lambda$ = 0.69 and $\lambda$ = 1.06 $\mu$m) in a haze is shown in Figure 6.3.3. Power-stabilized radiation from cw Argon-ion or He-Ne lasers was used in this study for recording the dynamics of the nonlinear process. The cw laser beam passed through the interaction zone ($\Delta x$ = 80 cm) and was then focused onto the photometric slit, behind which was the film of a high-speed camera. An alternative version of the recording system used a dissector as a photodetector. In the latter case the successive registrograms of the cw laser beam's angular structure were displayed on the oscilloscope screen.

Figure 6.3.4 presents typical oscillograms of the nonlinear distortions observed in the angular distribution of the sounding beam's intensity $I_n(\kappa)$ during the interaction between a Nd-glass laser pulse and sooty aerosols [20]. It is easily seen from this figure that the angular structure of the sounding beam undergoes broadening (for comparison, the lower oscillogram shows the angular structure of the initial, undisturbed beam).

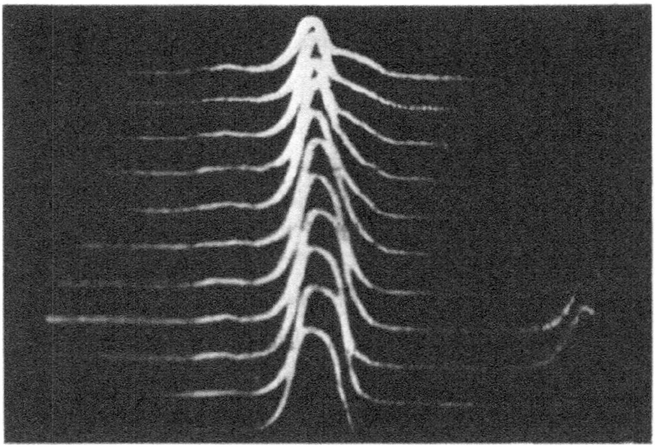

Fig. 6.3.4. Typical oscillograms of the nonlinear distortions of the beam's intensity along the diameter of the beam's cross-section obtained in a sooty aerosol irradiated by pulsed radiation from a Nd-glass laser. The mean intensity of the laser pulse is 10 $kW/cm^2$; $\tau_0 = 0.4$; $t_p \simeq 1$ ms. The lower oscillogram corresponds to an unperturbed angular structure of the sounding beam. The succeeding oscillograms were obtained over 0.2 ms intervals.

In order to separate the contributions due to the nonlinear effects of the scattering of light by thermal aureoles and thermal defocusing, the sounding beam ($\lambda = 0.48$ μm) was narrower ($R_S = 0.25$ cm) than the high-power radiation beam ($\lambda = 1.06$ μm), $R_0 = 0.65$ cm [2]. This allowed us to obtain information on the nonlinear refraction through a thermal lens by measuring the angular deflections of the sounding beam when it made its parallel displacements relative to the axis of the main beam. On the other hand, it was possible to measure the angular self-broadening caused by the scattering of light by the beam-induced thermal aureoles when the sounding beam was fixed in space. The fine structure of the sounding beam's angular distortions, when its optical axis was displaced at a distance $R_0/2$ relative to the axis of the high-power beam, in cement dust can easily be seen from the registrogram of the high-speed camera presented in Figure 6.3.5. The wavelength of the high-power radiation was 1.06 μm, pulse duration $t_p \approx 1$ ms, and pulse energy $W_0 \approx 45$ J. Quite a slow relaxation of the beam channel's optical properties takes place over about 10 ms and longer after the laser pulse of 1 ms duration terminates. The relaxation is due to the blurring of the thermal aureoles and turbulent mixing.

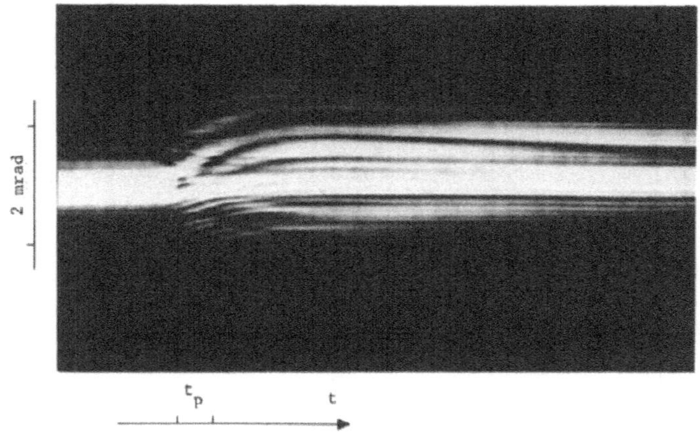


Fig. 6.3.5. High-speed photosweeping to determine the angular structure of a narrow sounding beam shifted with respect to the high-power beam's axis. $\lambda$ = 1.06 μm; $w_0$ = 120 j/cm$^2$. The initial optical depth of the cement dust in haze was $\tau_0$ = 0.1. Vertical lines in the figure denote the invervals of laser pulse propagation.

The experimental profiles $I_n(\kappa)$ where then used for the determination of the root-mean-square angle of beam divergence by integrating according to the formula

$$\theta_{sq}^2(t) = \frac{1}{F_0^2} \frac{\int_{-\infty}^{\infty} \kappa^2 I_n(\kappa, t)\, d^2\kappa}{\int_{-\infty}^{\infty} I_n(\kappa, t)\, d^2\kappa}, \tag{6.3.34}$$

where x is the distance from the centre of the point of focus; and $F_0$ is the focal length of the objective (24) in Figure 6.3.3. The angular correction $\theta_N$ for nonlinear effect was found as follows: $\theta_N = (\theta_{sq}^2 - \theta_0^2)^{1/2}$, where $\theta_0$ is the angle of the initial divergence of the sounding beam.

Figure 6.3.6 shows the measured temporal vehavior of changes of the weighted mean angular divergence of the sounding beam occurring in different model media (wood smoke, a suspension of cement particles in air, soot, and talc) during the laser pulse due to nonlinear effects. Two kinds of experimental points representing one medium were obtained during two laser shots with very similar pulse energies. As the observations showed, the process of the nonlinear angular broadening of the laser beam is less efficient in wood smoke. The observations were made with the Nd-glass laser emitting about 150 J per pulse with $\lambda$ = 1.06 μm. The optical depth was $\tau_0$ = 1. The wide scatter of experimental points in Figure 6.3.6 is probably due to the non-reproducibility of the exact laser pulse energy and aerosol parameters.

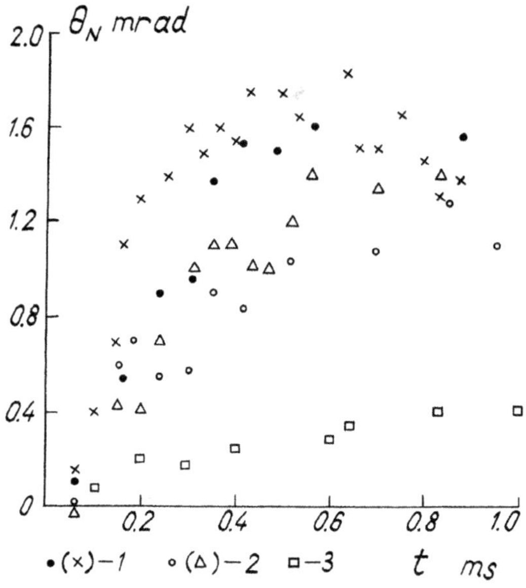


Fig. 6.3.6. Weighted mean nonlinear angular divergence of the sounding beam, measured in various polydispersed systems of particles. (1) - Suspension of cement particles ($CaSO_4 \cdot 2H_2O$ - 5%; ballast - about 10%, $Ca_2\ SiO_4$; $CaMgSi_2O_6$; and $Ca_2Mg[OH]_2Si_8O_{22}$ - 80%). Root-mean square radius $a_{sq} = 2\ \mu m$; $\tau_0 = 0.9$; $\lambda = 0.69\ \mu m$; pulse energy $W_0 = 40$ J. (2) Suspension in air of talc particles; $M_gSi_4O_{10}[OH]_2$; $\lambda = 0.69\ \mu m$; $W_0 = 40$ J, $a_{sq} = 3\ \mu m$; $\tau_0 = 0.1$. (3) Wood Smoke; $a_{sq} \approx 10^{-4}-10^{-5}$ cm; $\tau_0 = 1.0$; $\lambda = 1.06\ \mu m$, $W_0 = 150$ J.

The characteristic time in these experiments is the time of averaging the thermal perturbations of the medium's density over the gaps between particles, $t_5 \approx (4N_0^{2/3}\chi_T)^{-1}$. In the case of talc and cement aerosols with a number density of $2\text{-}6 \times 10^3\ cm^{-3}$, $t_5$ greatly exceeds the pulse duration $t_p$. It is also characteristic for this experiment that $t_p \gg R_0/C_s$, where $R_0$ is the radius of the beam's cross-section and $C_s$ is the speed of sound in the medium. Taking into account the above peculiarities of the experiment, one can arrive at the conclusion that the effect of laser beam self-broadening is mainly due to the scattering of light by local thermal inhomogeneities of the medium. The temporal behavior of the nonlinear correction $\theta_N$ (measured experimentally) has a maximum at t = 0.5 to 0.8 ms; this is in good agreement with the theoretical calculations [1].

The number density of particles in woody smokes, assessed from the measured optical depth, was between $10^6$ and $10^8$ $cm^{-3}$ for an $a_0$ assumed to be of the order of $10^{-4}$ to $10^{-5}$ cm. In this case, the characteristic time of averaging of the thermal aureoles is about $10^{-5}$ to $10^{-4}$ s, and is much shorter than the pulse duration $t_p \approx 10^{-3}$ s. Thus, in woody smoke, laser beam self-action appears as the effect of thermal self-defocusing in the mean profile of the medium's dielectric constant.

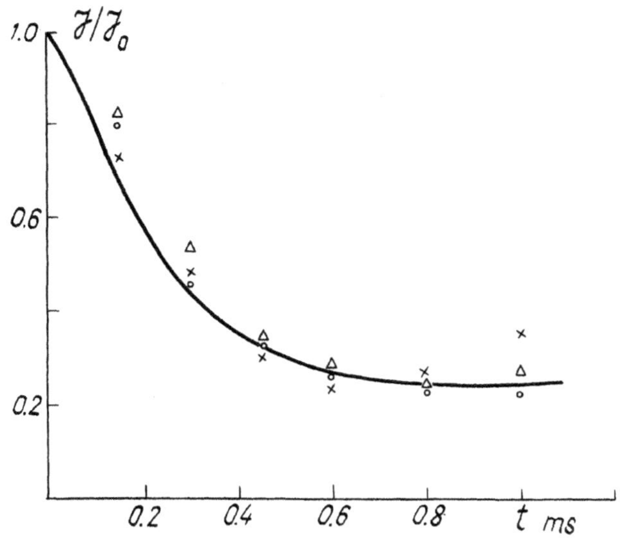


Fig. 6.3.7. The behavior of the sounding beam's intensity during the interaction of high-power laser radiation ($w_0$ = 100 J/cm$^2$; $\lambda$ = 1.06 μm) and aerosol particles composed of $Ni_2O_3$. Points represent experimental data averaged over 3 laser shots; the solid line is the theoretical curve [2].

Figure 6.3.7 presents a quantitative comparison of the experimental and theoretical data relating to the dynamics of the beam's intensity $I_n$ ($\kappa$ = 0) of the sounding beam's radiation normalized by the value of the nonperturbed beam intensity. The case presented in this figure is when $Ni_2O_3$ aerosol particles are irradiated with a laser pulse of 1 ms duration and a pulse energy of $W_0 \approx 50$ J at a wavelength $\lambda$ = 1.06 μm. The size spectrum of the aerosol ensemble is described by a single peaked curve with a modal radius $a_m$ = 2.6 μm and a r.m.s. radius $a_{sq} \approx 3.6$ μm. As seen from this figure, the beam channel becomes significantly turbid during the laser pulse, and this results in the scattering of light on the beam-induced inhomogeneities of the medium. The theoretical model is in satisfactory agreement with the experimental data.

Laboratory measurements of the transmission of a gas-dispersed medium [14, 19] as a function of incident radiation showed that, in the region where the energy distribution varies from $10^1$ to $10^3$ j/cm$^2$, one can observe a decrease of the on-axis intensity caused by the scattering of light by aureoles and defocusing of the beam. It was observed in the focused beam, when $w_0 \gtrsim 10^3$ Jcm$^{-2}$, that the turbid medium clears. This fact can be explained as being a result of the effects of the radial photophoresis of the particles and of the non-uniform heating of particles and their photo-reactive acceleration in the field of high-power radiation observed in [22].

It should be noted that some new effects can appear in a medium with laser-induced thermal aureoles which need further experimental verification. The effect of thermo-acoustic self-focusing of light [1, 11] should be mentioned as being among these effects. It can occur during the interval of the characteristic times (6.1.5) of pulsed heating during particle vaporization or gas-dynamic explosion. The 'aureole' mechanism of turbid medium self-clearing can also take place [1]. The latter mechanism is possible because there exists the possibility of compensating, under certain heating conditions, for the phase changes of a light wave in a particle and on its thermal aureole.

## 6.4. LASER RADIATION TRANSFER IN COMBUSTIBLE AEROSOLS

It must be said that the problem of laser beam self-action in a combustible aerosol assumes that a common solution of the nonlinear parabolic equation for the complex amplitude of the light field and the system of aerothermochemistry equations (see §2.4) must be found. Basic mechanisms of aerosol nonlinearity in this case are: (1) the decrease of the geometrical cross-section of burning or splitting particles which leads to changes in the medium's transmission; (2) light scattering by thermal and mass aureoles around absorbing particles; and (3) regular refraction of the laser beam based on the statistically mean profile of the medium's refractive index occurring due to exothermic chemical reactions and dissipation of the laser beam's energy.

The up-to-date theoretical results relating to high-power radiation transfer in such media, obtained for carbon aerosol particles [23], take into account the effect of clearing of the medium due to the regular burning of the particles. The applicability of such a model is limited by the case of a wide cw laser beam propagating through an aerosol cloud of relatively short length. In this case, the scattering of light by thermal aureoles and thermal defocusing are the effects of the next order of magnitude ($t \gg (4N_0^{2/3}\chi_T)^{-1}$; $\delta T(x/R_0)^2 10^{-6} \ll 1$, where $\delta T$ is the mean temperature of the superheated medium within the beam; $R_0$ is the beam's radius; and $x$ is the distance along the path of propagation). The description of the process is based on the one-dimensional transfer equation for beam intensity $I(x, t)$, written following the single-scattering approximation:

$$\left(\frac{d}{dx} + \frac{1}{c}\frac{d}{dt}\right) I_n(x, t) = -I_n(x, t) \int_0^\infty da\, N_a(a_0(a)) a^2 \left|\frac{da_0(a)}{da}\right| \times$$

$$\times K^{(0)}(a)[1 + O(K^N)]; \qquad (6.4.1)$$

where $N_a(a)$ is the size-distribution function; $a_0(a)$ is the inverse functional dependence of the current radius of a burning particle on the initial one, $a_0$; and $K^{(0)}(a)$, $K^N$ are the extinction efficiency factors of an individual particle of radius $a(t)$ and of its thermal aureole, respectively.

The peculiarity of the process of the interaction between laser radiation and a combustible aerosol, compared with the case of water-droplet aerosols, is the threshold character of particle inflammation. This means that the process begins when the intensity of incident radiation at a point x inside the medium reaches a threshold value $I_i$. The value $I_i$ can be estimated according to (2.4.14). If $I \gtrsim I_i$, then one can neglect, in the first-order approximation, the dependences of the surface temperature $T_a$ and of the vaporization rate (2.4.12) on the intensity I. In this case the simplified model of the diffusion-limited process of particle burning can be described using relationships derived from (2.4.15):

$$a(t) \approx \begin{cases} a_0(1 - t/t_i(a_0))^{1/2}[1 + O(A_d/A_k)], & t_c \leqslant t < t_i \\ 0, & t \geqslant t_i. \end{cases} \qquad (6.4.2).$$

where $t_c$ is the moment of time at which the intensity of incident radiation at the point x reaches the value $I \geqslant I_i(a_0)$.

According to [23], the stationary speed of the burning front can be estimated as follows, $v_i = dx/dt_c$:

$$v_i(I_0) \approx \frac{2/\pi}{\int_0^\infty da\, a^2 K N_a(a) t_i(a)} \frac{\int_0^\infty da\, a^2 K N_a(a) \ln(I_0/I_i(a))}{\int_0^\infty da\, a^2 K N_a(a)}, \qquad (6.4.3)$$

where $I_0$ is the intensity of incident radiation at the boundary of the medium. This expression is valid for $t \geqslant t_i(a_{sq})$.

In the case of a monodispersed aerosol,

$$v_i(I_0) \approx \frac{2 \ln(I_0/I_i)}{t_i \pi a^2 N_0 K}. \qquad (6.4.4)$$

Thus, for $a_0 = 2$ µm; $K = 2$; $N_0 = 10^3$ cm$^{-3}$; $\ln(I_0/I_i) \approx 1$, one finds that $v_i \approx 6 \times 10^6$ cm/s. Under these conditions, the aerosol layer of length $x_0 \approx \ln(I_0(I_i)/(\pi a^2 N_0 K)$ burns up practically simultaneously (time delay $t = x_0/c$ is neglected).

Paper [14, 23] presents the results of the experimental investigations into the nonlinear distortions of the sounding beam ($\lambda$ = 0.63 μm) taking place in the beam of a high-power Nd-glass laser ($\lambda$ = 1.06 μm, W ≈ 1000 J, $t_p$ ≈ 1 ms). A suspension in air of sooty particles (exponential size spectrum from 2 to 50 μm radii) and wood smoke were used in the experiments. The experimental set-up and measurement technique were analogous to those described in §6.3. The diameters of the sounding and the high-power beams were 0.5 and 4.0 cm, respectively. The useful length of the aerosol chamber x was 73 cm. Figure 6.4.1 shows the data relating to the dynamics of the aerosol's optical depth (curves 1 and 2) obtained from the oscillograms of the Nd-glass laser pulses. The envelope of the laser pulse has a maximum at t ≃ 0.6-0.7 ms.

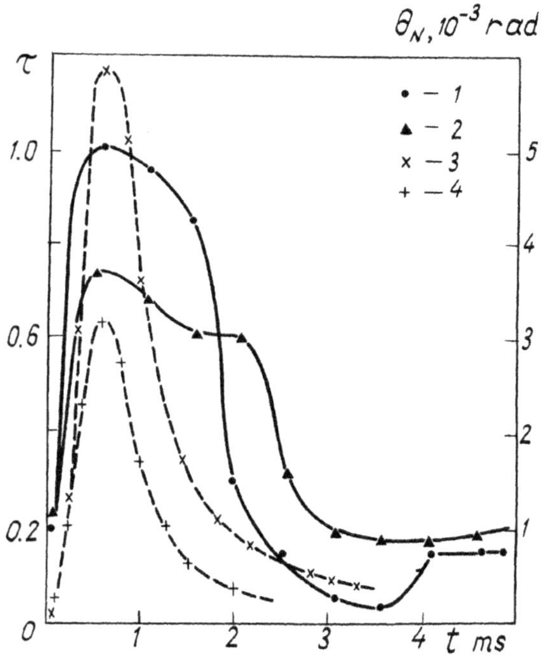


Fig. 6.4.1. Dynamics of the optical depth of an inflammable sooty aerosol $\tau$ (cases 1, 2) and of the angular divergence of the sounding beam $\theta_N$ (cases 3, 4) in a high-power Nd-glass laser beam channel, when $t_p$ ≈ 1 ms; $\lambda$ = 1.06 μm. 1 - w = 150 J/cm$^2$; 2 - w = 100 J/cm$^2$; 3 - w = 7.9 J/cm$^2$, $\tau_0$ = 1.4; 4 - w = 2.3 J/cm$^2$, $\tau_0$ = 0.7.

As seen from this figure, 0.1 to 0.2 ms after laser firing the effect of aerosol turbidity is observed, resulting in a 4- to 5-fold increase in the optical depth relative to the initial one ($\tau_0$ ≈ 0.2). Then, in 1.5 to 2 ms, the turbidity is replaced by the partial clearing of the medium. The relaxation of the optical properties in the laser beam channel took place during several tens of milliseconds. The decrease in aerosol transmission occurring during the laser pulse is due to the joint effect of the scattering of light by thermal and mass aureoles around the burning particles and the fragmentation of large particles into smaller ones with radii of about $a \approx 0.1\ a_0$. In practice, nonlinear divergence of the sounding beam (experimental points of the 3rd and 4th types) follows the high-power laser pulse shape over time, that means that this divergence is caused by the effect of light scattering by thermal aureoles. The partial 'clearing' of the medium occurring just after the cessation of the high-power pulse can be related to the relaxation of the thermal and mass aureoles, which makes, as a consequence, the effect of particles burning the dominating factor.

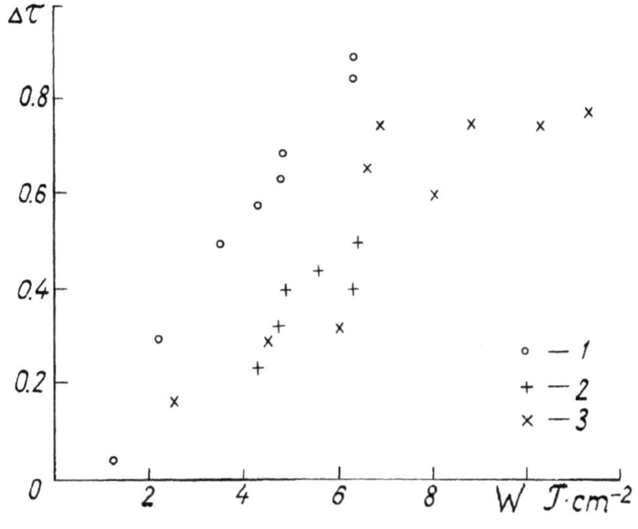


Fig. 6.4.2. The dependence of the maximum optical depth of a turbid sooty aerosol, for a sounding beam, on the energy density of incident radiation from a Nd laser pulse of 1 ms: in a nitrogen atmosphere ($N_2$ - 97%; $O_2$ - 3%) - case 1, and for air - cases 2, 3. Cases 1, 2 - $\tau_0 = 1.1$; case 3 - $\tau_0 = 1.5$.

Figure 6.4.2 shows data from [23] illustrating the dependence of the maximum optical depth of sooty aerosols, observed during the laser pulse, on the pulse energy in the range 0 to 12 J/cm$^2$. Curve 1 represents data obtained in a nitrogen atmosphere, while curves 2 and 3 refer to air. It can be seen from this figure that combustion compensates a little for the turbidity effect ($\Delta\tau \approx 0.1$ to 0.2 of the optical depth due to the turbidity). According to [23], the maximum turbidity also increases monotonously with an increase in the pulsed laser energy in the range $W = 100$ to 700 J/cm$^2$ and $\tau_0 \approx 0.2$, reaching the values $\Delta\tau \approx 0.7$ to 1.3. The experiments which were carried out with smoke having only fine particles, whose radii ranged from $10^{-6}$ to $10^{-5}$ cm, and an initial optical depth of $\tau_0 \approx 1$, showed that, in every case with Nd-glass laser pulses of energy $W \lesssim 10^3$ J/cm$^2$, only an increase in smoke transmission was observed. The corresponding decrease in the optical depth was $\Delta\tau = -(0.1$ to $0.8)$.

Thus, the results presented above of the first investigations of high-power laser beam interaction with model combustible aerosols reveal the possibility of observing nonlinear optical effects in chemically reactive atmospheric hazes.

## 6.5. THERMAL BLOOMING OF THE CW AND QUASI-CW LASER BEAMS DUE TO LIGHT ABSORPTION BY ATMOSPHERIC AEROSOLS AND GASES

### 6.5.1. General Discussion of the Problem

Consider the propagation of high-power laser radiation through a slightly turbid atmosphere. In this practically important case the extinction of radiation by aerosols and by gases is of the same order, i.e., the volume extinction coefficients of aerosols $\alpha$ and gases $\alpha_g$ have similar values. For simplicity, the duration of optical action on a fixed volume of atmosphere is assumed to be much longer than the characteristic times of averaging the thermal and acoustic perturbations over the spaces between the light absorbing centers: $t_2 = (N_0^{1/3} c_s)^{-1}$; $t_5 = (4N_0^{2/3}\chi_T)^{-1}$ (see Figure 6.1.1).

The propagation of a slightly divergent laser beam through a medium composed of light absorbing aerosols and gases, in the above case, can be described by a nonlinear parabolic equation for the complex field amplitude $E(x, \vec{\ell}, t)$ which is derived from (6.1.6). This equation describes the beam's energy losses following the single-scattering approximation by the total volume extinction coefficient $\alpha_\Sigma = \alpha + \alpha_g$, while the thermal perturbations of the gaseous medium are described by the nonlinear correction $\bar{\varepsilon}_N$ of the dielectric constant averaged over the space between particles. The equation is written

$$2ik\,\frac{\partial E}{\partial x} + \Delta_\perp E + ikE\alpha_\Sigma + k^2 E\bar{\varepsilon}_N = 0. \tag{6.5.1}$$

The concrete forms of the material equations for $\alpha_\Sigma$ and $\bar{\varepsilon}_N$ are considered below. These, together with (6.5.1), form the closed system of equations describing nonlinear propagation of high-power radiation. A description is presented of the models of both conservative and nonconservative light-absorbing components of the atmosphere.

### 6.5.2. The Effects of Laser Beam Interaction with a Conservative Light-Absorbing Component

A gas-aerosol medium containing particles of mineral or organic origin can serve as an example of a conservative absorbing mixture for incident radiation in the intensity range over which no changes in the phase state or chemical composition of the particulate matter are observed ($I_{ab} \leqslant 10^2,\ldots,10^3$ Wcm$^{-2}$).

The mechanism of the thermal self-action of radiation, in the case of a conservative dispersed admixture, does not differ, in principle, from that observed in the case of a homogeneously absorbant gas medium. It is connected with the changes of air density in the region of laser beam heating, which leads, consequently, to the appearance of gaseous lenses which defocus the beam. Parameters of the gaseous lenses depend on the distribution of the intensity of the incident radiation over the beam's cross-section, as well as on the heat transfer regime.

The description of thermal blurring of the beam in this case can be based on the results of numerous investigations carried out in homogeneously-absorbant media (see, e.g., the reviews [24-32]).

The linearized thermohydrodynamic equation for the nonlinear correction $\bar{\varepsilon}_N$ of the dielectric constant is written as follows [30-31]:

$$(d^3/dt^3 - \gamma X_{eff}\nabla_\perp^2\, d^2/dt^2 - c_s^2\nabla_\perp^2\, d/dt + X_{eff}C_s^2\nabla_\perp^4)\varepsilon_N(x, \vec{r}, t) =$$

$$= (d\varepsilon/d\rho)(\gamma - 1)(\nabla_\perp^2 I(x, \vec{r}, t))\alpha_\Sigma^{ab}[1 + O(t_3/t) + O(t_c/t)]. \quad (6.5.2)$$

where $d/dt = (\partial/\partial t) + \vec{V}_\perp\nabla_\perp$ is the full substantial derivative with respect to time and the radial coordinates $\vec{r}_\perp = (y, z)$; $I(x, \vec{r}_\perp, t) = (c\varepsilon_0^{1/2}/8\pi)|E(x, \vec{r}_\perp, t)|^2$ is the function of intensity distribution over the beam's cross-section, which is assumed to be a doubly differentiable function with respect to $\vec{r}_\perp$; $\alpha_\Sigma^{ab} = \alpha_{ab} + \alpha_g$ is the total volume coefficient of both aerosol and gas absorption; $t_3 = (a^2/3X_T)(c_a\rho_a/c_p\rho)$ is the characteristic time delay for the transfer of heat from particles to the surrounding medium; and $t_c$ is the characteristic time of the thermal decay of the excited vibration states.

Besides the characteristic times of the processes (see §6.1.1), the qualitative analysis of the conditions under which the formation of gaseous lenses occurs requires information on some additional parameters. These parameters are: Mach number $M = V_\perp/C_s = t_4/t_6$; the effective Peclet number $Pe = 4X_{eff}/R_0V_\perp = t_6/t_{eff}$, and the pulse repetition period $t_r$ (for quasi-cw radiation). Here, $t_{eff} = t_7$ with $\beta \ll 1$ and $t_{eff} = t_9$ for $\beta \gg 1$.

The following asymptotic cases can occur in practice, depending on the values of M, Pe, and on the relationship between the characteristic times of the mass and heat transfer and the duration of optical action:

$$t \lesssim t_4. \quad M \ll 1. \tag{6.5.3}$$

$$t_4 \ll t \ll t_6. \quad M \ll 1. \tag{6.5.4}$$

$$t \gtrsim t_{eff}. \quad M \ll 1. \quad Pe \ll 1. \tag{6.5.5}$$

$$t \gtrsim t_6. \quad M \ll 1. \quad Pe \ll 1. \tag{6.5.6}$$

$$t \gtrsim t_6. \quad M \ll 1. \quad Pe \gtrsim 1. \tag{6.5.7}$$

$$t \gtrsim t_6. \quad M \gtrsim 1. \quad Pe \gg 1. \tag{6.5.8}$$

Here, the duration of optical action of quasi-cw radiation on the medium, in the cases (6.5.3)-(6.5.8), means: (1) the pulse duration, if $t_r \gg \min\{t_6, t_{eff}\}$, or (2) the total time of a series of pulses, if $t_r \ll \min\{t_6, t_{eff}\}$.

In situations like (6.5.3)-(6.5.6) the thermal gaseous lens induced by the cylindrically-symmetric high-power beam will also be axially symmetric. The formation of thermal lenses in the regions limited by (6.5.3) and (6.5.4) is determined by the thermoacoustic perturbations of the medium ($P \neq$ const) and by the processes of isobaric thermal expansion ($P = P_0 =$ const), respectively. In these cases, the diffusion and convective mechanisms of heat transfer can be neglected. The situations (6.5.5) and (6.5.6) can occur for quasi-cw beams with a low off-duty factor, or for a cw beam, and are characterized by the presence of 'rest' zones if the beam's axis is oriented strictly along the wind vector, or if the beam's rotation speed is equal to the speed of the side wind. In these two cases there are no heat losses caused by the relative movement of the beam and the medium, and the mechanisms of forming the gaseous lenses are of the next order of magnitude down. Under real atmospheric conditions this mechanism corresponds to turbulent heat transport, and is characterized by the time $t_{eff}$.

In the regions (6.5.6) and (6.5.7), the thermal gaseous lenses are asymmetric in the direction of the side wind that controls the self-deflection of the beam as a whole. The sign of deflection depends on the intensity profile across the beam's cross-section and on the Mach number. Thus, for $M < 1$ a beam with its intensity maximum along its axis is deflected towards the air stream. In this case, the defocusing of the beam along the perpendicular axis also takes place. For $M > 1$, defocusing along the perpendicular axis again occurs, while along the direction of the beam's scan the thermal gaseous lens focuses symmetrically. When passing from subsonic to supersonic scanning ($|M - 1| \to 0$), the anomalous accumulation of hydrodynamic perturbations of the dielectric constant occurs, this leads to an increase in the nonlinear distortions of the beam structure, as in the case of 'subsonic rest zones' of the type (6.5.6).

It should be noted that asymmetry of thermal gaseous lens can occur due to convection caused by light absorption, even if the side wind is absent or scanning is not carried out.

For a beam propagating along a horizontal path, the velocity of laminar convection $V_g$ can be assessed, according to [31], using

$$V_g \approx \begin{cases} c_1 \nu c_p (\gamma - 1) g R_0^4 \alpha_\Sigma^{ab} I/(c_s^2 \lambda_T), & Pe^* \lesssim 1; \\ c_2 [(\gamma - 1) g R_0^2 \alpha_\Sigma^{ab} I/(c_s^2 \lambda_T \rho_0)]^{1/3}, & Pe^* \gg 1, \end{cases}$$

where g is the acceleration due to gravity; $c_1$ and $c_2$ are constants of the order of unity; $\nu$ is the coefficient of kinetic viscosity of air; and $Pe^* = t_6(V_g)/t_8$. The convection due to the absorption of light in the beam channel is negligible in the majority of real atmospheric situations. Thus, for normal atmospheric conditions and for $\pi R_0^2 I \alpha^{ab} \approx 1$, one finds that $V_g(Pe^* \gg 1) \approx 10\ \mathrm{cms}^{-1}$.

The above qualitative analysis of the equation for the perturbations of the dielectric constant in an aerosol-gas medium (6.5.1), made by comparing the characteristic times of the heat transfer processes, shows that there exist two groups of self-action effects for cylindrically-symmetric laser beams. The first group of effects, occurring under the conditions (6.5.3)-(6.5.6), results in the symmetric self-braodening of the beam and in the stratification of the aberration beam. In addition to the nonlinear distortions of the beam's structure, the second group of effects defined by (6.5.7) and (6.5.8) results in self-deflection of the beam taking place along the axis coinciding with the side wind or with the direction of scanning.

We will discuss the basic quantitative relationships characteristic of these two groups of self-action effects below. The discussion is based on the results of experimental studies carried out in gas medium having a constant absoprtion coefficient [24-32].

If the medium is irradiated by a series of pulses of duration $t_p$ and pulse separation time $t_r$ then, according to the principle of superposition of small perturbations, one can write the solution of (6.5.1) at the time of the j-th pulse as follows [29]:

$$\varepsilon_{Nj}(\vec{r}, t) = \sum_{i=1}^{j} \varepsilon_N^0(\vec{r}_\perp, t - (i-1)t_r) - \sum_{i=2}^{j} \varepsilon_N^0(\vec{r}_\perp, t - (i-2)t_r - t_p), \tag{6.5.9}$$

where $\varepsilon_N^0(\vec{r}_\perp, t)$ is the perturbation of the dielectric constant induced by the first pulse of the train acting during the interval [0, t].

Under the conditions (6.5.3)-(6.5.4), the solution of (6.5.1) for $\varepsilon_N^0$ in quadratures has the form [31]

$$\varepsilon_N^0(\vec{r}_\perp, t) = \frac{4\pi\varepsilon_0^{1/2}}{W_n k^2}\begin{cases} \frac{c_s^2}{2}\int_{-\infty}^{t_0} (t - t')^2 \nabla_\perp I \, dt'[1 + O(t_0^2/t_4^2)]; & (6.5.10) \\ \int_{-\infty}^{t_0} I \, dt' \, [1 + O(t_4^2/t_0^2)], & (6.5.11) \end{cases}$$

where $W_n$ is the threshold energy of the thermal self-action defined by (6.3.9), in which $\alpha_{ab}$ is substituted for by the total volume (aerosol and gas) absorption coefficient $\alpha_\Sigma^{ab}$; and $t_0 = t - x/c$.

The estimates of $\varepsilon_{N_j}(0, t)$ in 'rest' zones, corresponding to the situations (6.5.5) and (6.5.6), can be obtained by using (6.5.9) and (6.5.11) and changing the upper limit of integration $t_0 \to t_{eff}$.

The analysis of the methods for solving the problem of the thermal blurring of the beam, based on the use of the nonlinear parabolic equation (6.5.1), can be found in a number of publications [26-32].

The simplest solutions were obtained by following the nonaberrational (paraxial approximation and solving the differential equations for the beam parameters which can be derived from (6.5.1). In the case of a Gaussian beam of the type (6.3.7) entering the medium, the solution can be sought also for the form of a Gaussian beam whose dimensionless width $f(x, t) = R(x, t)/R_0$ and weighted mean radius of the phase front curvature $F(x, t)$ (see (6.3.28)) depend on x and t. The equations for $f(x, t)$ and $F(x, t)$ are as follows [31]:

$$F(x, t) = f(df/dx)^{-1}; \tag{6.5.12}$$

$$d^2f/dx^2 = L_d^{-2}[f^{-3} + fD_j], \tag{6.5.13}$$

where $L_d = kR_0^2$ and

$$D_j = (L_d^2/(4\varepsilon_0^{1/2}R_0^2)\varepsilon_{N_j}(x, 0, t) \exp(-\alpha_\Sigma x). \tag{6.5.14}$$

The function $\varepsilon_{N_j}(x, \vec{r}, t)$ is described by expressions analogous to (6.5.9)-(6.5.11).

In the case of weak absorption by aerosols and gases ($\alpha_\Sigma x \ll 1$), the solution of (6.5.13) can be written in the form

$$f(x, t) = (1 + \frac{x}{F_0})\begin{cases} (1 + \xi)^{1/2}, & \xi < 1; \\ (1 + \xi)^{-1}\xi^{3/4}/\sqrt{2}, & \xi \gtrsim 1; \end{cases} \tag{6.5.15}$$

$$\xi = (x^2/L_d^2)(1 + x/F_0)^{-1}(1 + L_d^2/F_0^2)^{-1}D_j^{-1/2} \quad (x = 0).$$

In the opposite case of a strongly absorbant medium ($\alpha_\Sigma x \gg L_d, F_0$), and for situations described by (6.5.4)-(6.5.5), one can obtain a simple expression for the nonlinear angular divergence $\theta_N$ of a Gaussian cw laser beam based on the results of [30]:

$$\theta_N(t) = R_0(df/dx) = R_0/F_0 + \frac{2P_0R_0}{W_nL_d\alpha_\Sigma} \times$$

$$\times [1/t_{eff} + \tfrac{1}{2}t]^{-1}[1 - \exp(-\alpha_\Sigma x)]. \qquad (6.5.16)$$

Here, $P_0$ is the power of the incident beam at the boundary of the medium and $R_0/F_0$ is the initial beam divergence. So, if $t \ll t_{eff}$, beam self-defocusing is proportional to $\sim t$, while for $t \gg t_{eff}$ the beam divergence reaches its maximum value, which is proportional to the effective time of turbulent mixing, $\sim t_{eff}$.

A numerical solution of a simplified form of (6.5.13) was obtained in [15, 33]. Numerical simulations of the reference parabolic equation (6.5.1) in combination with (6.5.2) have been carried out in [29-32]. The calculations showed that, under the conditions of strong nonlinearity ($D_j \gg 1$), the aberrations of the gaseous lenses become significant and restrict the application of the paraxial approach. The influence of the aberrations is strongest in the case of a thin lens located at the beginning of the propagation path, and is caused by the deviation of the beam-induced profile of the dielectric constant from the ideal parabolic one.

Figure 6.5.1 illustrates the peculiarities of the self-action of short ($t \ll t_4$) and long ($t \gg t_4$) laser pulses. The data were obtained from laboratory experiments which were carried out under a wide range of interaction durations t, relative to the characteristic relaxation time of thermo-acoustic perturbations, $t_4 = R_0/c_s$. The experimental parameters were [34]: $\lambda = 1.06$ µm, $W = 57$ J, $t_p = 3$ to 100 µs; $R_0$ $(x = 0) = 0.74$ cm; $R_0$ $(x = 100$ cm$) = 0.24$ cm. The laser beam was focused onto a sample cell 100 cm in length. The cell was filled with a concentrated, light absorbing component of ammonia. The absorption coefficient $\alpha_\Sigma^{ab} \approx 8 \times 10^{-4}$ cm$^{-1}$. As can be seen from this figure, the effect of the beam's intensity at the axis is weaker for short pulses compared with that for long ones. It can also be seen from the figure that the action of long pulses results in isobaric changes of the air density which strictly follow the heating of the medium by the laser pulse.

The specific feature of high-power laser beam self-action in the case where there is relative movement of the beam and the absorbing medium is the beam's self-deflection from the initial axis in situations like those described by (6.5.7) and (6.5.8), as was pointed out above in the qualitative analysis of the problem. The paraxial description of the deflecting gaseous lens correponds to the truncation of the series expansion of the solution $\bar{\varepsilon}_N$ for a near-axial region of the beam at the second order term:

$$\varepsilon_B(x, \vec{r}, t) \approx \varepsilon_y(y/R_0) + \varepsilon_{yy}(y/R_0)^2 + \varepsilon_{zz}(z/R_0)^2 + \cdots .$$

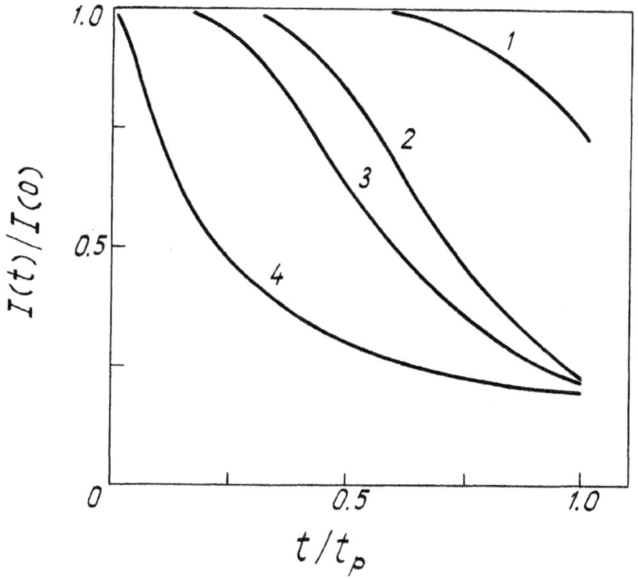


Fig. 6.5.1. Nonlinear variations of the relative radiation intensity along the Nd laser beam axis during laser pulse propagation through a cell containing energy-absorbing gas. $\lambda$ = 1.06 μm; W = 57 J; $t_p$ = 3-100 ms; $R_0$ = 0.24-0.74 cm; $\alpha_\Sigma^{ab}x \approx 0.08$. The curves correspond to variations for different values of the ratio of pulse duration $t_p$ to characteristic time of pressure relaxation in the beam, $t_4 = R_0/C_s$. (1) - $t_p/t_4$ = 0.3; (2) - 1.0; (3) - 2.0; (4) - 10.0.

For convenience, the direction of scanning (the same as for the side wind) is chosen as lying along the Y-axis. The coefficients of expansion $\varepsilon_y$, $\varepsilon_{yy}$, $\varepsilon_{zz}$ are functions of the Peclet number $Pe = 4x_{eff}/(V_\perp R_0)$. These coefficients determine both the bulk beam deflection and its blurring along the corresponding axes, respectively. When the Mach number $M < 1$, maximum deflection takes place for $Pe \approx 0.3$, after this the deflection decreases with increasing Pe as $\approx(Pe)^{-1}$. The defocusing effects also decrease monotonously with increasing Pe. Over the range of Peclet numbers $Pe \gg 1$, we find that $\varepsilon_{yy} \sim (Pe)^{-2}$ and $\varepsilon_{zz} \sim (Pe)^{-1}$, that means that during the process of wind deflection the beam becomes elongated across the wind.

The stationary ($t \gtrsim R_0/V$) distribution of the perturbations of the dielectric constant over the beam channel is described, for $Pe \gg 1$, by expressions (6.5.9)-(6.5.11), in which the variable t is substituted for by the variable $(y/R_0)$.

A quantitative estimate of Gaussian beam self-deflection toward the side wind, taking place within the limits defined by (6.5.6) and for

$t_r \ll t_6$, can be made using the following formula [31]:

$$\theta_y(x) \approx \frac{2R_0F_0}{L_T^2}\Big[1 - (1 + \frac{X}{F_0})^{-1} + \exp(-\alpha_\Sigma F_0) \times$$
$$\times (E_i^*(\alpha_\Sigma(x + F_0)) - E_i^*(\alpha_\Sigma F_0))\Big],$$

where $E_i^*$ is the exponential integral function; $L_T$ is the effective length along which thermal self-action takes place in the moving medium; and

$$L_T = (kR_0^2)(P_n/P_0)^{1/2}. \qquad (6.5.18)$$

$P_n = W_n/t_6$ is the threshold power for isobaric thermal self-action. The expression (6.5.17) has been derived following the approach of a weakly nonlinear medium, and does not account for blurring of the diffraction beam ($x \ll L_d$).

In the case of a collimated beam ($F_0 \to \infty$), one can obtain from (6.5.17) that

$$\theta_y \approx (2R_0/(\alpha_\Sigma L_T^2))[1 - \exp(-\alpha_\Sigma x)]. \qquad (6.5.19)$$

Figure 6.5.2 presents the results of calculations [27] of the stationary wind deflection of Gaussian collimated and ring beams, carried out for the gravity centers $\Delta Y$ of the beams in a homogeneously-absorbing gaseous medium.

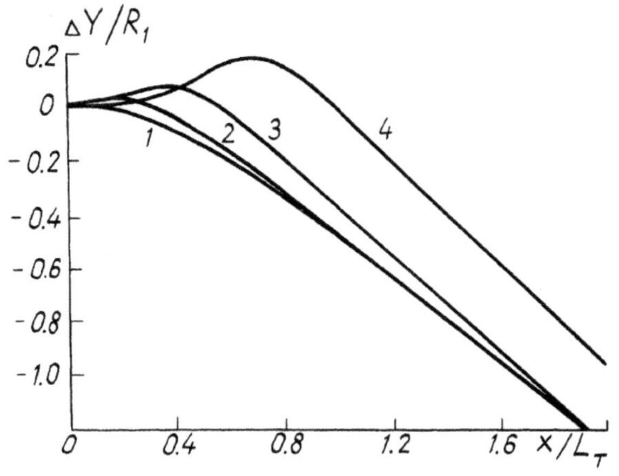


Fig. 6.5.2. Nonlinear energetic center deflections of the $CO_2$ laser collimated beams calculated for a standard model of summer ground-based atmosphere. Curve 1 - Gaussian beam with $L_T$ = 0.12 km; curves 2-4 are ring beams with $L_T$ = 0.12 and $R_1/R_2$ = 0.125, 0.25, 0.50, respectively. The positive values $\Delta Y/R_0$ characterize the beam's deflection due to the wind.

The intensity distribution of ring beams was taken to be: $E(\vec{r}_\perp, t) = E_0[\exp(-r^2/2R_1^2) - \exp(-r^2/2R_2^2)]$, where $R_1$ and $R_2$ are the radii of the inner and the outer circles of the ring-beam's cross-section, respectively.

$$\Delta Y(x, t) = \int_{-\infty}^{\infty} |E(x, y, z = 0)|^2 y\, dy/(\int_{-\infty}^{\infty} |E|^2\, dy).$$

For weak energetic losses ($\alpha_\Sigma x \ll 1$), the calculations are also valid for beam self-action in an atmosphere of a conservative gas-aerosol mixture. As seen from the figure, ring beams undergo a weaker deflection compared with that for collimated Gaussian beams, and the sign of deflection changes along the path. Thus, at the beginning of the path, $\Delta Y > 0$ (deflection along the wind) and then vice versa ($\Delta Y < 0$) towards the end of the path.

An example of the calculated [32] effective lengths of thermal self-action $L_T$ of beams propagating through the absorbing medium, when scanning at a speed close to the speed of sound, is presented in Figure 6.5.3. In Figure 6.5.3 $q_0 = \alpha_\Sigma^{ab} IR_0(\gamma - 1)/c_s^3\rho_0)$; and $\beta_0 = |M^2 - 1|^{1/2}$. The expression expressions describing the intensity of beams travelling close to the speed of sound, as well as numerical results, can be found in [32].

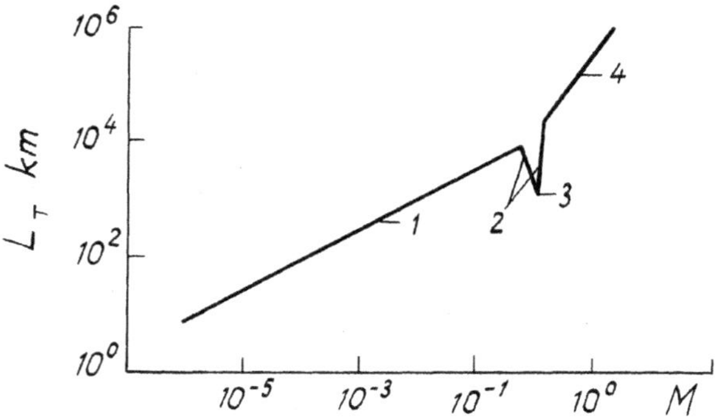


Fig. 6.5.3. The calculated dependence of the effective lengths of the thermal self-action of a beam propagating through an absorbant medium on the Mach number, when scanning is carried out at speeds close to the speed of sound: $M = V_\perp/C_s$. The calculated parameters are $t_4 = 10^{-3}$ s; $q_0 = 10^{-6}$. 1 - $L_T \sim (M/q_0)^{1/2}$, 2 - $L_T \sim (\beta_0 M/q_0)^{1/2}$, 3 - $L_T \sim (1/q_0)^{1/2}$, 4 - $L_T \sim (M^3/q_0)^{1/2}$.

The data presented in this figure show that the dependence of the parameters of the beam-induced lenses on the scanning speed are quite different in the subsonic ($M < 1$) and the supersonic regions. At speeds $M \approx 1$, anomalous refraction is observed, as mentioned in §6.5.2, in the channel of the high-power beam, due to the effect of the accumulation of hydrodynamic perturbations of the density of the air within the beam channel.

The analysis of thermal self-action of quasi-cw radiation when $t_r \sim t_6 = R_0/V_\perp$ shows that the effect of additional focusing of the leeward part of the beam by the thermal profile of the preceding pulse can occur [29, 32]. As a result, the maximum intensity of the radiation within the beam can increase compared with that in the linear medium. This is clearly demonstrated by the results of numerical simulations of this problem carried out in [32] (see Figure 6.5.4).

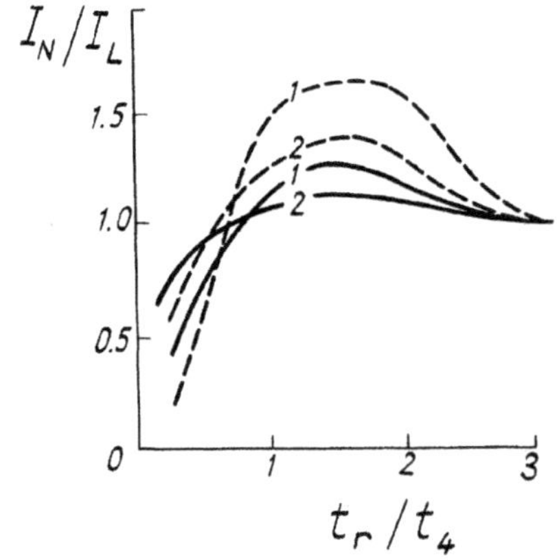


Fig. 6.5.4. The relative variation of the main frequency-pulsed radiation intensity in the atmosphere with thermal nonlinearity as a function of the number of pulses for the time the wind blows across the beam's cross-section. 1 - $W_0/W_n$ = 27; 2 - $W_0/W_n$ = 13. The solid curves correspond to $x/L_d$ = 0.25. The dashed curves correspond to $x/L_d$ = 0.5.

### 6.5.3. Thermal Self-Action of Laser Beams in Water-Droplet Hazes

Water droplet hazes are characterized by relatively low threshold intensities ($IK_{ab} \gtrsim 1 - 10$ Wcm$^{-2}$) of incident radiation, above which their microstructure undergoes modifications (that is, the condition of conservatism is broken). This essentially complicates the picture of laser beam propagation through such media. The expression for the volume extinction coefficient of the aerosol and gas mixture, as well as for the nonlinear correction for the dielectric constant of the air entering (6.5.1), are written as

$$\alpha_\Sigma = \alpha_g + \pi\int_0^\infty a^2 N_a(a, \vec{r}, t) K(a, m_a, \lambda)\, da; \qquad (6.5.20)$$

$$\varepsilon_N(\vec{r}, t) \approx (\partial\varepsilon/\partial\rho)\delta\rho + (\partial\varepsilon/\partial\rho_n)\delta\rho_n,$$

where $\alpha_g$ is the absorption coefficient of the gaseous component; $N_a(a, \vec{r}, t)$ is the size-distribution function of aerosols normalized w.r.t. $N_0$; and $\delta\rho$ and $\delta\rho_n$ are the deviations of the air and water vapor densities from their corresponding equilibrium values.

The averaged transfer equations for $N_a$, $\delta\rho$, and $\delta\rho_n$ corresponding to isobaric processes (situations analogous to (6.5.4)-(6.5.7)) have the following universal form:

$$(\partial/\partial t) + v_\perp\nabla_\perp - D_j\nabla_\perp^2)S_j = q_j, \tag{6.5.21}$$

where $j = 1, 2, 3$; $S_1 = N_a$; $S_2 = \delta\rho$; $S_3 = \delta\rho_n$; $D_j$ is the diffusion coefficient, and $q_j$ is the source function of the j-th component:

$$q_1 = (da/dt)N_a; \tag{6.5.22}$$

$$q_2 \approx (\gamma - 1)I(\vec{r}, t)[\alpha_g + \alpha_{ab}(1 - \bar{\beta}_T)]; \tag{6.5.23}$$

$$q_3 \approx \alpha_{ab}\bar{\beta}_T I(\vec{r}, t)/Q_e. \tag{6.5.24}$$

The vaporization rate of the haze droplets can be satisfactorily described by

$$\frac{da}{dt} \approx -\beta_T \frac{IK_{ab}(a, m_a, \lambda)}{4\rho_a Q_e}[1 + O(\frac{a^2}{4tD_{2,3}})], \tag{6.5.25}$$

where $\rho_a$ and $Q_w$ are the density and the specific heat of vaporization of the liquid water phase, respectively; $\beta_T$ is the vaporization efficiency which determines the fraction of the absorbed optical energy expendend on the liquid-to-vapor phase transition, except for losses due to thermal conductivity and thermal re-emission [35]; $\bar{\beta}_T$ is the mean value of this efficiency, taken over the range of I and a changes; and it was found that $\bar{\beta}_T \approx 0.5$ to 0.9.

The main features of the dissipation process (or clearing) of water droplet aerosols by laser radiation have been discussed in the preceding chapters. Thus, in particular, changes in the intensity of the laser pulse ($t \sim t_4$) in a water droplet haze composed only of fine fractions ($ak|m_a - 1| \ll 1$) can be described by the following simple relationship [2]:

$$I(x, t) = I_0 \exp\{-\alpha_g x - \ln\{1 + (\exp(\tau_0) - 1) \times$$

$$\times \exp[-3\beta_1 w_0(t)]\}\}[1 + O(x/ct)], \tag{6.5.26}$$

where $w_0(t)$ is the power density of the radiation at the boundary of the medium, and

$$\beta_1 = \bar{\beta}_T(\rho_a Q_e)^{-1} \kappa_a k \exp(-0.2|m_a - 1|),$$

where $\kappa_a$ is the absorption coefficient entering the complex refractive index $m_a = n_a - i\kappa_a$; and $\tau_0$ is the initial optical depth of the haze. The stationary solution ($t \gtrsim t_4$) is analogous to (6.5.26), in which the variable t is substituted for by $y/v_\perp$, where y is the coordinate in the beam's cross-section that is oriented parallel to the side-wind.

The overall effect of the clearing of the haze and the defocusing occurring in the beam channel due to absorption of laser radiation by aerosols and gases have been discussed in [32, 36].

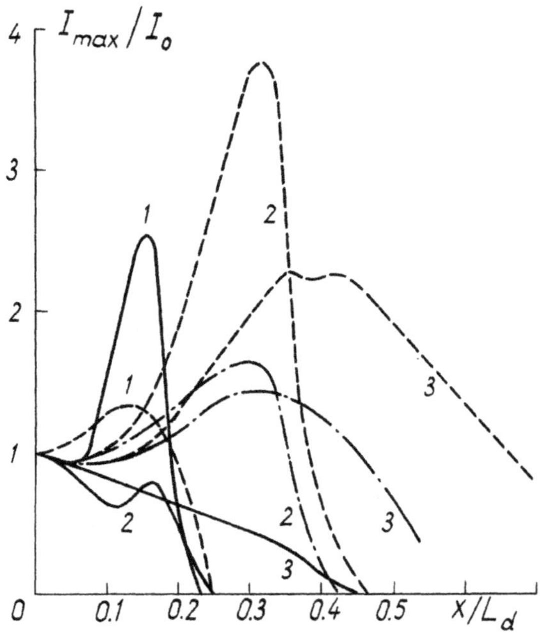


Fig. 6.5.5. Variations of cw $CO_2$ laser peak-power radiation propagating in a water droplet fog for different initial optical depths and with a diffraction length $\tau_d = \alpha_\Sigma L_d$. The dashed line represents consideration of the effect of radiative fog clearing. The dot-dash line is the calculation of the joint influence of the effects of dissipation and thermal lens formation due to aerosol absorption. The solid curve is the result of solving the total problem of thermal beam self-action in the ground layer of the atmosphere together with the consideration of aerosol-gas absorption 1 - $\tau_d$ = 180; 2 - $\tau_d$ = 90; 3 - $\tau_d$ = 45.

The results of numerical simulations of the full system of equations (6.5.1), (6.5.20)-(6.5.25) obtained in [36] are presented in Figures 6.5.5 and 6.5.6.

The curves in Figure 6.5.5 represent the relative changes of the peak power of the cw collimated Gaussian beam ($\lambda$ = 10.6 μm) as a function of the relative distance $x/L_d$ inside the water droplet fog at different initial optical depths $\tau_0$, other parameters are: $\bar{\beta}_T = 0.7$; $\kappa_0 = \gamma_0/t_4 = 0.043$; $\gamma_0 = \rho_a Q_e (\bar{\beta}_T bI)^{-1}$ is the characteristic time of the droplet vaporization; and $b = 0.75 \times 10^3$ cm$^2$ is the ratio of the volume absorption coefficient of the fog droplets to the fog's water content $q_0$.

As follows from this figure, a correction for the molecular absorption of a slightly turbid atmosphere($\tau_g = \alpha L_d$ = 45, 90) results in a decrease in the peak power of the beam, due to the thermal blooming of the beam which is not compensated for by the effect of aerosol dissipation.

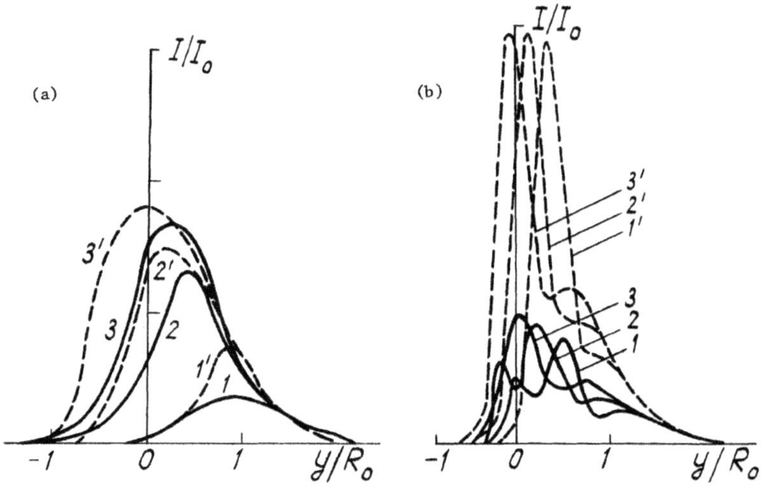


Fig. 6.5.6. Dynamics of the pulse intensity profile of quasi-continuous radiation propagating through a water haze. $\gamma_0$ = 89; $\kappa_0$ = 0.086; (1) $t/t_p$ = 0.2; (2) 0.6; (3) 1.0. (a) $x = 0.1\ L_d$; $\beta_s = t_r/t_p = 1.5$. (b) $x = 0.2\ L_d$; $\beta_s = 0.5$.

The calculated results presented in Figure 6.5.6 represent the intensity profiles over the beam's cross-section at Z = 0. These profiles correspond to different moments after the beginning of a regular pulse in the pulse train of quasi-continuous radiation. The calculation used the values $\alpha_g = 10^{-7}$ cm$^{-1}$; $\alpha = 1.8 \times 10^{-4}$ cm$^{-1}$; $b = 0.75 \times 10^3$ cm$^2$; $\bar{\beta}_T = 0.7$. Broken curves represent the solution of the problem of haze dissipation when the effects of thermal defocusing are neglected. The solid lines represent the solution which takes this effect into account. It is seen that, for the Striel number $\beta_S = t_r/t_p = 1.5$ (where $t_r$ is the time interval between succeeding pulses of duration $t_p$), the leading edge of the pulse undergoes self-focusing, while the influence of the thermal lens on the pulse's 'tail' is quite weak.

The increase in the overlapping of the temperature fields of succeeding pulses ($\beta_S = 0.5$) results in an increase in beam blurring. This, in turn, tends to lead to the alignment of the beam's side-wind direction and the aerosol stream, and to the defocusing of the beam in the direction perpendicular to the stream. As a result, the beam undergoes aberrational destruction. For a constant mean power of the quasi-continuous beam, the depth of clearing (at a level of $0.1\alpha$) only weakly depends on the value $\beta_s$ above $\beta_s = 0.5$ (about 10 to 15%) and reaches its maximum value at $\beta_s \approx 0.5$.

## REFERENCES: CHAPTER 6


[1] V.E. Zuev and Yu.D. Kopytin: 'Nonlinear high-power light propagation in a gaseous medium with solid micro-inclusions', Izv. Vyssh. Uchebn. Zaved. Fiz. 1, 79-105 (1977), in Russian.

[2] V.E. Zuev, Yu.D. Kopytin, and A.V. Kuzikovskii: Nonlinear Optical Effects in Aerosols (Nauka, Novosibirsk, 1980), in Russian.

[3] L.T. Matveev: Foundations of General Meteorology (Gidrometizdat, Leningrad, 1965), in Russian.

[4] V.I. Klyatskin: Statistical Description of Dynamic Systems with Fluctuating Parameters (Nauka, Moscow, 1975), in Russian.

[5] O.A. Volkovitskii, Yu.S. Sedunov, and L.P. Semenov: High-Power Laser Radiation Propagation in Clouds (Gidrometizdat, Leningrad, 1982), in Russian.

[6] Yu.D. Kopytin and S.S. Khmelevtsov: 'High-Power Light Pulse Propagation in an Absorbing Randomly-Heterogeneous Medium' in Optical Wave Propagation in the Atmosphere, ed. by V.E. Zuev (Nauka, Novosibirsk, 1975) pp. 84-94, in Russian.

[7] G.A. Askarian: 'The decrease of intensive light penetrability due to scattering on refracting aureoles of medium optical perturbation', Zh. Eksp. Teor. Fiz. 45, 810-814 (1963), in Russian.

[8] A.A. Manenkov: 'Nonlinear light scattering by small particles', Dokl. Akad. Nauk. SSSR 190, N6, 1315-1317 (1970), in Russian.

[9] Yu.K. Danileiko, A.A. Manenkov, V.S. Nechitailo, and V.Ya. Khaimovmal'kov: 'Nonlinear light scattering in inhomogeneous media', Zh. Eksp. Teor. Fiz. 60, 1245-1250 (1971), in Russian.

[10] V.I. Bukaty, Yu.D. Kopytin, and A.S. Khmelevtsov: 'Intensive Light Pulse Propagation in an Absorbing Heterogeneous Medium', in VIth Conf. Nonlinear Optics Abstracts (Minsk, USSR, 1972) p. 38, in Russian.

[11] Yu.D. Kopytin and S.S. Khmelevtsov: 'New mechanism of light self-focusing in a gaseous medium in the presence of absorbing centers', Zh. Eksp. Teor. Fiz. Pis'ma Red. 21, 45-48 (1975), in Russian.

[12] Yu.D. Kopytin and S.S. Khmelevtsov: 'Thermal self-broadening of intensive light pulses propagation in an absorbing aerosol', Kvant. Elektron. 1, 806-811 (1974), in Russian.

[13] V.B. Kutukov, Yu.K. Ostrovskii, and Yu.I. Yalamov: 'Nonlinear scattering of optical radiation in a medium with absorbing centers', Zh. Eksp. Teor. Fiz. Pis'ma Red. 5, 585-587 (1975), in Russian.

[14] V.I. Bukaty, Yu.D. Kopytin, S.S. Khmelevtsov, and D.P. Chaporov: 'Thermal self-action of optical pulses in simulated aerosol media', Izv. Vyssh. Uchebn. Zved. Fiz. 3, 33-39 (1976), in Russian.

[15] Yu.D. Kopytin: 'Optical Radiation Fluctuations in a Medium with Induced Thermal Inhomogeneities', in 1st Meeting on Atmospheric Optics Abstracts, Part II (Institute of Atmospheric Optics, Tomsk, U.S.S.R., 1976) pp. 222-226, in Russian.

[16] P.N. Svirkunov: 'The effect of thermal aureoles on the scattering properties of particles', Trans. Inst. Exp. Meteorology 18, 110-115 (1978), in Russian.

[17] V.B. Kutukov, Yu.K. Ostrovskii, and Yu.I. Yalamov: 'Nonlinear Scattering of Laser Radiation on an Absorbing Particle Moving in Gas', 1st Meeting on Atmospheric Optics Abstracts, Part II (Institute of Atmospheric Optics, Tomsk, U.S.S.R., 1976) pp. 194-197, in Russian.

[18] G.A. Askarian, V.G. Mikhalevich, and G.P. Shupilo: 'Aureole scattering of high-power light under the effect of radiation on a macroparticle', Zh. Eksp. Teor. Fiz. 60, 2170-2174 (1971), in Russian.

[19] S.V. Zakharchenko, S.D. Pinchuk, and L.M. Skripkin: 'Laser radiation interaction with a solid aerosol'. Trans. Inst. Exp. Meteorology 18, 103-109 (1978), in Russian.

[20] V.I. Bukaty, N.I. Mishenko, S.M. Slobodyan, and D.P. Chaporov: 'Variation of radiation intensity distribution over a laser-beam's cross-section', Prib. Tekh. Eksp. 2, 166-168 (1976), in Russian.

[21] P.N. Svirkunov: 'Radiation propagation in a medium with thermal spots', Trans. Inst. Exp. Meteorology 13, 34-43 (1976), in Russian.

[22] V.I. Bukaty, Yu.D. Kopytin, and S.S. Khmelevtsov: 'Experimental investigation of the optical characteristics of mist in a light beam channel in the explosion regime of droplet vaporization', Izv. Vyssh. Uchebn. Zaved. Fiz. 1, 113-116 (1974), in Russian. Also: Abstracts 2nd Symposium on Laser Radiation Propagation in the Atmosphere (Tomsk, U.S.S.R., 1973) pp. 322-324.

[23] V.I. Bukaty, Yu.D. Kopytin, and V.A. Pogodaev: 'Laser radiation induced combustion of carbon particles', Izv. Vyssh. Uchebn. Zaved. Fiz. 2, 14-22 (1983), in Russian.

[24] S.A. Akhmanov, A.P. Sukhorukov, and P.V. Khokhlov: 'Light self-focusing and diffraction in a nonlinear medium', Usp. Fiz. Nauk 93, 10-70 (1967), in Russian.

[25] G.A. Askarian: 'Light self-focusing effect', Usp. Fiz. Nauk 111, 249-260 (1973), in Russian.

[26] F.G. Gebhardt: 'High-power laser propagation', Appl. Opt. 15, 1479 (1976).

[27] V.V. Vorob'ev: 'Thermal laser beam self-action along inhomogeneous atmospheric paths', Izv. Vyssh. Uchebn. Zaved. Fiz. 11, 61-77 (1977), in Russian.

[28] D.S. Smith: 'High-power laser propagation: thermal blooming', Proc. IEEE 65, N12, 1679-1714 (1977).

[29] S.F. Clifford, M.E. Gracheva, A.S. Gurvich, A. Ishimaru, S.S. Kashkarov, V.V. Pokasov, I.H. Shapiro, I.W. Strohbehn, P.B. Ulrich, and I.L. Walsh: 'Laser Beam Propagation in the Atmosphere', ed. by J.W. Strohbehn, Topics in Applied Physics, Vol. 25 (Springer, Heidelberg, 1978).

[30] M.P. Gordin, L.V. Sokolov, and G.M. Strelkov: 'High-Power Laser Radiation Propagation in the Atmosphere', in Results in Science and Technology. Radiophysics, Vol. 20 (VINITI, Moscow, 1980) pp. 206-289, in Russian.

[31] V.E. Zuev: Laser Radiation Propagation in the Atmosphere (Radio i Svyaz, Moscow, 1981), in Russian.

[32] V.E. Zuev (ed.): 'Nonlinear Atmospheric Optics, Thematic Issue', Izv. Vyssh. Uchebn. Zaved. Fiz. 2, 66-104 (1983).

[33] V.A. Petrishchev and V.I. Talanov: 'On nonstationary light self-focusing', Kvant. Elektron. 6, 35-42 (1971).

[34] K. Kleiman and R.W. O'Neil: 'Thermal blooming of pulsed laser radiation', Appl. Phys. Lett. 23, N1, 43-44 (1973).

[35] V.A. Vysloukh and V.P, Kandidov: 'Dynamics of Water Aerosol Dissipation Under Thermal Self-Action', in 5th Symposium on Laser Radiation Propagation in the Atmosphere, Abstracts, Part III (Tomsk, U.S.S.R., 1979) pp. 62-64, in Russian.

CHAPTER 7

# IONIZATION AND OPTICAL BREAKDOWN IN AEROSOL MEDIA

The propagation of high-power laser radiation through the atmosphere is accompanied by a great variety of nonlinear phenomena, among which the effects caused by ionization and optical breakdown are of particular importance. This is, first of all, due to the threshold character of these two phenomena, which means that they are very sensitive to the combined effect of the three fundamental factors involved in the linear interaction between radiation and media, viz. molecular absorption, aerosol absorption, and the scattering of light by aerosol particles and by atmospheric turbulence

The essential influence of aerosols on the processes of ionization and optical breakdown should be noted since the changes of the aerosol's microphysical and/or optical parameters can result in corresponding changes in the thresholds of these two processes by some orders of magnitude. The particles of condensed matter play important roles as prime centers of ionization and centers of initiation of the shock wave following the optical breakdown [2-14, 25-28].

Optical breakdown causes nonlinear energetic attenuation of light and provides the principle limitation to the beam power that is transportable through the atmosphere [5-11].

On the other hand, the effects of optical breakdown and partial ionization are interesting independently in connection with the problem of the creation of plasma formations in the atmosphere; this field is aimed at various applications, including monitor antennas, reflectors of electromagnetic waves, and streamers for electric discharges [1, 2]. Other promising fields are the applications of laser-beam-induced plasma formations to the remote emission spectral analysis of the atomic composition of aerosol substances and noble gases [4-5], as well as to the opto-acoustic sensing of meteorological parameters [12, 29]. All this attracts a lot of attention to the investigations of optical breakdown in model aerosol media and in natural turbulent atmospheres.

## 7.1. PHYSICAL AND MATHEMATICAL FORMULATIONS OF THE PROBLEM

The theoretical and laboratory studies of the optical breakdown of aerosol media have been thoroughly reviewed in [7-9]. The papers [5-7] summarized the first results of field investigations into pulsed optical breakdown and the nonlinear attenuation of radiation in the atmosphere.

Consider the mathematical formulation of the problem of pulsed optical breakdown in dusty air.

The low-threshold optical breakdown of air initiated by quickly melting particles was first observed in [3, 9, 14]. Paper [3] gives the

the estimations of the threshold intensity of light $I_{th}$ made using the 'thermal explosion' model of an absorbing aerosol particle, based on the condition that the rate of energy consumption from the electromagnetic field by the prime thermoelectrons exceeds the energy losses of these electrons due to elastic electron-atomic collisions or diffusion. As shown in [5] and [7], a rigorous theoretical treatment of the problem should involve the joint solution of the system of equations describing the processes of radiation-induced heating and vaporization of an absorbing particle, the appearance of the prime thermoelectrons for ionization in the stream of evaporated substance, as well as the equations describing the energy spectrum of the electrons and the thermohydrodynamic equations for the vapor-gas mixture in the vicinity of a particle.

Low energies corresponding to IR quanta, as compared with the ionization energies of atoms (molecules) and the mean energy of the electron gas, allow one to neglect the multiphoton ionization effects and the effect of the photoionization of atoms and molecules previously excited by electron impact. For the same reason it is possible, when describing the energy spectrum of electrons $n_e(\varepsilon)$, to pass from the quantum finite differences equation to a corresponding differential equation (the so-called diffusion approximation) [1, 5].

The reference system of equations for the optical ionization of atoms in the vapor aureole of a particle (in the case of a monatomic vapor) is written as

$$\frac{dn_e}{dt} - \nabla(D_e \nabla n_e) + \nabla(\vec{V} n_e) = \frac{d}{d\varepsilon}\left[A\varepsilon^{3/2} \frac{d}{d\varepsilon} \frac{n_e}{\varepsilon^{1/2}} + 2\nu_m(\varepsilon)\varepsilon n_e M_e/M_a\right] + + Q_i + Q^* + Q_r + Q_a; \quad (7.1.1)$$

$$\frac{d}{dt}(\rho(c_v T + V^2/2)) + \nabla(\vec{V}\rho(C_p T + V^2/2)) - \nabla(X_T \nabla T) = = \alpha_g I + \langle\varepsilon\nu_m\rangle 2N_e M_e/M_a - \langle Q_1^*(\varepsilon)\rangle\alpha^*; \quad (7.1.2)$$

$$d\rho/dt + \nabla(\rho\vec{V}) = 0; \quad (7.1.3)$$

$$d\vec{V}/dt + (\vec{V}\nabla)\vec{V} = -\rho^{-1}\nabla p; \quad (7.1.4)$$

$$p = \rho KT/M_a. \quad (7.1.5)$$

The initial and boundary conditions are:

$$t = t_b; \quad N_e = N_{e0}; \quad \rho = \rho_n; \quad \vec{V} = \vec{V}_n; \quad T = T_n; \quad (7.1.6)$$

$$R = +a; \quad \rho = \rho_n^a; \quad T = T_n^a; \quad \vec{V} = \vec{V}_n^a; \quad (7.1.7)$$

$$R \to \infty; \quad \rho \to \rho_0; \quad V \to 0; \quad N_e \to 0; \quad T \to T_0, \tag{7.1.8}$$

where $\rho$, p, T and $\vec{V}$ are the density, the pressure, the temperature, and the velocity of the hydrodynamic motion of the medium, respectively; $\rho_n$, $\vec{V}_n$, $T_n$ and $\rho_n^a$, $\vec{V}_n^a$, $T_n^a$ are the density, the velocity, and the temperature of the vapor in the vapor aureole and near the aerosol particle, respectively, these are obtained from the equations for the gas-dynamic vaporization of a particle (2.3.1)-(2.3.4) or (2.3.11)-(2.3.13), solved without taking ionization into account; $t_b$ is the characteristic time necessary for the particle's surface to be heated to a temperature corresponding to well-developed vaporization; $N_e$ is the total number density of electrons; $N_{e0}$ is the number density of the electrons that have appeared due to thermal ionization, calculated using the Saha formula [15] with $\rho = \rho_n$ and $T = T_n$; $D_e$ is the coefficient of ambipolar electron diffusion [9]; $M_e$ and $M_a$ are the electron and atomic masses, respectively; $\nu_m(\varepsilon)$ is the frequency of elastic electron-to-atom collisions, which depends on the electron's kinetic energy $\varepsilon$ and, for the initial stage of ionization ($N_e \ll N_a$), this is determined by the approximate formula [1]

$$\nu_m(s^{-1}) \approx 2.2 \times 10^{24} \varepsilon^{1/2}\,(\mathrm{eV}) \cdot p\,(\mathrm{torr}) \cdot \sigma(\mathrm{cm}^2) \cdot (1 - \overline{\cos\theta}), \tag{7.1.9}$$

where $N_a$ is the number density of the uncharged vapor particles; $\overline{\cos\theta}$ is the mean cosine of the electron scattering angles (normally this value tends towards zero); $\sigma_m(\varepsilon)$ is the elastic scattering cross-section of the electrons, which depends on the energy of the electrons and on the kind of atoms (extensive experimental and theoretical information on this parameter can be found elsewhere in the literature, see, e.g.,[15]). A is the source function of the electrons, and is related to the intensity of incident radiation I as follows:

$$A = 16\pi e^2 I \nu_m(\varepsilon)\,[3M_e C(\omega^2 + \nu_m^2(\varepsilon))]^{-1}; \tag{7.1.10}$$

where e is the electron charge. The terms in (7.1.1) ($Q_i$, $Q^*$, $Q_r$ and $Q_a$) describe the birth and annihilation of electrons due to cascade ionization, and atomic and molecular excitation due to both electron impact and the process of recombination and sticking to heavy molecules. The function $Q_i$ is determined as follows [1]:

$$Q_i = -n_e(\varepsilon)\nu_m(\varepsilon) + 2\int_{\varepsilon+\phi_i}^{\infty} d\varepsilon' q(\varepsilon, \varepsilon')\nu_i(\varepsilon')n_e(\varepsilon'), \tag{7.1.11}$$

where $\phi_i$ is the energy necessary to ionize an atom (molecule) from the ground state; $q(\varepsilon, \varepsilon')$ is the probability density of the event that, after the ionization collision of an electron with energy $\varepsilon'$, the energy of one of the electrons is $\phi$ ($q \neq 0$ if $0 < \varepsilon < \varepsilon' - \phi_i$); the normalization condition for q is

$$\int_0^{\varepsilon' - \phi_i} d\varepsilon q(\varepsilon, \varepsilon') = 1. \qquad (7.1.12)$$

$\nu_i(\varepsilon)$ is the ionization frequency. The term $Q^*$ in (7.1.1) accounts for acts of excitation caused by non-elastic collisions between electrons and atoms (molecules), and

$$Q^* = Q_1^* + Q_2^* = -\sum_{j=1}^{J} n_e(\varepsilon)\nu_j^*(\varepsilon) + \sum_{j=1}^{J} n_e(\varepsilon + \phi^*)\nu_j^*(\varepsilon + \phi_j^*), \qquad (7.1.13)$$

where $\nu_j^*(\varepsilon)$ and $\phi_j^*$ are the frequency and the excitation energy of the j-th energetic level and J is the total number of levels $(\phi_j^* \quad \phi_i)$. The values $\nu_i(\varepsilon)$ and $\nu^*(\varepsilon)$ are determined by expressions similar to (7.1.9), but with the term $\sigma_m$ substituted for by the corresponding cross-sections of ionization $\sigma_i(\varepsilon)$ and excitation $\sigma^*(\varepsilon)$.

The functions $Q_r$ and $Q_a$ for the case of a single-component medium can be represented as follows: $Q_r = \lambda_r N_i n_e(\varepsilon)$ and $Q_a = \lambda_a N_a n_e(\varepsilon)$, where $N_i$ is the number density of ions $(N_i \approx N_e)$; $\lambda_r$ and $\lambda_a$ are the frequencies of electron recombination and electron capture per unit number density of electrons and heavy particles of every kind.

The right-hand-side of the equation of energy conservation for ionized gas (7.1.2) involves terms that account for thermal sources due to light absorption by heavy particles (absorption coefficient $\alpha_g$), the dissipation of the electron's kinetic energy in elastic electron-atom collisions, as well as the nonradiative (inelastic collisions between atoms) deactivation of the excited atomic levels with the termalization coefficient $\alpha^*$. The angle brackets < > denote the operation of averaging over the energy spectrum of the electron gas.

The plasma aureole appearing around the particle can function as an opaque screen for the laser radiation, being at the same time an additional heat source for the particle with plasma thermal radiation. These processes can be taken into account by introducing the particle's effective factor of light absorption into the heat balance equation and by correcting the term in this equation that describes the radiative heat exchange [5]. However, the estimates made in [5] showed that the above effects are significant only in the final stage of cascade ionization and heating of the vapor aureole, when the role of the initial aerosol particles in the further development of the breakdown wave in the surrounding air is neglibile. The system of equations (7.1.1)-(7.1.13), together with (2.3.11)-(2.3.13), describes the thermodynamic and ionization processes in a single-component vapor. In the case of a physical mixture, or a chemical combination of different substances, the corresponding generalization of the solution can be made by summing, where necessary, the partial contributions of different atoms (molecules) in (7.1.1)-(7.1.15).

Below we present the results of the approximate analysis of the problem of the low-threshold optical breakdown of aerosols, including calculated data for the case of corundum aerosol particles ($AL_2O_3$).

## 7.2. THEORETICAL ANALYSIS OF PULSED OPTICAL BREAKDOWN OF SOLID AEROSOL PARTICLES

### 7.2.1. Evaluations of the Order of Magnitude

As it follows from the above discussion, low-threshold optical breakdown of solid light-absorbant aerosols can occur if at least two fundamental conditions are fulfilled. The first one requires the particles to be heated to a temperature corresponding to well-developed vaporization, $T_b$. The corresponding threshold intensity of the laser beam, in the case of small particles ($k\kappa_a a \ll 1$), can be estimated according to [8] by:

$$I_b \approx T_b \lambda_T(T_b)(4k\kappa_a a^3)^{-1}[1 + O(t_3/t_p)],$$

where $\lambda_T(T_b)$ is the coefficient of molecular thermal conductivity of air, at a temperature $T_b$; a is the particle's radius, $\kappa_a$ is the absorption coefficient of the complex refractive index of particulate substance; $t_p$ is the laser pulse duration, and $t_3$ is the characteristic time of the particle's heating to the maximum temperature (see §6.1.1).

The second condition demands that the rate of energy consumption by electron gas from the field of intense light waves be higher than the rate of energy loss due to elastic collisions of electrons with the uncharged particles of the vapor. The threshold intensity $I_{av}$ for avalanche ionization to take place in the atomic vapor can be expressed, according to [1, 9] as

$$I_{av} \approx \frac{cM_e^2}{2\pi e^2 M_a^*} \cdot \left\langle \frac{\omega^2 + \nu_m^2}{\nu_m} \right\rangle \langle \varepsilon \nu_m \rangle, \tag{7.2.1}$$

where $M_a^*$ is the reduced mass of the multicomponent vapor particles;

$$M_a^* = \sum_k c_k M_k / (1 + 0.5 \sum_i \sum_k c_i c_k (M_i - M_k)^2 / (M_i M_k));$$

where $c_k = N_k/N_a$ is the relative concentration of the k-th vapor component of the mass $M_k$.

In this case, the breakdown threshold $I_{th}$ in the vapor aureole around the particle can be defined as

$$(I_{th})_{vap} = \max\{I_b, I_{av}\}.$$

The characteristic breakdown time is determined only by the process of cumulative ionization and elastic collisions, and its evaluation can be made using

$$t_{th}(I) \approx \ln (N_a/B_{e0}) \left[ (I - I_{av}) \frac{4\pi e^2 \langle \nu_m \rangle}{cM_e \omega^2 \phi_i} \right]^{-1}, \tag{7.2.2}$$

where $\phi_i$ is the ionization energy of the atoms; $N_{e0}$ is the initial number density of the prime thermoelectrons in the vapor aureole, as determined by the Saha formula.

For making the estimate, one can take the reduced energy of the lowest excited states of vapor atoms as the characteristic energy of the electron gas ($\varepsilon^* \approx \sum_k c_k \phi_k^* \simeq 3$ eV). Then the estimations made for $\lambda = 10.6$ μm, $\phi_i = 6$ eV, $(N_a(N_{e0}) \approx 10^6$, $T_b \approx 2.7 \times 10^3$ K, $\langle\nu_m\rangle \approx 1.6 \times 10^{-7}$ $N_a$ $(cm^{-3}) \approx 2 \times 10^{12}$ $s^{-1}$, give $I_b \approx 10^6$ to $10^7$ W/cm$^2$, $I_{av} \approx 10^7$ W/cm$^2$, $t_{th}(2I_{av}) \approx 5 \times 10^{-6}$ s.

The threshold conditions for the breakdown wave's escape into the ambient air, $(I_{th})_{air}$, can be formulated by taking the action of two mechanisms, viz., the avalanche ionization of the gas molecules at the threshold intensity $I_{av}^m$ or by the heating and compression of the region of prime breakdown of the vapor aureole to such a value of temperature T and pressure p at which the plasma becomes a strongly-absorbant medium. The corresponding threshold intensity is denoted by $I_{ab}$. In this case the escape of the discharge into the air has a thermal nature (the regime of a subsonic thermal conductivity wave or a light-induced detonation shock wave).

The region of prime breakdown is the source of vacuum UV radiation and soft x-rays, which initiate pre-ionization and dissociation of air molecules. These effects can play an important role in increasing the speed of the optical discharge [25] in the case of short, high-power pulses (pulse duration $t_p \sim 10^{-7}$ to $10^{-8}$ s).

The threshold intensity of laser radiation necessary for the initiation of the light-induced breakdown of air in the vicinity of aerosol particles can be defined as

$$(I_{th})_{air} \approx \min\{I_{av}^m, I_{ab}\}.$$

Normally, the value of the threshold intensity of avalanche ionization in air $I_{av}^m$ exceeds that in atomic vapors by one or two orders of magnitude. This is due to the fact that the main atmospheric gases, ($N_2$, $O_2$, and $CO_2$) have low-energy vibration-rotation levels; this leads, as one consequence, to significant energy losses for the electrons on the excitation of these levels, and this suppresses the avalanche. Moreover, the capture of electrons by heavy molecules ($O_2$, $CO_2$, and others [1]) also hinders the avalanche of electron generation in 'cold' gas. Thus, for $t_p \gtrsim 10^{-5}$ s, the value of $I_{av}^m$ is about $2 \times 10^9$ W/cm$^2$ when $\lambda = 10.6$ μm [2].

Now, we will give the method for estimating the threshold intensity of the beam $I_{ab}$ for the thermal mechanism of air breakdown.

The plasma appearing as a result of the initial optical breakdown of the vapor aureole is essentially not in equilibrium. In such a plasma the effective temperature of the electron gas $T_e = 2\langle E\rangle/3K$ strongly differs from the temperature of the heavy particles [4]:

$$T \approx (t - t_{th}(I))\langle\varepsilon\nu_m\rangle N_e M_e (c_{p(v)} \rho M_a)^{-1}, \tag{7.2.3}$$

where $\rho$ is the gas density; $c_{p(v)}$ is the specific heat of air (at constant) pressure or volume, depending on the character of the process; $t_{th}(I)$ is the time of the avalanche ionization of the vapor, determined by (7.2.2).

For this mechanism of air breakdown to work it is necessary for the gas temperature to rise to the critical value during the action of the laser pulse ($T_{cr} \simeq (1.5\text{-}2.0)\times 10^4$ K), at which the absorption coefficient of the plasma reaches significant values, e.g., $\alpha_g \approx 1$ to $10^2$ cm$^{-1}$; this corresponds to pressures $p \approx 1$ to $10^2$ atm.

The threshold intensity $I_{ab}$ is estimated using an expression derived from (7.2.2) and (7.2.3):

$$t_p - t_{th}(I_{ab}) = T_{cr} c_{p(v)} \rho M_a (N_e M_e \langle\varepsilon\nu_m\rangle_{vapor})^{-1}.$$

In order to obtain estimates for the isobaric process, one assumes that $\rho \approx \rho(T_{cr})$, while, in the case of an isochoric process, one can take $\rho \approx \rho_0$ ($\rho_0$ is the density of undisturbed air).

As follows from the above relationship, the threshold intensity of light necessary for air breakdown $(I_{th})_{air}$ can only slightly exceed the value $(I_{th})_{vap}$ if the laser pulse duration $t_p$ greatly exceeds the value $t_{th}$.

The simplified expression for the threshold intensity $I_{ab}$ necessary to form the plasma in air under the regime of quasi-stationary shock wave propagation was suggested in [30]:

$$I_{ab} \approx (2p^{3/2}/\rho_0^{1/2})[(\gamma + 1)/(\gamma - 1)][1 - \exp(-\tau)]^{-1}, \tag{7.2.4}$$

this can be approximated by the following expression:

$$I_{ab}(\text{MW/cm}^2) \approx 10^3[1 + 6.6 \log (r_p\ (\mu\text{m})\ /\tau)]^{-1}.$$

Here, p is the mean pressure of the plasma, determined by its temperature T and its density $\rho = \rho_0$ according to the equation of state (7.1.5); $r_p$ is the radius of plasma formation; $\tau = d_g r_p$ is the optical depth of absorption of the plasma; and $\gamma$ is the adiabatic exponent.

The estimates made in [30] gave the following values for the parameters of plasma in vapor aureoles and the corresponding parameters of the threshold intensity necessary for air breakdown: (a) $I_{ab} = 2.7\times 10^7$ W(cm$^2$; $T = 0.9\times 10^4$ K; $p = 54$ atm; $(\gamma - 1) = 0.177$; $r_p/\tau = 1.1$ cm ($\alpha_g = 0.9$ cm$^{-1}$). (b) $I_{ab} = 1.1\times 10^8$ W/cm$^2$, $V \approx 2$ km sec$^{-1}$; $T = 1.5\times 10^4$ K; $p = 126$ atm. $(\gamma - 1) = 0.148$; $r_p/\tau = 16$ μm ($\alpha_g = 624$ cm$^{-1}$).

Note the conclusion that there is another possibility for lowering the threshold value of the laser beam intensity $I_{th}$ necessary for the breakdown of air containing large, fast-melting aerosol particles in high concentrations ($N_a \gtrsim 10^4$ cm$^{-3}$) [26].

The thermal interaction of gas-vapor aureoles of individual absorbing centers can result in cooperative temperature instabilities and, hence, breakdown can start when the beam's intensity reaches a value sufficient for the well-developed vaporization of the particulate matter, thus omitting the stage of avalanche ionization. The duration of laser action must be longer than the characteristic time of overlapping of the vapor aureoles in the gaps between particles, $t \gtrsim (4N_0^{2/3}D_n)^{-1} \approx 10^{-3}$ s. The mass of the particles must initially be sufficient to fill the gaps between them with the products of vaporization ($\bar{a} \gtrsim (4 \times 10^3\ N_0)^{-1/3} \approx 10^{-3}$ to $10^{-2}$ μm).

Such conditions can hardly be met in the atmosphere, however, they can be successfully realized in laboratory conditions or technological installations.

### 7.2.2. The Analysis of Avalanche Ionization Processes in the Vapor Aureoles of Light-Absorbing Particles

The joint solution of the kinetic equation (7.1.1), for the energy spectrum of the electrons, and equations (7.1.2)-(7.1.18) cannot be obtained in an analytical form if no simplifications are made. The basic simplification is in the assumption of a quasi-stationary character of the ionization process. In this case, the energy distribution function of the electrons can be represented as follows [1, 5]:

$$N_e(\varepsilon, \vec{r}, t) = N_e(\vec{r}, t) f_e(\varepsilon).$$

The assumption of a quasi-stationary character of the ionization process can then be considered to be valid if the characteristic time of the electrons' number density changes exceeds the relaxation time $t_e$ of the normalized energy spectrum of electrons, $f_e(\varepsilon)$. Here, it is also assumed that $f_e(\varepsilon)$ depends on T and r parametrically. Estimates show that $t_e \approx \max\{\langle\nu_m\rangle, \langle\nu^*\rangle\} \approx 10^{-10}$ to $10^{-7}$ s.

The equation for the number density of the electrons $N_e(\vec{r}, t)$ can be obtained by integrating the kinetic equation (7.1.1) over the range of energies, and has the form [5]

$$\frac{dN_e}{dt} - \nabla(D_e \nabla N_e) + \nabla(\vec{V} N_e) = [\langle\nu_i\rangle - \langle\nu_r\rangle - \langle\nu_a\rangle] N_e. \qquad (7.2.5)$$

Then, neglecting the derivative $df_e/dt$, one obtains

$$[\langle\nu_i\rangle - \langle\nu_r\rangle - \langle\nu_a\rangle] f_e(\varepsilon) = \frac{d}{d\varepsilon}[A\varepsilon^{3/2} \frac{d}{d\varepsilon} \frac{f_e(\varepsilon)}{\varepsilon^{1/2}}] + $$

$$+ 2\varepsilon f_e(\varepsilon)\nu_m(\varepsilon) M_e/M_a + Q/N_e, \qquad (7.2.6)$$

where $Q = Q_i + Q^* + Q_r + Q_a$; $\nu_r = \lambda_r N_i$; $\nu_a = \lambda_a N_a$.

Equation (7.2.2) can be solved analytically if some additional simplifications are made, while a solution of (7.2.6) can be found numerically.

In [5] one can find an example of the calculation of $f_e(\varepsilon)$ for the particular case of two-component vapor of corundum ($Al_2O_3$). As is known, the vaporization of $Al_2O_3$ follows the scheme $Al_2O_3 \rightarrow 2Al + 3O$. Small amounts of other compounds can also be present in the vapor, but their influence can be neglected. Thus, as a result, we have a two-component gas with the components having quite different properties. The atoms of aluminium have low ionization and excitation thresholds ($(\phi_i)_{Al}$ = 5.984 eV and $(\phi^*)_{Al}$ = 3.14 eV), while the oxygen atoms are more stable ($(\phi^*)_O$ = 9.15 eV and $(\phi_i)_O$ = 13.614 eV). Taking into account the rapid fall of the function $f_e(\varepsilon)$ at high energies, one can neglect the ionization of oxygen atoms due to electron impact. Thus, atomic oxygen can be considered to be a buffer gas which affects the shape of $f_e(\varepsilon)$ but takes no part in the process of avalanche ionization.

Taking into account the above, one can write the functions of electron sources in (7.2.6) as follows:

$$Q_i/N_e = [-f_e(\varepsilon)(\nu_i(\varepsilon))_{Al} + 2\int_{\varepsilon+\phi_i}^{\infty} d\varepsilon'\, q(\varepsilon, \varepsilon')f_e(\varepsilon')(\nu_i(\varepsilon'))_{Al}];$$

$$Q^*/N_e = \sum_{J=1}^{J_O,J_{Al}} [-f_e(\varepsilon)\nu_j^*(\varepsilon) + f_e(\varepsilon + \phi_j^*(\varepsilon))\nu_j^*(\varepsilon + \phi_j^*(\varepsilon))];$$

$$Q_r/N_e = -f_e(\varepsilon)(\nu_r(\varepsilon))_{Al}.$$

The frequency of striking $(\nu_\alpha)_O$ can be neglected, owing to the fact that the vapor's temperature is high enough to cause significant initial thermal ionization [5].

All the frequencies $\nu(\varepsilon)$ can be expressed in terms of cross-sections $\sigma(\varepsilon)$ of the relevant processes as $\nu(\varepsilon) \approx N_a\sigma(\varepsilon)(2\varepsilon/M_e)^{1/2}$, where $N_a$ is the number density of the atoms involved in this process.

For solving these problems one needs knowledge of the ionization cross-section of aluminium atoms in their ground state, the excitation cross-sections for both atoms $(\sigma_j^*)_{Al,O}$, the recombination cross-section $(\sigma_r)_{Al}$, and the cross-section of elastic collisions $\sigma_m$ between all particles, including both atoms and ions. Both experimental and theoretical [15] estimates of the cross-sections were used. The energy distribution between an incident (ionizing) electron of energy $\varepsilon'$ and the electron appearing as a result of the ionization process is described by the function $q(\varepsilon, \varepsilon') = (\varepsilon' - \phi_i)^{-1}$.

It was also assumed that the decay of the excited atomic states is very rapid, due to the great efficiency of spontaneous emission and inelastic colissions between the atoms, therefore the concentration of the excited atoms can be neglected. Thus, the gas can be divided into three components, viz., uncharged atoms of aluminium ($N_{Al} - N_i$), neutral oxygen atoms, and singly-charged ions of aluminium $N_i \quad N_e$.

As a result, (7.2.6), in the case of a two-component system, takes the form

$$f_e(\varepsilon)\int_0^\infty d\varepsilon'(2\varepsilon'/M)^{1/2}[\sigma_i(\varepsilon')(N_{Al} - N_i) - \sigma_r(\varepsilon')N_i] =$$

$$= \frac{\partial}{\partial\varepsilon}[A\varepsilon^{3/2}\frac{\partial}{\partial\varepsilon}\frac{f_e(\varepsilon)}{\varepsilon^{1/2}} + (2M_e/M_a)\varepsilon f_e(\varepsilon)(2\varepsilon(M_e)^{1/2} \times$$

$$\times [(\sigma_m)_{Al}(N_{Al} - N_i) + (\sigma_m)_{Al}N_i + (\sigma_m)_O N_O]] + Q/N_e. \qquad (7.2.7)$$

The procedure for solving this integro-differential equation is described in [5].

This procedure is based on iteration. The normalization condition is $\int_0^\infty f_e(\varepsilon)\, d\varepsilon = 1$. The initial iteration was taken in the form of Maxwell or Margenow distributions [1].

The results of the calculations are the energy-distribution function of electrons $f_e(\varepsilon)$, the mean frequencies of ionization and excitation of aluminium and oxygen atoms, and the mean frequency of recombination, i.e. $\langle\nu_i\rangle$, $\langle\nu^*\rangle_{Al}$, $\langle\nu^*\rangle_O$, $\langle\nu_r\rangle_O$. Also, the calculations furnished the value of $\langle\varepsilon\nu_m\rangle$ which determines the rate of heating of the gas due to elastic collisions.

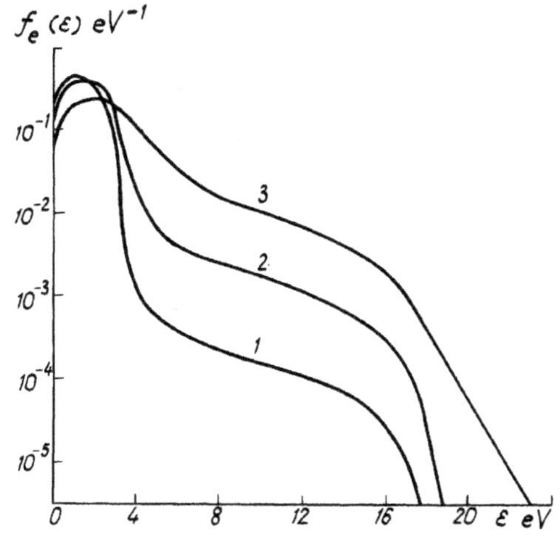


Fig. 7.2.1. The energy-distribution functions of electrons $f_e(\varepsilon)$ (normalized by 1), calculated for different intensities of incident radiation: (1) $I = 10^7$ W/cm$^2$, (2) $I = 10^8$ W/cm$^2$, (3) $I = 10^9$ W/cm$^2$.

Figure 7.2.1 presents the function $f_e(\varepsilon)$ calculated for different intensities of incident radiation. These calculations took into account the change of the initial number density of the electrons due to the increase in vapor temperature. As can be seen from this figure, an increase in the beam's intensity results in a decrease in the maximum of $f_e(\varepsilon)$. This can be explained by the growth of rate of the energy consumption by electrons.

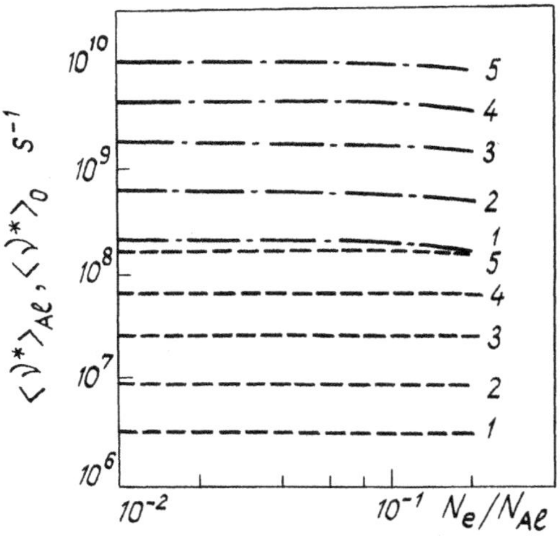


Fig. 7.2.2. Mean frequencies of excitation of Al and O atoms calculated as functions of the degree of plasma ionization $N_e/N_{Al}$ at different intensities of incident radiation: (1) $I = 10^7$ W/cm$^2$, (2) $I = 3 \times 10^7$ W/cm$^2$, (3) $I = 10^8$ W/cm$^2$, (4) $I = 3 \times 10^8$ W/cm$^2$, (5) $I = 10^9$ W/cm$^2$.

Figure 7.2.2 presents the mean excitation frequencies $\langle\nu^*\rangle_O$, $\langle\nu^*\rangle_{Al}$ as functions of incident radiation. As was to be expected, the value $\langle\nu^*\rangle_{Al}$ decreased with an increase in the degree of ionization, since it depends on the number density of neutral Al atoms. The fall in $\langle\nu^*\rangle_O$ can be explained by the decrease in the number of high-energy electrons caused by the growth of energy losses due to elastic collisions, caused by an increase in $N_i$ due to the higher values of the cross-section of elastic collisions $(\sigma_m)_{Al}$ at higher electron energies.

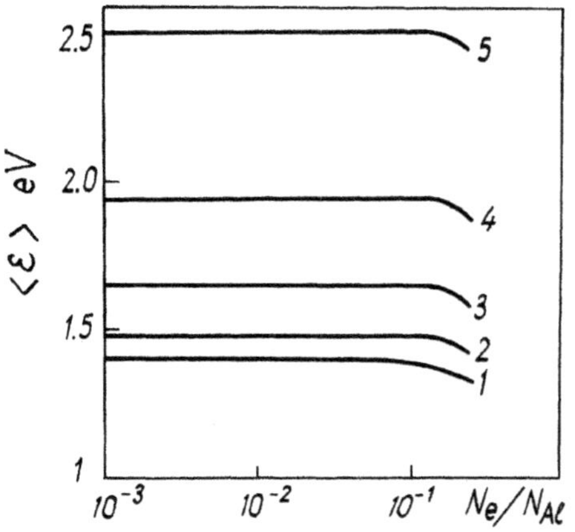


Fig. 7.2.3. The mean energy of electrons <ε> as a function of the degree of ionization of the plasma $N_e/N_{Al}$, calculated for different intensities of incident radiation: (1) $I = 10^7$ W/cm$^2$, (2) $I = 3 \times 10^7$ W/cm$^2$, (3) $I = 10^8$ W/cm$^2$, (4) $I = 3 \times 10^8$ W/cm$^2$, (5) $I = 10^9$ W/cm$^2$.

Figure 7.2.3 shows the analogous dependence of $\langle\varepsilon\rangle$ on $N_e/N_a$. The dependence of $\langle\varepsilon\rangle$ with $N_e/N_a$ can also be explained by the growth of energy losses due to elastic collisions.

It should be noted that knowledge of the frequencies of atomic level excitations can be useful for spectrochemical analysis of the composition of the aerosol particle vapor, as well as the ambient air. In principle, it is possible to make not only qualitative, but also semi-quantitative, analyses, especially of small admixtures which have only a weak effect on the shape of $f_e(\varepsilon)$. For a detailed discussion of this problem the reader is referred to Chapter 8 of this book.

The calculated results of $f_e(\varepsilon)$ and the weighted mean parameters of initial plasma formation around the aerosol particle can be used for solving the second part of the problem relating to the analysis of the dynamics of plasma evolution.

Let us seek a solution of (7.2.5), assuming that its coefficients are constant values independent of the ionization process. This assumption is quite justified because of the fact that the vapor cloud reaches a size of the order of the Debye radius $R_D(\text{cm}) = 6.9\, T_e^{1/2}\,(\text{K})\, N_e^{-1/2}\,(\text{cm}^{-3})$ sufficiently rapidly, this allows one to neglect the rate of ambipolar electron diffusion in comparison with the ionization rate (here, $T_e$ is the effective temperature of the electron gas). Further simplifications can be introduced if two limiting regimes of ionization are considered, viz., isobaric and isochoric regimes. In the former, the rates of ionization and heating of the plasma are so low that the pressure of the plasma is practically constant. The necessary condition for this is

$$r_p/C_s' \ll \langle\nu_i\rangle^{-1} \ln[N_{Al}/N_{e0}(T_b)],$$

where $r_p$ is the radius of the plasma formation and $C_s'$ is the velocity of sound in the plasma. The isochoric regime can prevail when the ionization and heating rates are so high that the plasma formation does not expand to any noticeable extent before it is completely ionized.

Since avalanche ionization is accompanied by plasma heating, the equation for the energy flux in the gas should account for the thermal sources.

Denoting the energy density obtained by the gas from external sources as w, and neglecting the conductivity of heat, one can obtain the following equation from (7.1.5):

$$\frac{d}{dt}(\rho(c_v T + v^2/2)) = -\nabla(\rho\vec{v}(c_p T + v^2/2)) + \rho\vec{v}\nabla w + \rho(dw/dt). \quad (7.2.8)$$

The last two terms in (7.2.8) describe the volumetric heat source. If plasma heating is caused by elastic collisions with electrons, then

$$\rho(dw/dt) \approx \langle\varepsilon\nu_m(\varepsilon)\rangle N_e 2M_e/M_a + \alpha_g I.$$

Also, in the case of a weakly absorbant, homogeneous plasma, the last term but one can be neglected.

Under the regime of isobaric plasma heating the speed of plasma expansion is small and (7.2.8) takes the form

$$c_p\rho(dT/dt) \approx \rho(dw/dt). \quad (7.2.9)$$

Under the isochoric regime one can neglect the derivatives of density and speed with respect to time, as well as the speed itself, and then (7.2.8) takes the form

$$c_v\rho(dT/dt) \approx \rho(dw/dt). \quad (7.2.10)$$

Now, write (7.2.5) for both regimes, taking into account the dependence of $\langle\nu_i\rangle$ on $N_{Al}$, and representing, in the isobaric case, the third term of (7.2.5) in the form $\nabla(\vec{V}N_e) = (N_e/T)(dT/dt)$.

The resulting simplified equations are:

$$dN_e/dt + (N_e/T)(dT/dt) = [\langle\lambda_i\rangle(\mu_{Al}\, p/(KT) - N_e) - \langle\lambda_r\rangle N_e]N_e; \quad (7.2.11)$$

$$dN_e/dt = [\langle\lambda_i\rangle(N_{Al} - N_e) - \langle\lambda_r\rangle N_e]N_e \quad (7.2.12)$$

for isobaric and isochoric regimes of ionization, respectively. The values entering these equations are as follows: $\lambda_i = \nu_i/N_{Al}$; $\lambda_r = \nu_r/N_{Al}$ are the ionization and the recombination coefficients, respectively; and $\mu_{Al}$ is the volumetric fraction of Al atoms w.r.t. the total number of particles ($\mu_{Al} \approx 0.4$).

It is convenient to write the solution for the system of equations (7.2.9) and (7.2.11) in the form of the dimensionless variables y, z, and $\tau$, assuming also that the coefficients at $N_e$ and T are independent of time:

$$y = N_e \frac{2KM_e\langle\varepsilon\lambda_m\rangle}{c_p M_a^2 p\mu_{Al}\langle\lambda_i\rangle}; \qquad z = T\,\frac{c_p M_a^2(\langle\lambda_i\rangle + \langle\lambda_r\rangle)}{2M_e\langle\varepsilon\lambda_m\rangle};$$

$$\tau = (t - t_b)\,\frac{c_p P\mu_{Al} M_a^2\langle\lambda_i\rangle[\langle\lambda_i\rangle + \langle\lambda_r\rangle]}{KM_e\langle\varepsilon\lambda_m\rangle}; \tag{7.2.13}$$

where p is atmospheric pressure; $\lambda_m = \nu_m/N_a$ is the coefficient of elastic electron-atom collisions; and $M_a$ and $N_a$ are the mean molecular weight of the mixture and the number density of the neutral atoms, respectively.

Equations (7.2.9) and (7.2.11), written in the form of dimensionless variables, are

$$2(dy/d\tau) = y/z - y^2 - y^2/z^2; \tag{7.2.14}$$

$$2(dz/d\tau) = y. \tag{7.2.15}$$

By excluding $\tau$ from (7.2.14) and (7.2.15), one can write the equation for y(z) as

$$(dy/dz) - z^{-1} + y(1 + z^{-2}) = 0.$$

The solution of this equation is

$$y = (y_0 z_0 - 1)\exp(z_0) + \exp(z)(z\exp(z))^{-1}. \tag{7.2.16}$$

Then substituting (7.2.16) into (7.2.15), we obtain the equation for $z(\tau)$;

$$2(dz/d\tau) = [(y_0 z_0 - 1)\exp(z_0) - \exp(z)](z\exp(z))^{-1}, \tag{7.2.17}$$

where $y_0$, $z_0$ are the values of y and z at $\tau = 0$. The initial value of $N_{e0}$ was calculated using the Saha formula [1].

Equation (7.2.17) can be solved using the iteration method, with the zero-order approximation taken in the form $z^{(o)} = (z_o^2 + \tau)^{1/2}$. Then, for the next iteration $z^{(1)}$, one has

$$z^{(1)} = [\tau + z_0^2 + 2(y_0 z_0 - 1)\exp(z_0)]\times$$

$$\times\left(\frac{z_0 + 1}{\exp(z_0)} - \frac{(\tau + z_0^2)^{1/2} + 1}{\exp(\tau + z_0^2)^{1/2}}\right)^{1/2}. \qquad (7.2.18)$$

As estimates show, the iterations rapidly converge. Therefore, the first iteration can be quite sufficient for making estimates. For large values of $\tau$, the asymptotic gives $z \approx \tau^{1/2}$.

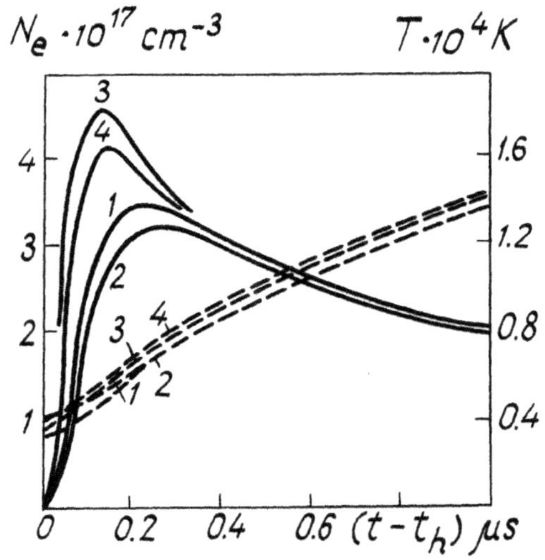


Fig. 7.2.4. Temporal behavior of the electron number density (solid curves) and gas temperature (dashed curves) in a plasma formation around a corundum particle, computed for different parameters of the particle and the incident radiation: (1) $a_0 = 10^{-6}$ cm, $I = 5\times10^7$ W/cm$^2$, (2) $a_0 = 10^{-5}$ cm, $I = 5\times10^7$ W/cm$^2$, (3) $a_0 = 10^{-6}$ cm, $I = 10^8$ w/cm$^2$, (4) $a_0 = 10^{-5}$ cm, $I = 10^8$ W/cm$^2$.

Figure 7.2.4 shows the behavior over time of $N_e$ and T under standard atmospheric conditions (values of atmospheric parameters at sea level and room temperature). The decrease in $N_e$, observed as beginning at a certain moment in time, is due to the thermal expansion of the plasma - the degree of ionization staying almost constant.

Now consider the system of equations (7.2.10) and (7.2.12). An approximate solution of this system is

$$N_e \approx N_{eo}\exp(\langle\nu_i^{(0)}\rangle t)[(\exp(\langle\nu_i^{(0)}\rangle t) - 1)\times$$

$$\times \frac{N_{eo}(<\lambda_i^{(0)}> + <\lambda_r^{(0)}>}{<\lambda_i^{(0)}>N_{Al}^{(0)}} + 1]^{-1}, \qquad (7.2.19)$$

$$T \approx T_0 + \frac{2M_e<\varepsilon\lambda_m^{(0)}>}{M_a^2\omega(<\lambda_i^{(0)}> + <\lambda_r^{(0)}>)} \times$$

$$\times \ln\left\{1 + (\exp(<\nu_i^{(0)}>t) - 1)\frac{N_{eo}(<\lambda_i^{(0)}> + <\lambda_r^{(0)}>}{N_{Al}^{(0)}<\lambda_i^{(0)}>}\right\}, \qquad (7.2.20)$$

where $\omega$ is the frequency of the incident radiation. The superscript '0' denotes the initial values of corresponding functions when $t = +t_b$; $t_b$ is the characteristic time necessary to heat the particle to its temperature of vaporization (see §2.3).

Asymptotically, as $t \to \infty$, the solutions take the form

$$N_e \to N_{Al}^{(0)}<\lambda_i^{(0)}>(<\lambda_i^{(0)}> + <\lambda_r^{(0)}>)^{-1}; \qquad (7.2.21)$$

$$T \to 2M_e t<\nu_i^{(0)}><\varepsilon\lambda_m^{(0)}>[\omega M_a^2(<\lambda_i^{(0)}> + <\lambda_r^{(0)}>)]^{-1}. \qquad (7.2.22)$$

Now consider the applicability of the two ultimate ionization regimes to the description of particle vaporization under the regime of metastable superheating (see (2.3.11)-(2.3.13)). For this it is necessary to take into account the characteristic times of the processes occurring in the vicinity of an aerosol particle during the action of the laser pulse. The characteristic heating time can be estimated as the time necessary to heat the particle to its maximum temperature $T_b$, taking the initial heating rate as

$$t_b \approx \min\{t_h, t_3\}; \qquad t_h = c_a\rho_a V_a(T_b - T_0)(\pi a^2 K_{ab} I)^{-1};$$

$$t_3 = (a^2/3X_T)(c_a\rho_a/c_p\rho_0), \qquad (7.2.23)$$

where $V_a$ is the volume of the particle. The time during which the radius of the vapor cloud reaches the value of the Debye radius can be estimated as the moment when avalanche ionization starts, under the above conditions and asymptotic situations, and can be represented as

$$t_D \approx R_D/C_S'. \qquad (7.2.24)$$

The values of the Debye radii are evaluated based on the vapor's parameters near the particle's surface at $t = +t_b$.

The next important parameter is the lifetime of the dense plasma formation close to the surface of a particle. This time determines whether or not it is possible to accomplish isochoric ionization in the dense vapor when the size of the vapor cloud exceeds $R_D$. This time $t_s$ can be estimated as the ratio of the characteristic vapor expansion distance to the expansion velocity, which is equal to the local speed of sound, i.e., $t_s \approx a/c_s'$.

A further two times are used as criteria of the breakdown process under both ionization regimes. Let us consider the time interval during which 50% of the aluminium atoms are ionized as the characteristic time of avalanche ionization (and, hence, of the breakdown), $t_{th}$. For the region of dense vapor we calculated the time of isochoric ionization t', while for the expanded vapor after the shock jump of isobaric ionization t" was calculated. By summing the durations of all the processes, one can estimate the time necessary for the breakdown in vapor with the consequent transition to the development of ionization in the ambient medium. Such estimates are presented in Figure 7.2.5 ($t_{th} \approx t_b + t_D + \min\{t', t''\}$).

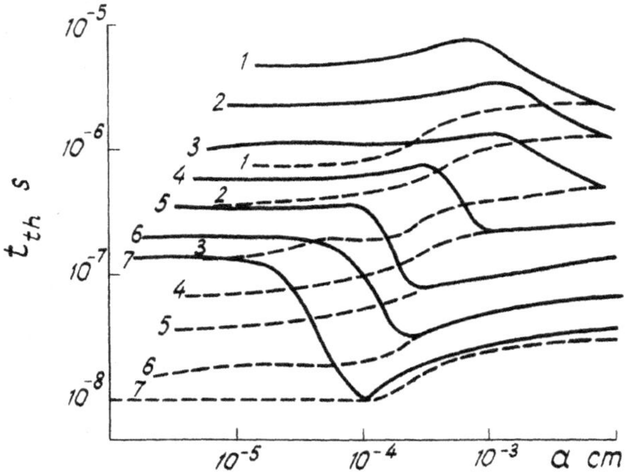


Fig. 7.2.5. The dependence of the characteristic times of prime optical breakdown of corundum particles on their radii, calculated for both isobaric and isohoric regimes of avalanche ionization induced by a $CO_2$ laser beam. Curves 1 to 7 represent the data calculated for beam intensities $I = 10^7$; $2 \times 10^7$; $5 \times 10^7$; $10^8$; $5 \times 10^8$; $10^9$; $5 \times 10^9$ W/cm$^2$, respectively. The actual time of breakdown is approximated by the solid line. The dashed curves represent the region of particle sizes in which the isohoric regime cannot occur.

The diagrams of characteristic times enable one to estimate the lower limit of the intensity of the laser beam necessary for breakdown to occur for different size particles and pulse durations t. Corresponding results were calculated for a corundum ($Al_2O_3$) particle, and are presented in Figure 7.2.6 in comparison with relevant experimental data from different authors [5, 9-11, 25]. The comparison shows that, within a spread of experimental data, the agreement between theory and experiment can be considered as being good enough.

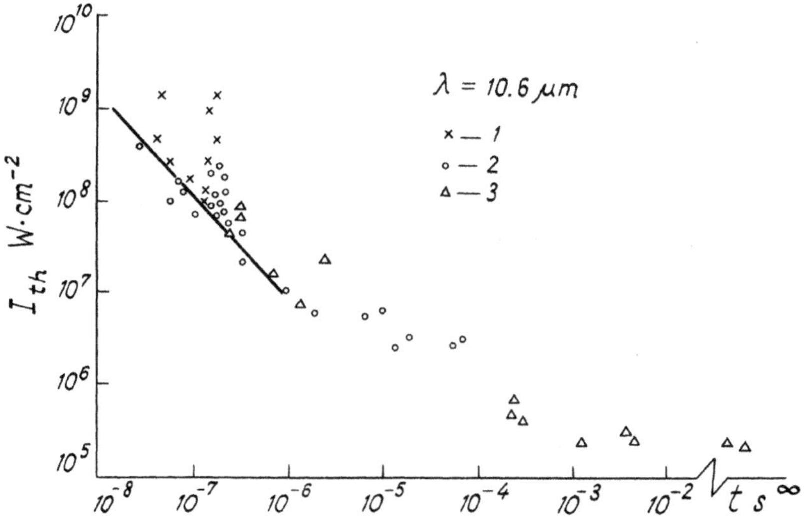


Fig. 7.2.6. The threshold intensity ($\lambda$ = 10.6 μm) necessary for optical breakdown in air as a function of laser pulse duration, according to [5, 9, 10, 11]: (1) represents data obtained in a room from air with an uncontrolled aerosol content; (2) solid aerosol particles of different chemical composition, with radii a ranging from 1 to 10 μm; (3) this curve represents data on the optical breakdown near the surface of macrotargets. The solid curve in this figure presents the theoretically-calculated data [5] for the case of corundum ($Al_2O_3$) particles.

The pattern of the evolution of the plasma formation during the second stage is the following. After vapor ionization, the resultant plasma cloud initiates the appearance of the so-called wave of 'burning', or wave of 'light detonation', depending on the intensity of the incident radiation. The former occurs at an intensity providing for low rates of plasma heating at which the plasma formation pressure is in equilibrium with the pressure of the surrounding medium. The intensities required for this case, $10^6$ to $10^7$ W/cm$^2$, are low enough and, as follows from the diagrams of characteristic times, the initial plasma cloud is also ionized under the isobaric regime.

At higher beam intensities, and for large particles, vapor heating and ionization can occur under the isochoric regime. The corresponding increase of pressure can cause values of the total pressure at the shock wave front such that the air surrounding the particle is heated to a temperature that provides for the development of a thermal wave of 'light detonation'.

The late stage of evolution of the ionization wave from the region of prime breakdown has been investigated numerically by several authors, see, e.g., [25, 26] and [30].

Numerical analysis is based on the solution of a system of thermohydrodynamic equations of the type (7.1.2)-(7.1.5), with the appropriate boundary and initial conditions. Thermodynamic parameters and the absorption coefficient of the plasma in the prime breakdown region of the vapor aureole form the initial conditions [25, 26] for the problem.

An approach to the determination of the boundary conditions has been suggested (in [26]); namely, modeling the inverse problem from the characteristics of undamped solutions for the plasma front's drift in air.

In [30] the authors suggested a simplified system of equations derived from the system (7.1.2)-(7.1.5), but written following the approach of uniform parameters over the plasma region and a quasi-stationary regime of plasma propagation:

$$\frac{d}{dt}(\rho v) = (p - p_0)4\pi r_p^2; \tag{7.2.25}$$

$$\frac{dU}{dt} = \frac{d}{dt}[\rho(q_0 + v^2/2)] = I\pi r_p^2[1 - \exp(4r_p\alpha_g/3)]; \tag{7.2.26}$$

$$\frac{dr_p}{dt} = v; \qquad p = \rho_0 KT/M_a, \tag{7.2.27}$$

where $\rho$ is the plasma density and $M_a$ is the mean mass of the plasma particles; $r_p$ is the radius of the plasma formation, p and $p_0$ are the mean pressure in the plasma and the pressure of non-perturbed air, respectively; U is the total energy; $q_0$ is the internal energy per unit mass; V is the plasma front speed; and K is the Boltzmann constant.

For calculating the functions $\alpha_g(P, T)$; $M_a(T)$; $q_0(T) = p[(\gamma - 1)\rho_0]^{-1}$, one can use tabulated data on air parameters at high temperatures.

In accordance with (7.2.25)-(7.2.27), it is assumed that the energy absorbed by the plasma from the laser beam is entirely spent on acceleration and heating of non-perturbed air captured by the shock wave front. The contribution of the plasma's self-emission to the heating of the air is neglected.

The estimation of the threshold intensity of laser beam necessary for breakdown, based on the solution of the system (7.2.25)-(7.2.27), can be determined by using an expression like (7.2.4). Typical results of the numerical solution of (7.2.25)-(7.2.27), in the case of a stationary regime of prime plasma formation growth in the light field of $CO_2$ laser radiation ($I = 10^8$ W/cm$^2$), are as follows [30]:

(a) $t = t_{th}$; $I_{ab} = 6.1 \times 10^7$ W/cm$^2$; $r_p = 10^{-2}$ cm; $T = 1.2 \times 10^4$ K; $p = 87$ atm; $\alpha_g = 41$ cm$^{-1}$; $U = 2.3 \times 10^{-4}$ J.

(b) $t = t_{th} + 10^{-7}$ s; $r_p = 0.38$ cm; $T = 1.42 \times 10^4$ K; $p = 120$ atm; $\alpha_g = 370$ cm$^{-1}$; $v = 3$ km/s; $U = 1.9 \times 10^{-2}$J.

It should be noted that, if the material of the particles has very weak absorption characteristics at the wavelength of incident radiation, then a situation can occur under which the intensity of the beam is insufficient for heating the particle's surface to the temperature of well-developed vaporization $T_b$ but, at the same time, it greatly exceeds the threshold intensity necessary for avalanche ionization in the presence of prime electrons in the medium. The initiation of the breakdown in this situation can be caused by electrons generated due to thermo-emission, microdischarges of static electricity accumulated due to the thermal splitting of a solid crystal, as well as by electrons appearing due to the multi-photon ionization of absorbed admixtures with low ionization energies. Some other effects can also contribute to the initiation of breakdown; the roles of these have not yet been studied.

## 7.3. THE INFLUENCE OF ATMOSPHERIC TURBULENCE ON THE CONCENTRATION OF OPTICAL BREAKDOWN CENTERS

Random redistribution of intensity over the beam's cross-section due to turbulence is characterized by the appearance of intensity spikes. The knowledge of the basic relationships determining the appearance of these spikes is very important in the study of the possibilities of initiating and carrying out the breakdown remotely. Below we give estimates of the concentration of breakdown centers for the case of a focused beam in a turbulent atmosphere [7].

Theoretical investigations into intensity spikes have been carried out in a number of papers, see, e.g., [16].

In [17] the authors derived an approximate formula for describing the random intensity spikes. This formula contains the empirical values, viz., the effective beam radius $R_e = (\langle S(I_1)\rangle/\pi)^{1/2}$, where $S(I_1)$ is the area of the beam's cross-section in which the intensity exceeds a certain preset level $I_1$ and the structural constant of the refractive index field $C_n^2$. The paper [16] presents the results of measurements of $R_e$ in the atmospheric ground layer. The measurements were carried out along atmospheric paths 180 and 650 m long and 1.5 to 2 m above the uniform plane surface of a steppe. The comparison made between theoretical and experimental results revealed the usefulness of the formula obtained in [17] for making numerical estimations of the beam's cross-sectional area, where the intensity of the spikes exceeds, on average, a preset value. In the case of a lognormal law of intensity probability distribution, the effective radius of the beam's cross-section can be found using

$$(R_e/R^*)^2 = \int_0^\infty dy_1 y_1 (1 - \phi \frac{\ln(g/f(y_1, \mu)) + (\chi^2/2)}{2^{1/2}\chi}), \qquad (7.3.2)$$

where g is the ratio of the preset intensity level $I_1$ to the intensity maximum of the ideal diffraction picture; $R^*$ is the radius of the ideal diffraction picture at half-maximum level; $y_1 = r_\perp/R_1$; $\chi$ is the variance of the logarithm of the beam's intensity; $\phi$ is the probability integral; $r_\perp$ is the radial coordinate; and $f(y_1, \mu)$ is the normalized mean intensity of the focused beam in the focal plane. The value of $f(y_1, \mu)$ was calculated for an initially Gaussian beam as follows:

$$f(y_1, \mu) = 2\int_0^\infty dt\ t\ J_0(2(\ln(2y_1 t))^{1/2}) \exp(-t^2 - (\mu^2 t^{5/3}/2)). \qquad (7.3.2)$$

Here, $J_0$ is the zero-order Bessel function; $\mu^2 \approx 1.1\ C_n^2\ xk^2(2R_0)^{5/3}$; x is the the path length; $k = 2\pi/\lambda$ is the wavenumber; and $R_0$ is the effective width of the Gaussian beam.

In order to carry out a comparison between the experimental results obtained with the uniform beam and the calculated results valid for a Gaussian beam, we have introduced the equivalent radius $R_0$ of a Gaussian beam, so that the radii of the ideal diffraction pictures of the focused beam operating at half-maximum level are equal. The effective $R_0$ and actual $R_1$ beam radii are related to each other as follows: $R_0 \approx 0.5\ R_1$.

Calculations were carried out, taking into account the above simplifications, of the probability of the occurrence of the breakdown centers in the region of the beam's caustics. As is known, the intensity distribution in the focal spot of a Gaussian beam in a vacuum has the form

$$I(x, r_\perp) = I_0(x/kR_0^2))^2 \exp(-r^2(x/kR_0))^{-2}). \qquad (7.3.3)$$

From this expression one can find $R^*$:

$$R^* = \ln 2(x/(kR_0^2))^2. \qquad (7.3.4)$$

Taking a certain value of $I_0$ and the breakdown threshold intensity $I_{th}$, one can determine the parameter g at which the breakdown can occur:

$$g_{th} = (I_{th}/I_0)(x/(kR_0^2))^2. \qquad (7.3.5)$$

Determining the ratio $(R_e/R^*)^2$ from (7.3.1), one then finds that the expression for the probability density of the appearance of $N_p$ breakdown centers per unit length is

$$N_p = \pi \ln 2N_0(R_e/R^*)^2(x/(kR_0^2))^2, \qquad (7.3.6)$$

where $N_0$ is the number density of the aerosol particles capable of initiating the breakdown ($N_0$ is assumed to be significant in the region occupied by the intensity spike ($I \gtrsim I_{th}$)). If the number density $N_0$ is too small, then one should take into account the joint probability of two events when calculating the value of $N_p$ viz., the appearance of a spike with intensity $I \gtrsim I_{thr}$ and the presence of an aerosol particle in the region of the spike.

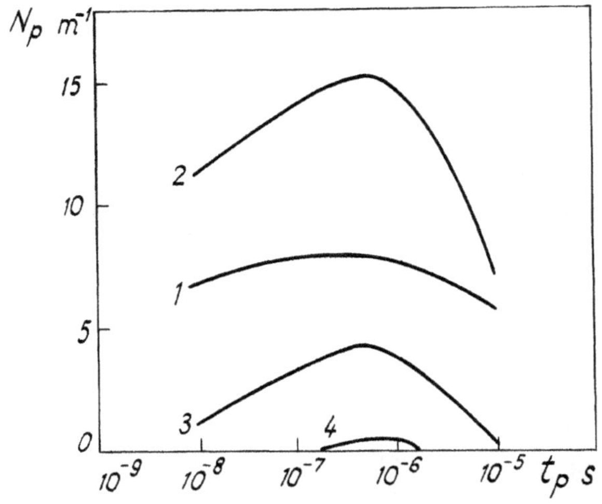


Fig. 7.3.1. Theoretically calculated number density of prime centers of optical breakdown as a function of pulse duration under different conditions of atmospheric turbulence. (1) $\mu = 0$; (2) $\mu = 2$; (3) $\mu = 10$; (4) $\mu = 25$.

Figure 7.3.1 presents the dependence of $N_p$ on the laser pulse duration $t_p$, calculated in [7] for the following beam parameters: beam energy $W_0 = 10$ J, $R_0 = 0.2$ m, $N_0 = 10^{-1}$ $cm^{-3}$, $x = 100$ m.

The calculations made used the experimentally-determined (based on the data presented on Figure 7.2.6) dependence of the threshold intensity for optical breakdown on the aerosol particles on the laser pulse duration.

As seen from Figure 7.3.1, atmospheric turbulence strongly affects the formation of plasma in a sharply focused beam. It should be noted also that, for a fixed pulse energy, there exists an optimal pulse duration of about 0.8 s at which the expected number of breakdown centers is at a maximum [7]. This is due to certain peculiarities of the dependence of $I_{th}$ on the pulse duration $t_p$.

## 7.4. LABORATORY EXPERIMENTS ON LASER SPARKING

In order to study the process of ionizing aerosols made of solid, strongly absorbant particles, we have carried out laboratory experiments.

An electro-ionization $CO_2$ laser ($\lambda$ = 10.6 μm) with tunable pulse duration (analogous to that used in [18]) was used in our experiments as a source of high-power radiation.

The tuning of the pulse duration in the laser was performed by varying the pressure of the working gas mixture within the range 1 to 10 atm. The pulse shape was asymmetric, with a short leading edge of 70 to 300 ns duration and a long trailing edge of 200 to 800 ns duration. The leading part of the pulse contained about 75% of the total pulse energy. The beam was focused into the aerosol chamber using a $BaF_2$ lens with a focal length of 80 cm.

The experiments were carried out with polydispersed aerosol ensembles of three different substances, viz., $Al_2O_3$, $Na_2CO_3$, and $SiO_2$, which significantly differ from each other in boiling point and dissociation temperature, as well as ionization and excitation potentials of the respective vapors. The size spectra were determined using the data from microphotography. The optical depth of the aerosol along the path of the high-power beam was $\tau \approx 0.08$ with an average number density $N_0 \approx 10^2$ cm$^{-3}$ which was kept constant during the series of experiments.

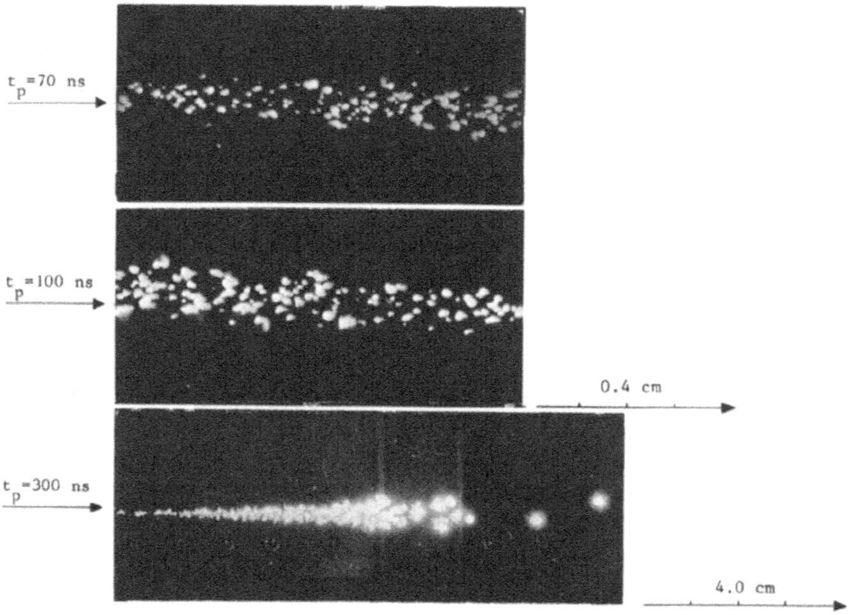


Fig. 7.4.1. Microphotographs of the beam channel ionized by pulsed $CO_2$ laser in a polydispersed aerosol of $Na_2CO_3$ particles.

The energy and spectral characteristics of the plasma formation were studied using data collected by microphotographing the glowing channel and from records of the plasma emission spectrum.

The microstructure of the regions of ionization was investigated using data collected by microphotographing the beam channel at different laser pulse durations. Figure 7.4.1 presents photographs of the beam channel with its plasma inhomogeneities, taken for the following durations of the leading edge of the laser pusle: 70, 200, and 300 ns. The aerosol particles were $Na_2CO_3$ with a number density of about $10^2$ $cm^{-3}$ and a r.m.s. radius $a_{sq}$ = 4 μm.

The mean (taken over the beam's cross-section) power density was kept constant (≈ 10 $J/cm^2$) for different pulse durations. As can be seen from the photo's, the number density of the plasma inhomogeneities increases with any decrease in pulse duration. This can be explained by the fact that, with a constant pulse energy, the critical size of aerosols initiating the plasma formation diminishes with diminishing pulse duration.

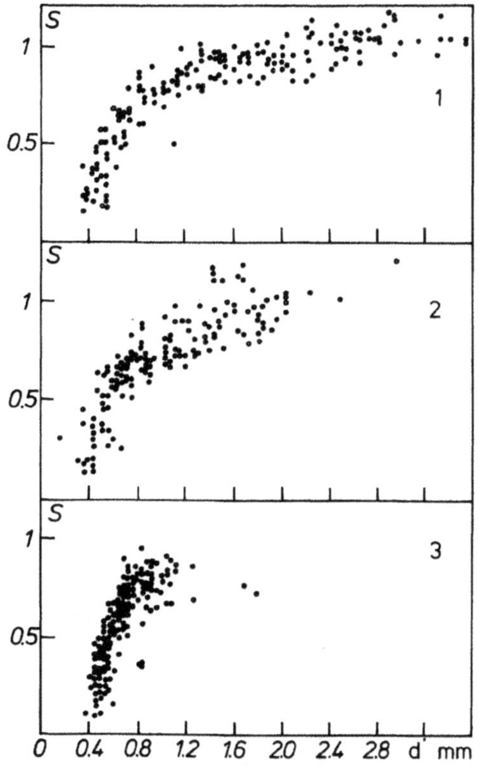

(a)

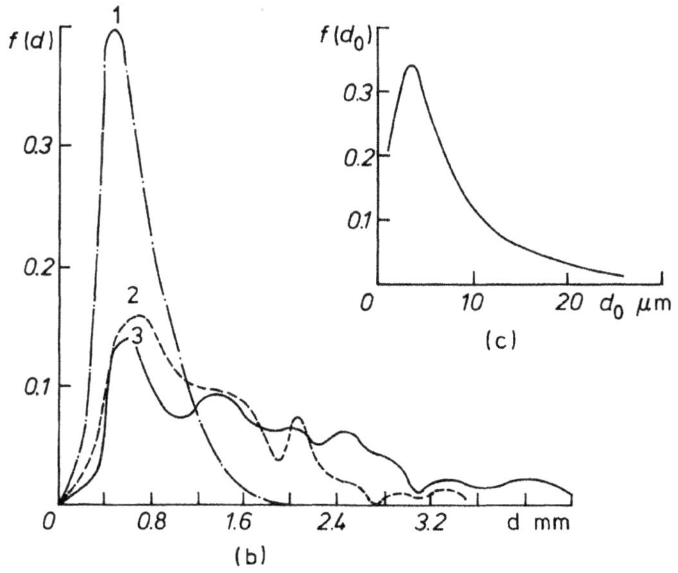


Fig. 7.4.2. The microstructure of the optical breakdown regions.

Figure 7.4.2 presents the results of a statistical analysis of the data concerning the dependece of the maximum brightness of the luminous region on its size. The experimentally-determined curves presented in Figure 7.4.2(b) are the size spectra of the localized plasma inhomogeneities (normalized w.r.t. the number density) observed in the aerosol at different pulse durations but with a constant average pulse power-density ($W = 12\ J/cm^2$). In comparison, Figure 7.4.2(c) presents the initial size-distribution function of $Al_2O_3$ aerosols before they are irradiated by the laser beam. Subscripts 1, 2 and 3 denote data corresponding to durations of the pulse's leading edge of $t_1 = 70, 100$, and 200 ns, respectively.

As follows from Figure 7.4.2, the value of the modal (the most probable) radius of the plasma inhomogeneities is independent (within the measurement error) of the pulse duration. The r.m.s. and maximum sizes monotonously increase with increasing pulse duration.

A minimum critical size of the plasma inhomogeneities of about $2 \times 10^{-2}$ cm was clearly observed in the experiments. Its value is almost independent of the pulse duration. According to estimates made, this size is one or two orders of magnitude greater than the radius of Debye screening in plasma, but it is in good agreement with the estimations of the sizes of the region of hydrodynamic spread of the vapor phase taking place due to particle evaporation during the action of the laser pulse. The increase in duration of the laser pulse results in a sharp increase in the sizes of luminous regions in the beam channel. The latter can be explained by a more complete vaporization of the large-size fraction of the aerosol ensemble and, that in turn, increases the probability (according to Poisson's law) of doubling the diameters of the inhomogeneities through their amalgamation, as $1 - \exp(-N_0^{-1/3}d)$.

The emission spectra of the plasma formations (Figures 8.4.2 and 8.4.3) contain intense spectral lines corresponding to the atomic composition of the aerosol substance and the surrounding gas. The lines are generated by neutral atoms or ions carrying a single charge. The spectral lines of molecular oxygen, nitrogen, and carbon dioxide observed in the emission spectrum of a laser spark had greater intensities than one could expect from the assumption of a thermal mechanism of excitation and a Boltzmann distribution of population over the vibration-rotation levels. This effect can be interpreted as the result of excitation due to electron impacts in the form of inelastic collisions.

At higher intensities of laser radiation, $\sim 10^9$ W/cm$^2$, molecular electron transitions can occur due to the essentially different electron and gas temperatures. The electron transitions can produce ultraviolet radiation and soft x-rays in vacuo. This can be used, in practice, for the chemical analysis of the aerosol surrounding the plasma formation using the methods of luminescent analysis, or the ionization of the substance around the plasma formations.

Thus, in particular, at a power density of incident radiation of about $5 \times 10^8$ W/cm$^2$, colour photos revealed a quite intense luminescent aureole around the plasma formations in the blue-green region. The diameter of this aureole was 0.5 to 1 cm. The appearance of this aureole can be explained by the action of hard radiation from the plasma on the surrounding medium [5].

The action of high-power laser pulses on aerosol media is accompanied by self-action effects. The possible self-action mechanisms occurring during breakdown are: (1) the decrease of the geometrical cross-section of solid particles due to beam-induced vaporization; (2) the appearance, due to laser action, of both thermal and mass aureoles around the aerosol particles, these scatter light, and the appearance of plasma formations around light-absorbing particles; (3) the formation, within the beam's cross-section, of the mean thermohydrodynamic profile of the density gradient.

Figure 7.4.3 illustrates the nonlinear behavior of laser pulse attenuation on the mean intensity of the incident beam which takes place during the passage of the leading edge of the pulse. The analysis of the results shows that the extinction of high-power beams in aerosols is caused by the joint effects of light absorption and scattering by localized plasma inhomogeneities appearing at the sites occupied by light-absorbing particles.

The pulsed optical breakdown of a discrete absorbing medium has two basic threshold characteristics, viz., the threshold intensity of one-fold ionization of the vapor and gas mixture around a vaporized particle $I_{th}$ and the threshold intensity $I'_{th}$ of the formation of a completely ionized channel within the beam scale caused, in turn, by the overlap of the propagating ionization fronts of individual plasma aureoles. The two threshold intensities can differ by several orders of magnitude. This difference is caused by the difference in ionization potentials and inelastic energy losses in the excitation (without ionization) of the vaporized particles matter on the one hand, and atmospheric gases, on the other. The functions of the speed of the drift of the ionization front also contribute to the difference in these intensities.

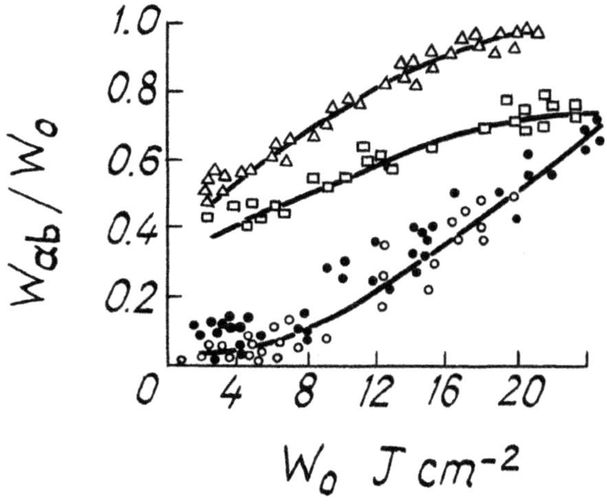


Fig. 7.4.3. Dependence of absorption $W_{ab}/W_0$ (with $\lambda$ = 10.6 μm) in the beam-induced plasma channel on the energy density of the laser pulse during the breakdown of particles of $Al_2O_3$ (□), $Na_2CO_3$ (Δ), and clear air (o) at the distance of 110 m from the laser source. The points (o) and (•) represent analogous data obtained under laboratory conditions during the spraying of water and after its termination.

As the measurements of the optical transmission of a beam channel ionized by a $CO_2$ laser pulse with a duration of microsecond and synchronously photographed showed, transmission blocking is already observed at the threshold intensity level $I_{th}$ .

Figures 7.4.4, 7.4.5, and 7.4.6 present a summary of the experimental data [5-14, 19, 27-28] relating to the intensity thresholds of the optical breakdown of air (obtained both from direct and indirect measurements) and the drift speed of the ionization front in the air surrounding the aerosol particles.

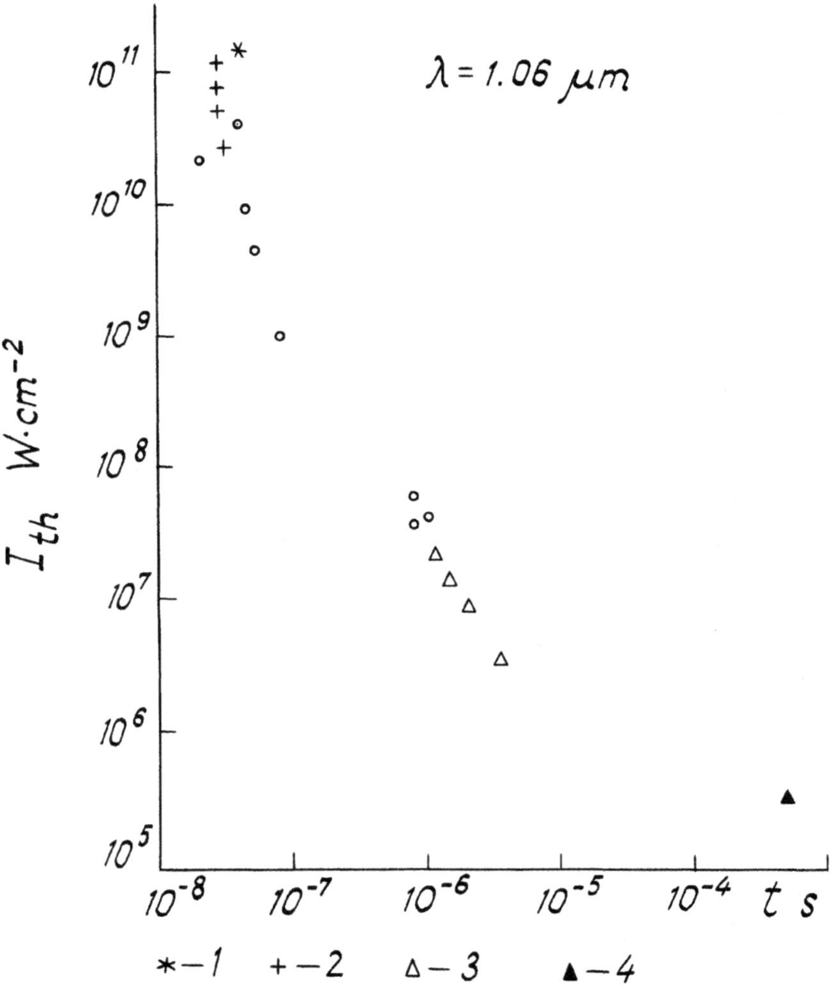


Fig. 7.4.4. Threshold intensities for optical breakdown in aerosols using Nd-glass laser radiation ($\lambda$ = 1.06 μm): 1 - technically pure air; 2 - 'room' air; 3 - solid aerosol particles of different chemical composition with radii from 1 to 70 μm [5, 9, 10, 13, 19]; 4 - the combined mechanism of optical breakdown [26].

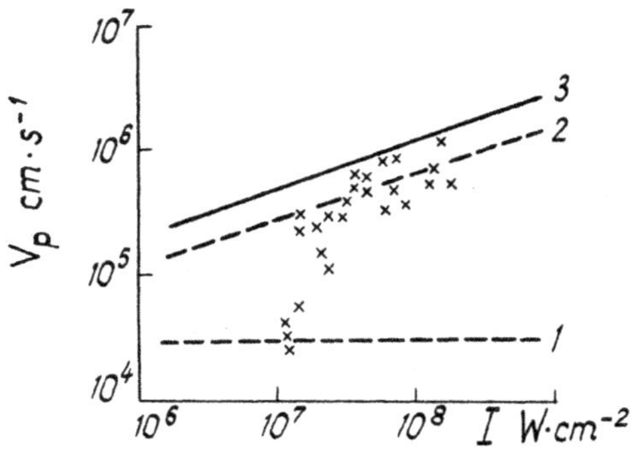


Fig. 7.4.5. Rate of plasma center growth as a function of $CO_2$ laser beam intensity, measured experimentally [9, 10]. Curve 1 is the speed of sound in air; curves 2 and 3 are the velocities of the transverse and longitudinal shock waves, respectively, calculated according to [1].

## 7.5. OPTICAL BREAKDOWN OF WATER AEROSOLS

Water aerosol droplets placed in a field of high-power laser radiation cause a significant decrease in the intensity threshold necessary for air breakdown. This effect was experimentally observed by several authors. In [14] it was observed for $CO_2$ laser radiation, and in [23] and [24] for laser pulses with $\lambda = 0.69$ and 1.06 μm.

The character of the process of the breakdown of air in the vicinity of water droplets depends on the radiation's wavelength, the pulse parameters, and the size of the droplets.

### 7.5.1. Optical Breakdown of Water Aerosols by a Pulsed $CO_2$ Laser

As was observed in [14], liquid aerosol particles cause a significant decrease in the intensity threshold necessary for breakdown; in the case of $CO_2$ laser pulses it is lowered to $10^9$ W/cm$^2$. The influence of a particle's surface on the process of optical breakdown is stronger for larger particles, and it reaches its maximum in the case of a plane liquid surface. The explanations of this effect suggested in [3] and [14] were based either on the assumption that the breakdown of air near a water droplet is caused by the presence of dense water vapor from an exploded droplet, or that it is due to the thermal ionization of air initiated by an intense shock wave produced by the explosion. In [20] it was shown that the breakdown of air in regions near water targets is, in fact, the process of sustaining and expanding the ionization produced by the shock wave from the explosive vaporization of liquid heated by the radiation.

The limiting values of the shock wave parameters achievable in the explosion can be found by solving the problem of a supercritical 'instantaneous' explosion ($t_p < a_0 / c_{so}$, where $a_0$ is the initial radius of a particle and, $c_{so}$ is the initial speed of sound in the liquid). In this problem there is no characteristic time scale of energy release and the amount of absorbed energy $W_{ab}$ is the only energetic parameter of the process.

The process of the heated region can be determined as follows:

$$p_H = (\gamma_E - 1)\frac{\rho_0 W_{ab}}{M} \simeq (\gamma_E - 1) k_{ab} W_p, \quad W_p = \int_0^t I \, dt,$$

where $\gamma_E$ is the integrated adiabatic exponent of water (under high pressures of $\sim 10^4$ bar, $\gamma_E \sim 2$); $\rho_0$ is the initial density of water; M is the mass of the heated region of liquid; and $W_p$ is the pulse energy. The calculations in [21] showed that, at values of the ratio $W_{ab}/M = 1.1\, Q_e^w$, $2Q_e^w$, and $3.3Q_e^w$, the corresponding pressures $p_H = 20$, 30, and 40 kbar. Here, $Q_e^w$ is the heat of vaporization of water at its normal boiling point. The initial (maximum) parameters of the shock wave in air were determined from the solution of the problem of the decay of an arbitrary explosion. Air pressure at the shock wave front $p_{sw}$ at the moment $t = \Delta t$, $\Delta t \to 0$ is determined by the following equation [22]:

$$\int_{p_H}^{p_{sw}} dp/\rho c_s = \frac{c_{s1}}{\gamma_1}\left(\frac{p_{sw}}{p_1} - 1\right)\left[\frac{\gamma_1 + 1}{2\gamma_1}\left(\frac{p_{sw}}{p_1} - 1\right) + 1\right]^{-1/2}, \qquad (7.5.1)$$

where $\rho$ and $c_s$ are the density and the speed of sound in the expanding substance; $c_{s1}$, $p_1$, $\gamma_1$ are the initial values of the speed of sound, air pressure, and the adiabatic exponent of air respectively.

The values of $p_{sw}$ corresponding to different models of the equation of state for water at high pressure are presented in Table 7.5.1. The calculations used the value $\gamma_1 = 1.4$ for air. With excess pressures at the shock wave front from 100 to 200 bar, the temperature changes from $5 \times 10^3$ to $10^4$ K, respectively. The number density of the electrons appearing due to thermal ionization $N_e$ (as calculated using the Saha formula) is from $10^7$ to $10^{13}$ $cm^{-3}$. The same values of the above parameters are also typical for the breakdown waves [1].

The experiments carried out in [20] were aimed at studying breakdown in the vicinity of large particles and plane targets exploded by laser radiation. The experimental set-up used a $CO_2$ laser delivering pulses with an output energy of 10 J as the source of high-power radiation. A high-speed camera and an ordinary 'Zenit-E' camera were used for the integral macrophotography of the process. The installation was also equipped with a power meter and a high-speed photodetector providing for a time resolution of $10^{-9}$ s.

TABLE 7.5.1. The initial values of air pressure on the shock wave front upon the 'instantaneous' explosion of a droplet.

| $P_H$, kbar | $\gamma_E$ | $P_{sw}/P_1$ |
|---|---|---|
| 20 | 1 + (PV/U) [31] | 60 |
| | 2 | 127 |
| | 2.5 | 89 |
| | 3 | 68 |
| 30 | 1 + (PV/U) [31] | 183 |
| | 2 | 189 |
| | 2.5 | 132 |
| | 3 | 111 |
| 40 | 1 + (PV/U) [31] | 309 |
| | 2 | 250 |
| | 2.5 | 176 |
| | 3 | 133 |

The field of investigation was illuminated using the laser's spark light reflected from a plane mirror. The laser spark was produced by $CO_2$ laser radiation focused by a lens onto a metallic pin-point. The time delay between the laser pulse and the laser spark burst-out was $10^{-8}$ s. The targets under investigation were introduced into the region between the lens and its focal plane.

A typical photoregistogram of the explosion and the breakdown processes in the air observed in the vicinity of a particle with a radius $r_0 \sim 100$ μm is presented in Figure 7.5.1(a). The light band observed in the beginning of the process shows that the temporal behavior of the intensity of the laser spark initiated on the metallic pin-point follows the shape of a high-power laser pulse. Therefore, the spark's intensity is at a maximum at the beginning of the process. The shadow image of the droplet cut out by the photocamera slit (slit width was ~10 μm) is in the middle of the band of illumination.

The size of the droplet increased insignificantly during the high-power laser pulse, due to the thermal expansion of water. Approximately 300 ns after the laser pulse action, droplet explosion took place, accompanied by an intense glow in the air near the particle. The presence of a dark region in the glowing zone means that the breakdown of the explosion products is not observed in the initial stage of the process. Thus, the same applies to the beginning of the explosive expansion of a particle and, hence, to the moment at which an intense shock wave appears. The comparison of photograms with the laser pulse oscillogram (Figure 7.5.1(b)) reveals the fact that the breakdown occurs just after the pulse peak. This, in turn, shows that the achievement of some intensity level is not quite sufficient for the breakdown to take place. It can also be seen from this figure that the spreadings of plasma has a wave-like character. Figure 7.5.1(c) presents the speed of this wave as a function of time.

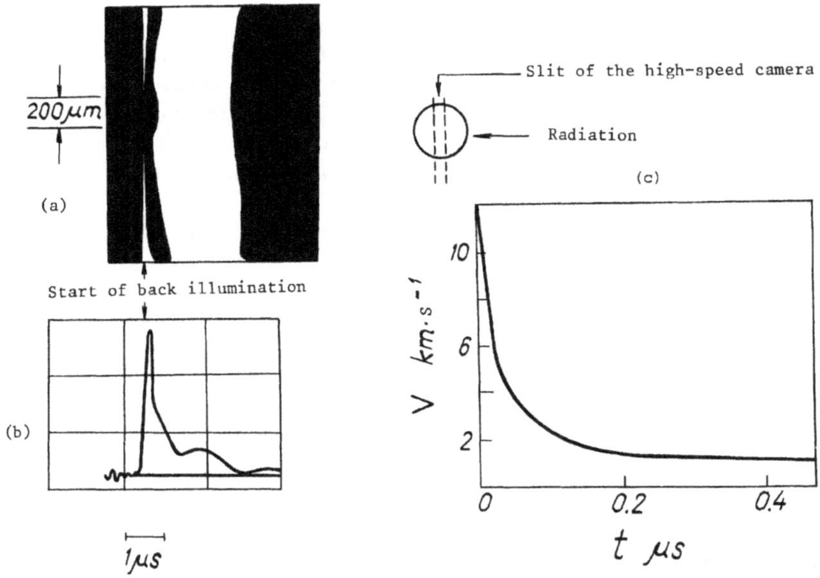


Fig. 7.5.1. (a) photogram of the breakdown process taking place during the explosion of a droplet 200 μm in diameter; (b) oscillogram of the $CO_2$ laser pulse intensity; (c) temporal behavior of the breakdown wave's speed.

Such results are in good agreement with the assumption of a light detonation nature of the plasma's expansion [1]. Since the velocity of wave propagation was a value measured experimentally, it is advisable to estimate the characteristic parameters of the plasma wave (temperature, pressure, and intensity of radiation) using their relationships with the wave speed, i.e.,

$$T = \frac{2}{\gamma_1 + 1}\,\frac{D^2}{2(\gamma_1^2 - 1)c_{v1}}; \tag{7.5.2.}$$

$$p = \rho_1 \frac{D}{\gamma_1 + 1}; \tag{7.5.3}$$

$$I = \frac{D^3 \rho_1}{2(\gamma_1 + 1)^2}, \tag{7.5.4}$$

where D is the velocity of the detonation wave. According to (7.5.4), a wave speed of 11.6 km/s corresponds to a beam intensity of $8 \times 10^7$ W/cm$^2$, this agrees quite well with the independent measurements of laser pulse intensity made at the moment of the appearance of breakdown in the air (~350 ns after the start of laser action). Corresponding estimates of the temperature and pressure are $6.2 \times 10^4$ K and 680 bar, respectively. The absorption length, evaluated using data on the absorption coefficients for radiation of $\lambda$ = 10.6 μm at the above temperature and pressure, was $10^{-4}$ cm. This allows one to arrive at the conclusion that we are dealing with a well-formed light detonation wave in the air even in the initial stage of the breakdown process.

Now, consider the problem of the prime breakdown. The above-mentioned values of shock wave parameters (pressure ~200 bar, air temperature $10^4$ K) and of the corresponding absorption coefficients (up to 1600 cm$^{-1}$ [1]) permit one to consider prime ionization mainly as the result of droplet explosion, but not of direct radiation action. The role of radiation is, therefore, reduced to the maintainance of the light detonation regime of the discharge. It should be noted, however, that the above values of shock wave parameters limited to the explosion of small droplets ($a_0 \sim k_{ab}^{-1} \sim 10$ μm) and pulse durations of $t_p > a_0/c_{s0}$.

In experiments on the laser-induced breakdown of air near a plane water surface [20], the targets used were pieces of thawing ice. The pieces of ice were 3 × 4 × 2 cm$^3$ in size and were introduced into the same region of the focused laser beam as an individual droplet, with the illuminated side perpendicular to the beam's axis. The irradiated area was 0.44 × 0.6 cm$^2$. The photographing of the front of the blocks was carried out through the back of the blocks (since the ice was transparent) using a normal camera (integral photographing) and a high-speed camera working in the same regime as the time magnifier. In the last case, the exposure time of a single shot was $0.44 \times 10^{-6}$ s.

The estimates of the speed of the luminescent front, made using the cinemagrammatic data, gave the value ~5 km s$^{-1}$. The time of the development of the breakdown was ~1 μs. This means that the breakdown kinetics in this case are the same as for droplets.

The experiments showed that several laser firings made at the same ice target led to a certain decrease (from ~15 to 12 J/cm$^2$) of the intensity threshold necessary for breakdown. This is due to a small cavity in the ice surface, appearing after laser firings, whose edges can focus hydrodynamic flows. On the other hand, it should be noted that, in the case of a solid target, there occurs an increase in the threshold due to the cleaning of the target's surface which becomes evident after several firings [10]. The laser spark was localized in the vicinity of the focusing edges of the cavity. If a cumulative cavity has been previously made in the target, then the breakdown is localized in it for the first firing at an energy density of the laser beam of $\sim 12\ J/cm^2$.

At higher (superthreshold) energy densities, the breakdown is observed in regions of small cracks in the target's surface, which, evidently, play the role fo focusing edges.

Thus, the above results confirm the assumption of the thermal nature of the prime ionization of air in the shock wave appearing near the water irradiated with pulsed $CO_2$ laser radiation. This prime ionization initiates the breakdown burst. Experiments carried out using plane targets distinctly showed that the surface inhomogeneities of the targets are prime centers of initiation of the breakdown wave. These inhomogeneities cause the appearance of corresponding inhomogeneities in the shock wave, where both temperature and pressure are much higher than the average values. In the physics of detonation [22] such regions are called ignition points. The ignition points appear at breaks in the shock wave, as well as when it is focused or when several shock waves collide. In the case of large droplets, the ignition points are evidently produced due to the nonuniform distortion of the shock wave's source over the droplet's surface. Probably, in the case of small aerosol particles, the collisions of shock waves from adjacent particles produce the ignition points. Therefore, optical action on aerosols that is aimed at the vaporization of the droplets at gas-dynamic rates should be performed with laser pulses of a short duration. The shape of the pulses must be such that the energy concentrated in the leading edge of the pusle is sufficient to produce the explosion of the droplet, while the energy in the trailing edge is insufficient to sustain the breakdown discharge.

### 7.5.2. Optical Breakdown Initiated at Weakly-Absorbing Water Aerosol Particles

The experiments [23-24] carried out using laser pulses with $\lambda = 0.69\ \mu m$ and $1.06\ \mu m$ revealed the fact that the presence of water aerosols in the beam channel leads to a significant lowering of the threshold intensity required to produce the breakdown of 'pure' air.

Since the imaginary part of the refractive index of water in this spectral range is small ($\kappa_a(\lambda = 0.69\ \mu m) = 3 \times 10^{-8}$ and $\kappa_a(\lambda = 1.06\ \mu m) = 3 \times 10^{-6}$) then, as a pose to the case of strongly absorbant particles, the mechanism of the breakdown is not related to thermal effects (i.e., to vaporization, phase explosion). The specific feature of the interaction between optical waves and droplets (whose diffraction parameter is large: $2\pi a_0/\lambda \gg 1$) is the focusing of the optical wave inside the droplet [32]. At experimentally attainable values of the optical field strength, the optical breakdown of water can occur, and this, in certain cases, can cause ionization of the air near the droplet.

Calculations of the optical field distribution inside weakly-absorbant particles carried out by different authors [32-35] using Mie's theory show that, at large diffraction parameters, it is extremely inhomogeneous. Two main maximums of the field inside the particle are localized near the droplet's diameter along the direction of light propagation, and it is characteristic that the maximum localized in the hemisphere in shadow is stronger than the one in the illuminated hemisphere (see Figure 5.2.1(b)).

For water droplet breakdown to take place (taking into account the effect of focusing) the following condition should be fulfilled:

$$I_0 \geqslant I^*/B_{max}, \tag{7.5.5}$$

where $I^*$ is the threshold value of the light intensity necessary for the breakdown of water, $B_{max}$ is the ratio of light intensity's ($I_a$) maximum inside the droplet to the intensity of incident radiation $I_0$.

The calculations made in [32-35] for $\lambda$ = 0.69 μm and $r_0$ = 1, 20, 60 μm gave $B_{max}$ = 25, 100, and 290, respectively.

At present, estimates of the threshold intensity for the breakdown of water can only be made when based on experimental data. The experimentally measured values of threshold intensity vary from $I^* \simeq 4 \times 10^8$ W/cm$^2$ [36] to $I^* \simeq 6 \times 10^{11}$ W/cm$^2$ [37]. The corresponding energy densities of the laser pulses $w_p^*$ in these cases were 8 J/cm$^2$ and $3 \times 10^4$ J/cm$^2$, respectively. Such a great difference between experimental data can be explained by the different purities of water used in the measurements. The higher value of threshold intensity was obtained for water of a higher purity. It is for this reason that we will take this value of theshold intensity for our estimates of the breakdown threshold in water droplets. Using the data in [32-35], one can ascertain that, with $\lambda$ = 0.69 μm and with droplet radii $a_0 \simeq$ 1, 20, 60 μm, the intensity of incident radiation necessary for the breakdown of water in a maximum of the light field inside a droplet is $2.4 \times 10^{10}$, $6 \times 10^9$, $2.1 \times 10^9$ W/cm$^2$ for a laser pulse of 50 ns duration.

The parameters of laser radiation necessary for the breakdown of water aerosol media are available in [23, 24], where they were used in the study of the breakdown both of air and aerosol particles.

Figure 7.5.2 presents the results of measurements of the breakdown thresholds for air and droplets of distilled water obtained using a ruby laser ($\lambda$ = 0.69 μm, $t_p$ = 50 ns, $W_p$ = 1 J). The results presented in this figure reveal a strong dependence of the threshold intensity values on the size of the particles.

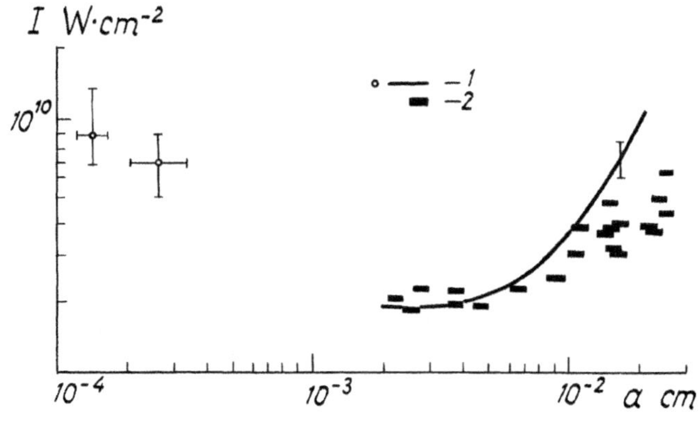


Fig. 7.5.2. Intensity threshold of optical breakdown in air in the vicinity of water droplets (curve 1), and inside the droplet, as a function of droplet size for one pulse of laser radiation with $\lambda = 0.69$ μm.

In the particle size region of about $10^{-3}$ cm, the decrease of the focusing effect of such droplets (the decrease of the factor $B_{max}(a_0)$) plays an important role. The increase of the threshold values for droplets of $a_0 \sim 10^{-2}$ cm and larger is connected with certain peculiarities of the breakdown process in the case of large droplets, as well as with the specific conditions involved in initiating the breakdown of the air when the energy required for this is transported through the water layer separating the prime source of breakdown in water from the air.

## 7.6. FIELD EXPERIMENTS ON THE NONLINEAR ENERGETIC ATTENUATION OF PULSED $CO_2$ LASER RADIATION DURING THE OPTICAL BREAKDOWN OF THE ATMOSPHERE

Field experiments aimed at studying the nonlinear effects on the propagation of high-power laser radiation through the natural atmosphere were carried out in a rural area, in order to avoid the influence of antropogenic factors. The measurements were taken, using a mobile installation, along the atmospheric path over the plane, uniform underlying surface. The measurement path was equipped with receiving points every 100 m. The optical investigations were accompanied by meteorological measurements, which provided the necessary information on air temperature, humidity, wind speed, and precipitation rate. The data on the structure constant of the atmospheric refractive index $C_n^2$ were also made available from measurements of the intensity fluctuations of the He-Ne laser radiation.

The $CO_2$ laser used in the mobile installation was capable of delivering, into the atmosphere, pulses of 500 J total energy and 1.5 μs total duration, with λ = 10.6 μm. The pulse shape had a main peak of 300 ns duration with about 75% of the total pulse energy concentrated in it. The 25% intensity inhomogeneities were characteristic for the distribution of intensity over the beam's cross-section. The high-power $CO_2$ laser beam could be focused at distances from 100 to 150 m using a Cassegrainian telescope with primary and secondary mirrors of diameters 500 and 120 mm, respectively [5-7].

A block-diagram of the experimental set-up and the scheme of measurements is depicted in Figure 7.6.1. The characteristics of the laser sparks were measured using data collected by panoramic photography (camera 6 in Figure 7.6.1), records of the spectral and integral luminosity of the sparks (refer to the elements on the scheme denoted by 9, 11, and 12), as well as measurements of the nonlinear transmission of high-power radiation. The apparatus used for measuring the transmission involves a mirror, 300 mm in diameter and with a focal length of 2500 mm, plane parallel plates 5 and 8, and power meters 2.

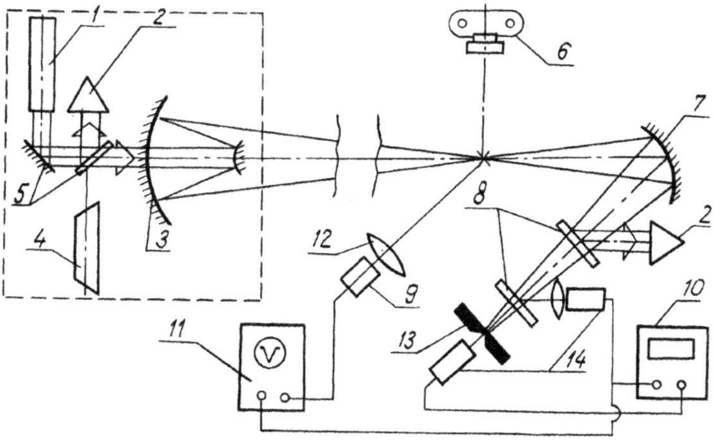


Fig. 7.6.1. The experimental set-up used in the field studies of $CO_2$ laser pulses propagating through the ground layer of the atmosphere. The components are: (1) the $CO_2$ laser; (2) a power meter; (3) the primary mirror of the receiving telescope; (4) an auxiliary He-Ne laser used in the alignment of the optical scheme of the set-up; (5) the reflecting plates; (6) a camera; (7) the focusing mirror; (8) beam splitters; (9) a PMT; (10) a voltmeter; (11) an oscilloscope; (12) a lens; (13) a slit diaphragm, (14) a PMT.

In order to avoid breakdowns in the optical systems of the measuring devices, the beams reflected from the plates 5 and 8 were attenuated by several layers of lavsan film. The structure of the high-power beam in its focused region was occasionally checked by analyzing the burn spots on the exposed photographic paper. The measurements of the values of $C_n^2$ were performed using a He-Ne laser in a measuring channel composed of the elements 5, 7, 8, 10, 11, 13, and 14 in the optical scheme.

A ST-1 spectrograph was used for recording the laser spark's spectrum. The entrance slit of the spectrograph was illuminated using the lens 12. The handling of the spectrograms was performed by a microdensitometer.

The first results of field experiments on the propagation of high-power $CO_2$ laser radiation [6, 7] revealed the heterogeneous spatial structure of the laser spark, as also have laboratory experiments. The spark occurs in the atmosphere as the result of the prime breakdown of solid aerosol particles of radius $a > 0.5$ μm

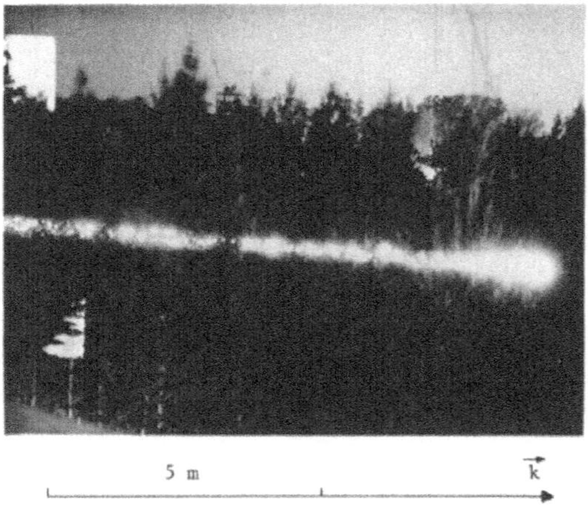


Fig. 7.6.2. (a) An example of a laser spark initiated by a $CO_2$ laser pulse of one microsecond duration in slightly dusted air at distance of 120 m from the laser ($W_0$ = 200 J; $C_n^2 = 2 \times 10^{-15}$ cm$^{-2/3}$; $F_0/R_0$ = 240). The length of the laser spark presented in this picture is about 15 m.

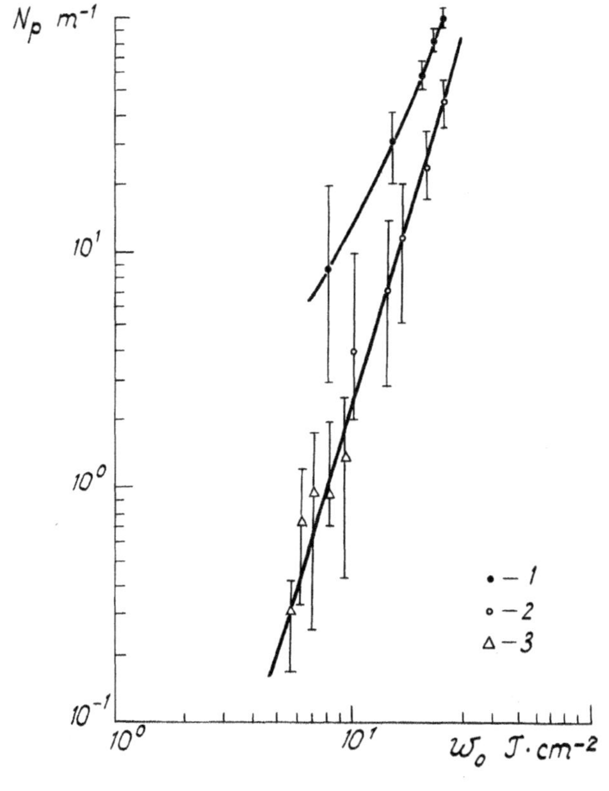


Fig. 7.6.2. (b) The dependence of the concentration of the prime centers of optical breakdown on the energy density of the $CO_2$ laser pulses, measured experimentally in the atmosphere. Curves 1 and 2 represent data obtained using a focused beam ($F_0/R_0 = 240$). Curve 3 represents data for a collimated beam. The data presented by curves 2 and 3 were measured in the atmosphere, while the curve 1 was obtained from cement dust introduced into the beam's focal plane. The parameters of the cement dust particles were: ($a_{sq} \approx 3$ μm, $N_0 \approx 1$ to $10^{-1}$ $cm^{-3}$).

Figure 7.6.2(a) shows a photograph of laser sparks generated at a distance of 100 to 120 m from the laser unit. Figure 7.6.2(b) presents data on the dependence of the linear concentration of breakdown centers on the density of the beam's energy at the focal point, $W_0$ J/cm$^2$, obtained both for artificially-dusted and natural atmospheres. The cement dust had a r.m.s. radius of particles $a_{rms} \approx 3$ μm and a mean volumetric number density $N_0 \lesssim 1$ cm$^{-3}$. The r.m.s. radius of natural aerosols (particles of soil) was about 1 μm and $N_0 \lesssim 10^{-2}$ cm$^{-3}$. It can be seen from this figure that the concentration of breakdown centers exceeds a value of $10^{-1}$ m$^{-1}$ at an energy

$W_0$ in the beam's focal plane of about 20 $J/cm^2$, both in natural and in artificially-dusted atmospheres. The microphotographs reveal a considerable spread of breakdown center size, i.e., from 0.1 cm to several centimeters. The r.m.s. diameter of the plasma formations was about 2 to 3 mm.

The presence of spectral lines of singly- and doubly-charged ions in the emission spectrum of plasma formations indicates that the temperature of the plasma is about 1.5-2 eV.

The time of the de-excitation of the spectral lines coincides with the duration of the laser pulse (1.5 μs), while the duration of continous emission from the plasma formations was about 10 to 15 μs. In the latter case, the emission time is determined by thermal relaxation. It was found in the experiments that the beginning of laser spark emission is delayed with respect to the moment of arrival of the laser pulse. The measured values of the time lag were between 0.2 and 0.4 μs. The observed emission lines of the aerosol's atoms are weakly broadened as compared with the strongly broadened lines of ions and atoms of atmospheric gases. This shows that, in the initial stage of avalanche ionization of atoms in the vapor aureoles of particles (until the moment of complete ionization), there exists a large difference between electron and gas temperature.

The estimates of the peak pressure in the light detonation wave, made using data from acoustic measurements, gave the values 70 to 80 dB at a distance of 0.3 to 0.5 m from the beam's axis. The experimentally estimated value of the threshold density of the beam's energy necessary for the initiation of the breakdown of solid aerosol particles was $w_{th} \approx 6$ $J/cm^2$. It should also be noted that, according to laboratory studies [9-11], the threshold values of the energy density necessary for the initiation of breakdown by $CO_2$ laser pulses of a duration of a few microseconds do not depend on the chemical composition of the solid fast-melting particles and their radii. This refers to particles with radii from 1 to 100 μm.

In the course of these field experiments we also studied the transmission in the atmosphere for high power $CO_2$ laser pulses in different meteorological conditions and we investigated different pulse energies necessary for optical breakdown.

Figure 7.6.3 presents the experimental results on the dependence of the integral (over the pulse duration) transmission of the beam channel path $T_{tr} = W/W_0$ on the pulse energy $W_0$. Here, W is the pulse energy at the end of the path. The curves presented in this figure illustrate two meteorological situations which differ in intensity of atmospheric turbulence by more than one order of magnitude.

In order to correctly interprete these results, we carried out a particular experimental study on the statistics of the intensity spikes in the beam's cross-section in the region of the beam's focal point. For this purpose we measured the areas of the spikes whose intensities exceeded some fixed level w for three values of w. The results of this study, averaged over 10 to 15 laser shots, are presented in Figure 7.6.4.

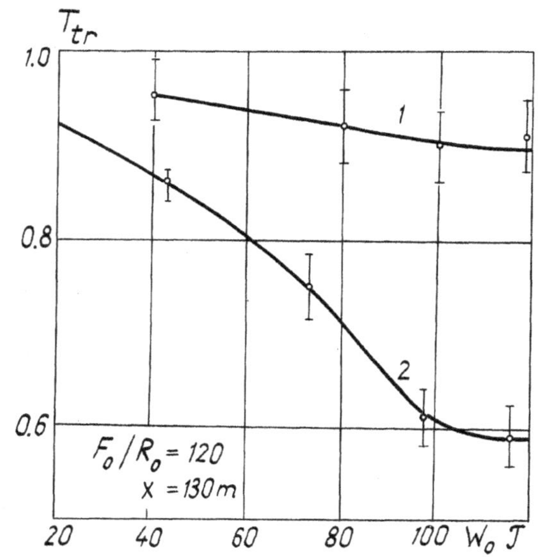


Fig. 7.6.3. Atmospheric transmission during optical breakdown, $C_n^2 = 10^{-14}$ cm$^{-2/3}$ (curve 1) and $C_n^2 = 4$ to $5.5 \times 10^{-15}$ cm$^{-2/3}$ (curve 2) as a function of laser radiation output.

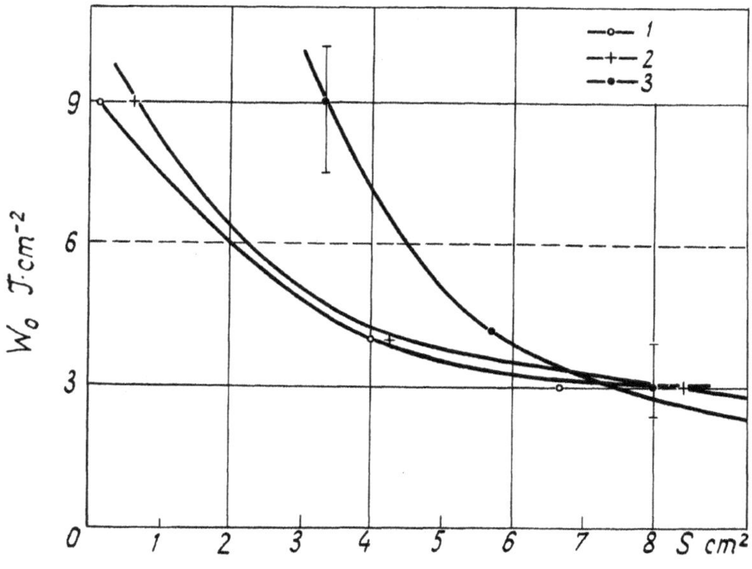


Fig. 7.6.4. The intensity of energy density spikes at the beam's focal point

as a function of the spike area, under the following conditions of atmospheric turbulence: (1) $C_n^2 = 10^{-14}$ $cm^{-2/3}$; (2) $C_n^2 =$ 2 to $3 \times 10^{-15}$ $cm^{-2/3}$; (3) $C_n^2 = 1$ to $4 \times 10^{-16}$ $cm^{-2/3}$.

The spike areas were assessed by analyzing the burnt areas on the sheets of exposed photographic paper. The degree of burning of the exposed photographic paper was calibrated with respect to the energy density of the incident $CO_2$ laser radiation beforehand. The results of these measurements allowed us to distinguish between three distinct degrees of burning, which correspondea to energy densities of incident radiation of 2-4, 4-6, and above 6 $J/cm^2$. As the results show, the strength of the nonlinear interactions between a high-power $CO_2$ laser beam with energy above the breakdown threshold and the atmosphere becomes significantly weaker in a strongly turbulent atmosphere, due to turbulent blooming of the beam. The effect of the turbulent blooming of the beam decreases the probability of reaching intensities above the breakdown threshold in beam regions occupied by large aerosol particles with radii from 1 to 100 μm. As a result, the probability of the appearence of breakdown centers also decreases.

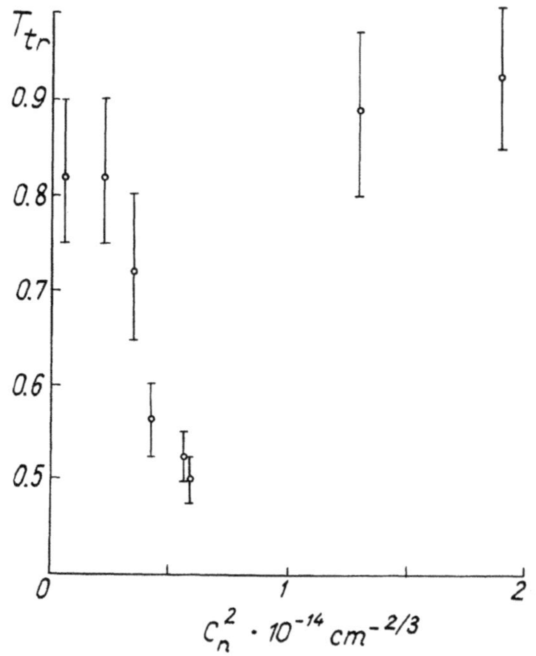


Fig. 7.6.5. Summary data of the measurements of atmospheric transmission in the ground layer at different values of $C_n^2$, with $W_0 = 100$ J, $F_0/R_0 = 200$, and X = 120 m.

Figure 7.6.5. presents measured values of the integral nonlinear transmission of the atmosphere $T_{tr}$ for $CO_2$ laser radiation versus measured values of the structure constant of atmospheric turbulence $C_n^2$. Vertical bars in the figure present r.m.s. deviations of the experimental data of $T_{tr}$ from the average (over 10 to 15 measurements) values. The results presented in Figure 7.6.5 reveal the interesting fact that, in conditions of both weak ($C_n^2 \lesssim 2 \times 10^{-15}$ cm$^{-2/3}$) and strong ($C_n^2 \gtrsim 1.5 \times 10^{-14}$ cm$^{-2/3}$) turbulence, the number of experimentally observed breakdown centers was lower than that under conditions of moderate turbulence ($C_n^2 \approx 0.5 \times 10^{-14}$ cm$^{-2/3}$). This effect can be explained by the apperence, under conditions of moderate turbulence, of intensity spikes in the sharply focused beam which make the probability of reaching the intensity threshold $I_{th}$ necessary for breakdown in certain beam regions higher. This result agrees qualitatively with the estimates of breakdown probability made in the case of coherent radiation in §7.3.

In order to study the influence of meteorological conditions on the transmission of high-power $CO_2$ laser radiation through the atmosphere we have carried out weekly cycles of field measurements. The measurements made in natural atmospheric aerosols at a relative humidity of about 90% showed that an increase in meteorological visual range from 1 to 12 km was accompanied by a weak increase in atmospheric transmission along the high-power beam channel (30%), i.e., $T_{tr}$ varied in the range from 0.5 to 0.8. The scatter of experimental data was about 60%. When the air's humidity changed from 60 to 90%, the transmission within the laser beam channel changed from 0.8 to 0.5, the meteorological visual range remaining the same. This clearly demonstrates the increase of nonlinear effects with increasing relative humidity. The atmospheric transmission for a low-power laser beam ranged, in this case, from 0.9 to 1.0.

Experiments on initiating optical breakdown in natural fogs and rains revealed a great difference in the behaviour of the process from that observed in laboratory experiments (see §7.4). A long spraying session of water aerosols in the laboratory experimental chamber led to a strong washing-out of aerosols that caused an increase in the breakdown threshold by one order of magnitude. However, in the atmosphere aerosol formations such as fog and rain caused weakening of the atmospheric turbulence, thus improving the conditions for laser beam focusing and, as a consequence, for laser sparking in the natural atmosphere. The washing-out of atmospheric aerosols was also observed in field experiments, but only after a long period of precipitation. In addition to plasma formations generated by the optical breakdown of large aerosol particles, a weak glow was observed in the beam channel. This glow was caused by the vaporization and partial ionization of the submicron aerosol fraction. The number density of the submicron particles in the atmosphere is normally quite significant ($N$ ($a$ = 0.1, 0.5 μm) = $10^1$–$10^3$ cm$^{-3}$).

The scattering of light by the breakdown centers, and by other irregularities in the air's refractive index induced by incident radiation, must be considered as one of the main mechanisms of energy losses of high-power beams propagating through the atmosphere, together with the breakdown process itself.

## REFERENCES: CHAPTER 7


[ 1] Yu.P. Raiser: Laser Sparking and Spreading of Discharges (Nauka, Moskow, 1974) p. 210, in Russian.

[ 2] I.P. Shkarofsky: RCA Review 35, 110-122 (1974).

[ 3] F.V. Bunkin, V.V. Savransky: Zh. Eksp. Teor. Fiz., 65, 2185-2191 (1973) (Sov. Phys.-JETP).

[ 4] E.B. Belyaev, A.P. Godlevsky, and Yu.D. Kopytin: Kvant. Electron., 6, 1152-1156 (1978) (Sov. J. Quantum Electron.).

[ 5] E.B. Belyaev, A.P. Godlevsky, V.E. Zuev, and Yu.D. Kopytin: 'Remote Spectrochemical Analysis of Atmospheric Aerosols Using Lasers', in Determination of Physico-Chemical Parameters of the Atmosphere Using High-Power Lasers, ed. by V.E. Zuev (IAO, Tomsk, 1979) pp. 3-56, in Russian.

[ 6] A.P. Godlevsky and Yu.D. Kopytin: Kvant. Elektron. 9, 1280-1283 (1982) (Sov. J. Quantum Elektron.).

[ 7] Yu.V. Akhtyrchenko, E.B. Belyaev, Yu.P. Vysotsky, et al.: Izv. Vyssh. Uchebn. Zaved. Fiz. 2, 3-13 (1983) (Sov. J. Phys.).

[ 8] V.E. Zuev, Yu.D. Kopytin, and A.V. Kuzikovsky: Nonlinear Optical Effects in Aerosols (Nauka, Novosibirsk, 1980) p. 180, in Russian.

[ 9] D.C. Smith: Appl. Phys. 48, 2217-2225 (1977).

[10] J. Reilly, P. Singh, and G. Weyl: AIAA, Paper N697, 11 (1977).

[11] N.N. Belov, N.P. Datskevich, F.V. Bunkin, et al.: Zh. Tekh. Fiz. 49, 333-338 (1979) (Sov. Phys.-Tech. Phys.).

[12] E.B. Belyaev, A.P. Godlevsky, Yu.D. Kopytin, N.P. Krasnenko, et al.: Zh. Tekh. Fiz. Pis'ma Red. 8, 333-337 (1982) (JTP Lett.).

[13] A.A. Boni and D.A. Meskan: Opt. Commun. 14, 115-118 (1975).

[14] D.E. Lencioni: Appl. Phys. Lett. 23, 12-14 (1973)

[15] A.V. Eletsky, L.A. Palkina, and B.M. Smirnov: The Transfer Phenomena in Low Ionized Plasma (atomizd., Moscow, 1975) p. 330, in Russian.

[16] V.L. Mironov: Laser Beam Propagation through the Turbulent Atmosphere (Nauka, Novosibirsk, 1981) p. 242.

[17] V.E. Zuev: Laser Beam in the Atmosphere (Plenum, New York, 1982).

[18] Yu.I. Bychkov, V.M. Orlovsky, and V.V. Osipov: Kvant. Elektron. 4, 2435-2441 (1977) (Sov. J. Quantum Electron.).

[19] V.A. Volkov, F.V. Grigor'ev, V.V. Kalinovsky, et al.: Zh. Eksp. Tecr. Fiz. 69, 115-121 (1975) (Sov. Phys.-JETP).

[20] A.A. Zemlyanov, A.V. Kuzikovsky, and L.K. Chistyakova: Zh. Tekh. Fiz. 51, 1439-1443 (1981) (Sov. Phys.-Tech. Phys.).

[21] A.A. Zemlyanov, A.V. Kuzikovsky, V.A. Pogodaev, and L.K. Chistyakova: 'A Macroparticle in a High-Power Optical Field', in Problems of Atmospheric Optics (Nauka, Novosibirsk, 1983) pp. 13-39, in Russian.

[22] F.A. Baum, L.P. Orlenko, K.P. Stanyukovich, V.P. Chelyshev, and B.I. Shekhter: Physics of Explosions (Nauka, Moscow, 1975), in Russian.

[23] V.A. Pogodaev and A.E. Rozhdestvensky: Zh. Tekh. Fiz. Pis'ma Red. 5, 257-261 (1979) (JTP Lett.).

[24] V.A. Pogodaev and A.E. Rozhdestvensky: 'Propagation of Laser Radiation in Dispersed Media', All-Union Symp., Conf. Abstracts Vol. II, (Obninsk, 1982) pp. 123-125, in Russian.

[25] P.K. Wu and A.N. Pirri: 'The dynamics of air plasma growth in a 10.6 μm laser beam', in Conf. Abstracts: 16th Aerospace Sciences Meeting AIAA (1978).

[26] V.A. Vdovin, S.V. Zakharchenko, S.M. Kolomiets, A.M. Skripkin, and Yu.M. Sorokin: 'The origin and evolution of plasma centers in an air-dispersed medium illuminated with laser radiation', in Abstracts, 12th Conf. on Propagation of Radio Waves (Tomsk, June 1978), pp. 159-161, in Russian.

[27] I.V. Aleshin, S.I. Anisimov, A.M. Bonch-Bruevich, et al.: Zh. Eksp. Teor. Fiz. 70, 1214-1223 (Sov. Phys.-JETP).

[28] S.V. Zakharchenko, G.A. Sintyurin, and A.M. Skripkin: Zh. Tekh. Fiz. Pis'ma Red. 6, 1065-1070 (1980) (JTP Lett.); also see Zh. Tekh. Fiz. Pis'ma Red. 7, 767 (1981).

[29] M.S. Sodha, R.L. Sawhney: Acustica 41, 139-142 (1978).

[30] Yu.V. Akhtyrchenko, A.A. Vaisljev, Yu.V. Vysotsky, and V.N. Soshnikov: 'Propagation of Laser Radiation in Dispersed Media', Conf. Abstracts: 2nd All-Union Symp. (Obninsk, 1982) pp. 86-97, in Russian.

[31] N.M. Kuznetsov: Prikl. Mekh. Tekh. Fiz. 1, 112-120 (1961) (Sov. J. Appl. Mechanics Tech. Phys.).

[32] A.P. Prishivalko: Optical and Thermal Fields inside Light-Scattering Particles (Nauka i Tekhnika, Minsk, 1983) p. 190, in Russian.

[33] N.V. Bukzdorf, V.A. Pogodaev, and L.K. Chistyakova: Kvant. Elektron. 2 (1973) (Sov. J. Quantum Electron.).

[34] V.S. Loskutov and G.M. Strelkov: 'Explosive vaporization of droplets by laser pulses', Preprint N12(295) (Institute for Radio Engineering and Electronics of the U.S.S.R. Ac. Sci., Moscow, 1980).

[35] V.N. Pozhidaev and A.I. Fatievsky: Kvant. Elektron 8, 119-123 (1981) (Sov. J. Quantum Electron.).

[36] A.I. Ioffe, N.A. Mel'nikov, K.A. Naugol'nykh, and V.A. Upadyshev: Prikl. Mekh. Tekh. Fiz. 10, 125-127 (1970) (Sov. J. Appl. Mechanics, Tech. Phys.).

[37] Ph. Roch and M. Davis: IEEE 58, 108-109 (1970).

CHAPTER 8

# LASER MONITORING OF A TURBID ATMOSPHERE USING NONLINEAR EFFECTS

## 8.1. BRIEF DESCRIPTION OF THE PROBLEM

The use of high-power lasers in atmospheric optics investigations allows one, on the one hand, to increase the potential of traditional lidar facilities and, on the other, provides for new possibilities for obtaining information concerning the composition of the atmosphere using the nonlinear effects of the interaction between laser radiation and atmospheric constituents.

The distortions of lidar returns, which can occur due to nonlinear interactions, are important factors limiting the power of lasers used in conventional lidar facilities, since they complicate the interpretation of experimental data. In this connection, investigations into lidar return distortions caused by nonlinear effects, and the determination of the applicability limits of the conventional 'linear' lidar equation for situations involving high-power sources of sounding radiation, are of extreme interest [1-3].

Another aspect of this problem is the determination of criteria for detecting the high-power laser beam by means of singly-scattered radiation under conditions of the thermal interaction of the beam and atmospheric aerosols [11-12]. One can expect that the results of such investigations may lead to the increase of the operational range of various laser navigation devices in bad weather conditions [10, 11, 13], as well as the development of a method for assessing the linear and angular sizes of high-power laser beams by analyzing the picture of the beam's thermal self-action [1].

State-of-the-art laser technology allows the great number of nonlinear optical effects in the atmosphere to be observed; these carry information on the physical and chemical properties of the atmosphere. The possibility of combining nonlinear and linear methods of remote sensing, aimed at obtaining multiparameter information concerning aerosols in the atmosphere sufficient for the correct solution of the relevant inverse problems using no *a priori* models of the aerosol media, as well as at measuring certain atmospheric parameters that cannot be investigated by conventional lidar techniques, should be mentioned here [1, 2, 6].

In this respect, the initiation of emission spectra of atmospheric aerosols by high-power laser radiation (due to vaporization, explosion, and ionization [1, 2, 4-7]), as well as the effect of resonance oscillations of droplets' shapes in the field of a modulated laser beam [6, 14], are important.

This chapter presents a discussion of the contemporary state of these problems. It should be noted that laser sensing of atmospheric aerosols using nonlinear interaction effects is quite a novel application of lasers to atmospheric studies, so only preliminary results of investigations are available for discussion.

## 8.2. DISTORTIONS OF LIDAR RETURNS CAUSED BY THE NONLINEAR EFFECTS OF THE INTERACTION OF HIGH-POWER RADIATION WITH AEROSOLS

Pulses of high-power radiation propagating through a turbid atmosphere affect the optical characteristics of aerosol particles. Taking into account the effect of the thermal action of laser radiation on aerosols, one can write the initial system of location equations, following the single-scattering approximation, as follows [2]:

$$P(x) = D(x)\alpha_\pi(x, (x/c))ct_p P(0, t - (2x/c)) \times$$
$$\times \exp[-\int_0^x \alpha(x', \frac{2x - x'}{c})dx' - \int_0^x \alpha(x', (x'/c))\, dx']; \qquad (8.2.1)$$

$$(\frac{\partial}{\partial x} + \frac{1}{c}\frac{\partial}{\partial t})P(x, t) = -\alpha(x, t)P(x, t), \qquad (8.2.2)$$

where $P_\pi(x)$ is the power of the light scattered at a distance x from the receiver; $P(x, t)$ is the beam's power at a distance x and moment of time t; $t_p$ is the laser pulse duration; $D(x)$ is the geometrical factor characterizing the overlap of the fields of view of the receiver and the transmitter; $\alpha(x, t)$ and $\alpha_\pi(x, t)$ are the volume extinction and backscattering coefficients of aerosol, respectively, where

$$\alpha_j = \pi \int_0^\infty a^2 K_j(a, [P])N(x, a_0(a))\frac{da_0(a, [P])}{da}, \qquad (8.2.3)$$

here $K_j$ is the extinction efficiency factor K, or the backscattering efficiency factor $K_\pi$, of a particle with radius a; $N(x, a)$ is the size-distribution function of an aerosol ensemble at a point on the sounding path x; and $a_0(a, [P])$ is the inverse function describing the dependence of the current particle radius on the initial one $(a_0)$. The brackets around P denote the functional dependence on the beam power P.

In the case of water aerosols, the physical mechanism governing the nonlinearity of the volume extinction $\alpha$ and backscattering $\alpha_\pi$ coefficients is regular vaporization or particle fragmentation, depending upon the radiation heating regime.

In the case of hazes, the processes of particle destruction and modification of the particles optical properties are determined by the physio-chemical properties of the particulate matter and by the energy of the incident radiation. From among the physical processes causing nonlinearity of the optical characteristics of aerosols, the following should be mentioned: radiation heating, vaporization, thermal dissociation, and the burning of particles resulting in the appearance of both thermal and mass aureoles around particles and the initiation of prime centers of optical breakdown on the aerosol particles. These effects that accompany the interaction between high-power laser radiation and aerosol particles have been considered in previous chapters. Taking into account the results contained in these chapters, we will now consider the calculational and experimental data on the nonlinear distortions of lidar returns.

Figure 8.2.1 presents data on the estimations of the nonlinearity parameter of the lidar equation made in [2, 15]:

$$\eta = P_\pi(x)/P_\pi^L(x),$$

where $P_\pi(x)$ is the power of the lidar return, calculated taking into account nonlinear effects, while $P_\pi^L(x)$ is the power of the lidar return, calculated assuming the linearity of the medium. The nonlinearity of water hazes is assumed to be caused by droplet vaporization. In the case of corundum dust, the nonlinearity was related to light scattering by thermal aureoles around the particles. The calculations were made for $\lambda$ = 10.6 μm, initial beam radius $R_0$ = 5 cm, and angular beam width $\theta$ = $10^{-4}$ rad.

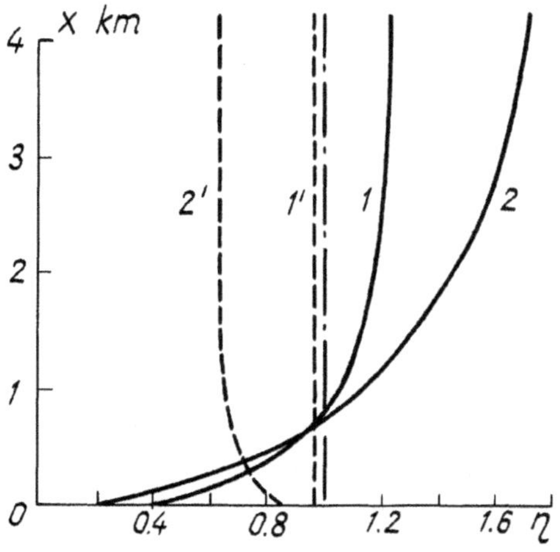


(a)

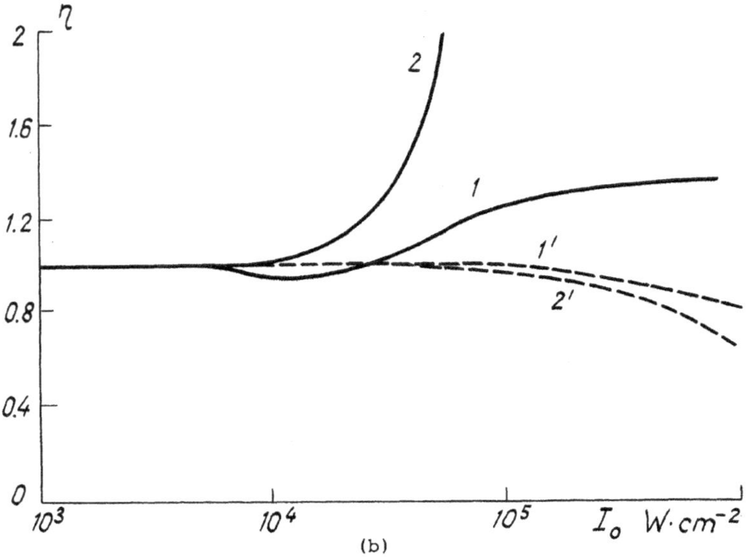


(b)

Fig. 8.2.1. The nonlinearity parameter η of the lidar return as a function of path length x (Fig. 8.2.1(a)) and of sounding beam intensity (Fig. 8.2.1(b)). Curves 1 and 2 represent the data for a moist haze and curves 1' and 2' for a dry haze. The laser beam's parameters and path lengths are as follows: $\lambda = 10.6$ μm; $t_p = 10^{-4}$ s. (a) $I_0 = 3 \times 10^4$ W/cm$^2$ for curve 1 and $5 \times 10^4$ W/cm$^2$ for curve 2. $I_0 = 5 \times 10^5$ W/cm$^2$ for curve 1' and $3 \times 10^5$ W/cm$^2$ for curve 2'. (b) x = 1km for curves 1 and 1', and z = 5 km for curves 2 and 2'.

Figure 8.2.1(a) shows (curves 1 and 2) that echoes from high-power lasers in water hazes can be weaker, as well as stronger, than those in linear media. This is due to the fact that two competing processes contribute to the power of the backscattered signal, viz., the increase in atmospheric transmission within the high-power beam channel and the decrease in the backscattering coefficient of a fixed volume. In the case of solid aerosol particles (curves 1' and 2'), the nonlinearity parameter η is always less than unity, due to additional extinction of radiation by thermal aureoles around aerosol particles heated by the laser radiation.

The deviation $|\eta - 1|$ shown in Figure 8.2.1(b) clearly demonstrates the possible errors (appearnce of false profiles) in estimating the behavior of aerosol scattering parameters with changing altitude, as obtained from the ordinary lidar equation without incorporating the nonlinear corrections.

The calculations in [3] pave the way for the method of lidar sensing of the high-power laser beam channel in a water aerosol. The numerical investigations are made by analyzing the energetics of successive lidar returns received from a media irradiated with a high-power cw laser beam of $10^2$ to $10^3$ W/cm$^2$ power with $\lambda = 10.6$ μm. The laser beam self-action in the forward direction was investigated numerically by solving the initial system of equations (8.2.1)-(8.2.3), (6.5.22), (6.5.25) for $V_T = 0$.

The calculations of lidar returns at $\lambda = 0.69$ μm were carried out using the Monte-Carlo method. The optical characteristics of the water droplets were calculated based on the Mie theory.

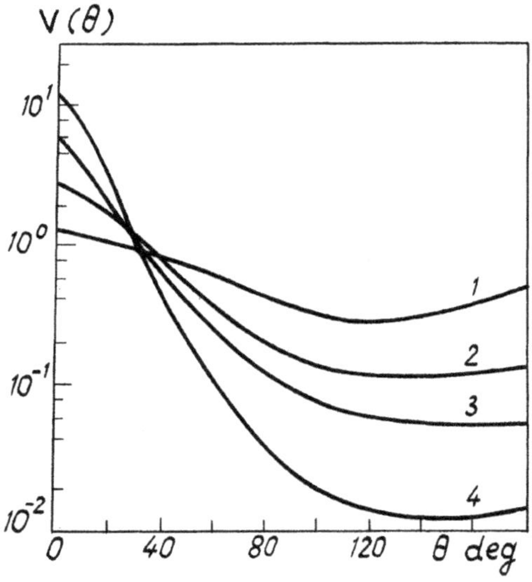


Fig. 8.2.2. Data illustrating the change of the scattering phase function of a water droplet fog ($\lambda = 0.694$ μm) at the moment 0.03 s from the beginning of irradiation by a high-power cw laser beam with $\lambda = 10.6$ μm ($I_0 = 1$ kW/cm$^2$). Curves 1, 2, 3, and 4 represent data for the heights 40, 60, 80, and 100 m, respectively.

Figure 8.2.2 illustrates the modification of the fog scattering phase function caused by the action of a high-power laser beam (angular width $\theta_0 = 20'$). The size spectrum of the aerosol droplets observed initially is described by (6.3.33), where $a_m = 4$ μm; $\theta_2 = 4$; $\theta_4 = 1$. As seen from the figure, the scattering phase function of the layer at the beginning of the beam channel (curve 1) has a form that is very close to a spherical one. This can be explained by the fact that the vaporization of the fog means that the small-size fraction of the droplets is dominant.

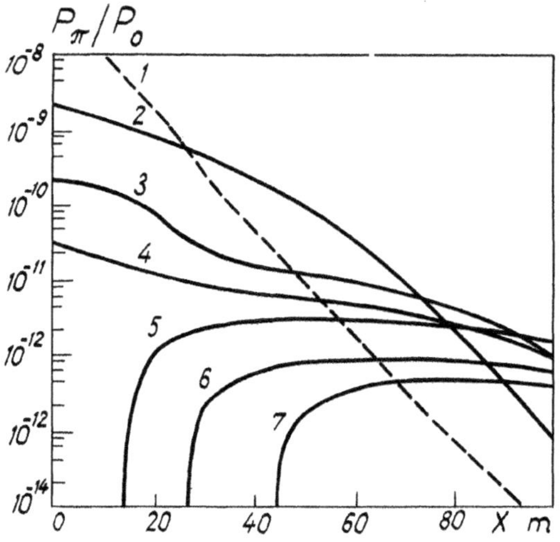


Fig. 8.2.3. Lidar returns from a cleared channel in a water aerosol at the moments from 0.01 s (curve 2) to 0.08 s (curve 7) in 0.01 s time steps from the beginning of high-power beam action. Curve 1 represents the lidar return in an undisturbed aerosol

Figure 8.2.3 presents the smoothed histograms of the temporal distribution of the intensities of lidar returns. The histograms are plotted in the coordinate system $x = ct$ ($c$ is the speed of light). The angular aperture of the lidar receiver $\theta_0 = 15'$.

Analysis has shown that, in this case, the optimal value of the receiver's angular aperture must not exceed 20', since any further increase results in a sharp growth in background intensity due to multiple light scattering outside the beam channel.

Table 8.2.1 presents the results of the numerical simulations of the reconstitution of the transmission profile $T_{tr} = \exp(-\tau)$ and of the backscattering coefficient $\alpha_\pi(x)$ within the high-power beam channel at $t = 0.04$ s. The lidar ratio $b_\pi(x, t) = \alpha_\pi(x, t)/\alpha(x, t)$ used in the calculations was established by using linear interpolation between the preceding and the current values of this ratio. Table 8.2.1 presents the model values (indexed by 'M' of the above optical characteristics, along with those obtained from the analysis of lidar returns as functions of the length of the sounding path. As seen from the table, the reconstitution errors $\delta_\alpha$ and $\delta_T$ do not exceed 3%.

TABLE 8.2.1. The results of numerical simulation experiments of the sensing of the beam channel in a water aerosol dissipated using a high-power $CO_2$ laser beam.

| x (m) | $\alpha_\pi^{(M)}$ ($m^{-1}$) | $\alpha_\pi$ ($m^{-1}$) | $(T_{tr}^{(M)})^2$ | $T_{tr}^2$ | $\delta_\alpha$ | $\delta_T$ |
|---|---|---|---|---|---|---|
| 10 | 0.00746 | 0.0191 | 0.999 | 0.978 | 0.00149 | 0.000152 |
| 20 | 0.0125 | 0.0222 | 0.998 | 0.977 | 0.00163 | 0.00166 |
| 30 | 0.0175 | 0.0253 | 0.993 | 0.974 | 0.0063 | 0.00641 |
| 40 | 0.0226 | 0.0284 | 0.985 | 0.967 | 0.0111 | 0.00112 |
| 50 | 0.0297 | 0.0315 | 0.972 | 0.956 | 0.0252 | 0.0256 |
| 60 | 0.0335 | 0.0346 | 0.952 | 0.937 | 0.0405 | 0.0411 |
| 70 | 0.0393 | 0.0377 | 0.926 | 0.911 | 0.0587 | 0.0596 |
| 80 | 0.0351 | 0.0408 | 0.894 | 0.881 | 0.0702 | 0.0712 |
| 90 | 0.0336 | 0.0439 | 0.854 | 0.849 | 0.0883 | 0.0889 |
| 100 | 0.0425 | 0.047 | 0.806 | 0.835 | 0.0134 | 0.0134 |

In [2, 11] the reader can find a discussion of experimental studies of the information content of scattered radiation required for assessing the necessary conditions for aerosol dissipation by high-power beams. Figure 8.2.4 presents the dependences of $\ln(P_\pi^N/P_\pi)$ and the decrease in optical depth $\Delta\tau$ on the initial optical depth $\tau_0$. Here, $P_\pi$ and $P_\pi^N$ are the powers of the backscattered radiation from a sounding beam ($\lambda$ = 0.63 μm, scattering angle ≈ 165°) in an undisturbed medium and in one activated with a high-power beam, respectively. The case studied in these experiments corresponded to a stationary pre-explosion regime of water fog dissipation using a $CO_2$ laser beam, 2.25 cm in radius and with a power of 50 W/$cm^2$. The fog was obtained by the adiabatic cooling of moist air in a water chamber of 8 m in length.

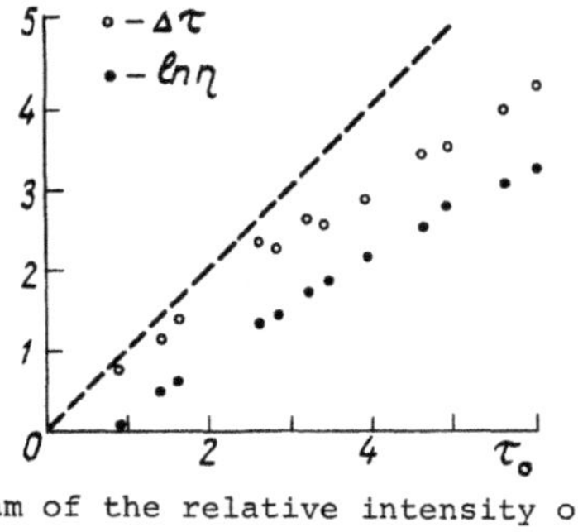


Fig. 8.2.4. Logarithm of the relative intensity of the sounding beam's backscattered radiation in the cleared channel produced by a $CO_2$ laser beam as a function of the initial optical depth of the fog $\tau_0$. The dashed line shows the boundary of complete aerosol dissipation.

As seen from this figure, the backscattered flux $P_\pi^N$ exceeds $P_\pi$ for $\tau_0 \gtrsim 0.9$, this is because of the prevailing role of the process described in (8.2.1) by means of the exponential factor. A dynamic analysis, starting from the moment of the switching-on of the high-power laser, allowed the estimation of the dissipation velocity, which in this experiment was found to be 50 m/s.

The distortions of the returned signals in a fog composed of droplets of aniline dye in water were investigated in [2]. Experiments in [2] were carried out using a ruby laser emitting free generation pulses which were focused into the fog chamber. The parameters of the fog microstructure were $a_{sq} \approx 8$ to 10 μm, $\kappa_a = 0.7 \times 10^{-3}$, $N_0 = 10^4\ cm^{-3}$. Radiation scattered at 150° from a region of size $\pi \Delta x R_0^2 \approx 0.1\ cm^3$ was recorded, using a photodetector. The scattered and the reference laser pulses were recorded with a dual-beam oscilloscope. Several steps of the interaction process can be seen from Figure 8.2.5. First, the vaporization of the fine aerosol fraction ('clearing') is observed, then follows the stage of fragmentation of the large particles, resulting in an increase in channel turbidity, finally, the vaporization of the fragments is observed, leading to further dissipation of the aerosols.

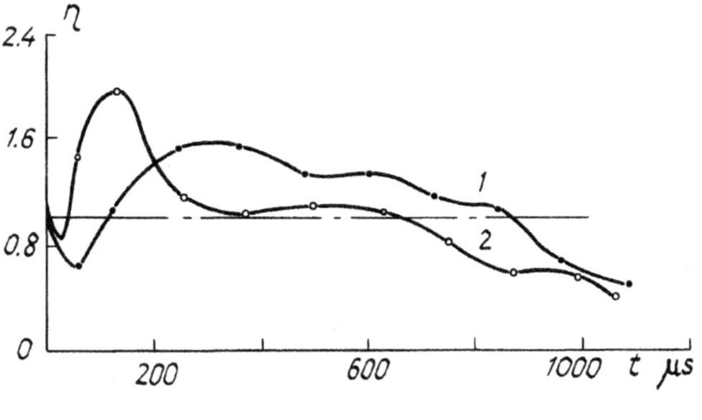


Fig. 8.2.5. Temporal behavior of the nonlinearity parameter of lidar returns at $\lambda = 0.694$ μm ($\eta = P_\pi(x)/P_\pi^L(0)$) under the explosion regime of fog droplet destruction (droplets of a weak solution of aniline dye in water). $W_0 = 1.1\ kJ/cm^2$ for curve 1, and $W_0 = 1\ kJ/cm^2$ for curve 2; $\kappa_a = 10^{-3}$.

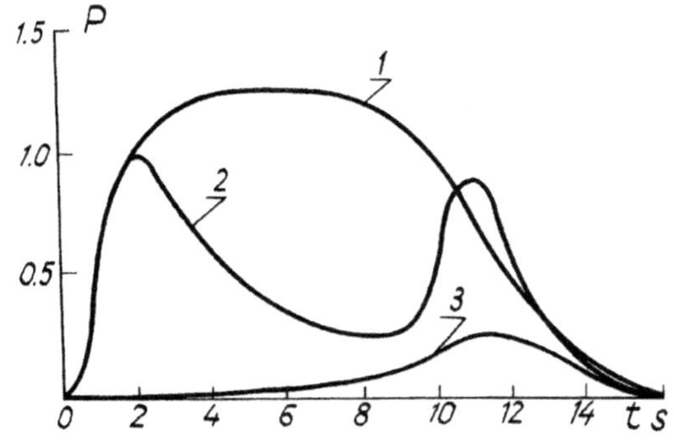


Fig. 8.2.6. Dynamics of the $CO_2$ laser backscattered radiation ($I \approx 50\ W/cm^2$) in the moist air in a room (curve 2). Curve 1 represents the power profile of the $CO_2$ laser radiation. Curve 3 shows the intensity of channel emission in the 2 to 7 μm spectral range.

Figure 8.2.6. shows the measured values of the relative intensity of the backscattered background radiation at $\lambda = 10.6$ μm (curve 2) within a high-power (30 to 50 $W/cm^2$) $CO_2$ laser beam propagating through moist air. The temporal behavior of the laser's output power is presented (in relative units) by curve 1. The measurements of the scattered background power were made using a HgCdTe photoresistor detector, whose threshold sensitivity at $\lambda = 10.6$ μm was $\approx 2 \times 10^{-7}$ W [16].

The FWHM of the filter used in the recording channel was 0.5 μm. Curve 3 in the figure presents the intensity of the thermal emission of the beam channel in the range 2 to 7 μm. The significant decrease of the background scattering signal observed during the few seconds after switiching the high-power laser beam on is caused by the evaporation of the water aerosol, while the braod-range maximum in the IR observed 10-12 s after switching on the high-power beam is related to thermal emission from the channel due to radiation heating and inflammation of the organic fraction of the aerosol substance.

The above results of preliminary experiments on the nonlinearity of lidar returns in aerosol media, including the atmosphere, enable one to evaluate the energetics of high-power lasers $I_n$ at which the nonlinearities discussed above become significant. Thus, for a cw-$CO_2$ laser, $I_n \approx 10^1$ to $10^2$ $W/cm^2$ and, for $CO_2$ laser pulses of a duration of a few microseconds, $I_n \approx 10^6$ $W/cm^2$; and $I_n \approx 10^3$ $W/cm^2$ for ruby laser pulses of a duration of a few milliseconds.

## 8.3. AN ANALYSIS OF THE CRITERIA FOR DETECTING A HIGH-POWER LASER BEAM PROPAGATING IN FOG WHEN THE BEAM POWER IS SUFFICIENT TO DISSIPATE THE FOG

The possibility of dissipating fogs and water hazes in the ground layer of the atmosphere using high-power laser beams can lead to the improvement of the operational range of laser navigation systems, whose working principle is the detection of the beam emitted from a laser beacon.

The main criterion for detecting the laser beam in a scattering medium follows from the fact that the singly-scattered light signal exceeds that from the background light due to multiple scattering, since the latter worsens the beam's image contrast [11-12].

Below, an approach to the solution of this problem is discussed which is based on the statistical modeling of radiation fluxes scattered in the direction of a photoreceiver (Monte-Carlo method) aimed at different points on the beam's axis, it being located off the beam's axis itself [12]. The experimental set-up is depicted in Figure 8.3.1(c). The laser source S emits a beam of angular width $\theta_0$ with $\lambda = 10.6$ µm. Scattered radiation is detected with a photoreceiver R, whose angular aperture is $\theta_d$. The length of the beam's path is denoted as $x_v$, $x_0$ is the shortest distance from the receiver to the laser beam, x is the distance from the source S to the point where the beam axis and the axis of the receiver's field of view intersect.

The estimates of the fluxes of scattered radiation, made for the case of a medium irradiated with a high-power beam, are presented in Figures 8.3.1 and 8.3.2. The fluxes presented are normalized w.r.t. the power of the incident radiation.

The values that were varied in the numerical simulations were the high-power laser beam path $x_v$, $x_0$, and the receiver's field of view $\theta_d$. The profiles of the nonlinear extinction coefficient along the beam's path were the same (as shown in Figures 8.3.1(b) and 8.3.2(b), curves 4 and 6, respectively).

In addition to the power of the optical signal, the contrast coefficient of the beam image,

$$b_c = \frac{F_{max} - F_{min}}{F_{max} + F_{min}},$$

is also very important for estimating the image quality and for separating the signal from the background noise by multiple scattering. Here, $F_{max}$ is the maximum of the beam image function in the plane of the receiver, and $F_{min}$ is its value at the boundary of the beam image calculated according to geomtrical optics laws. The values of $b_c$ calculated for the portions of the beam channel where linear light scattering occurs (that corresponds to an interval of $x \gtrsim 80$ m in Figure 8.3.2) are presented in Figure 8.3.3.

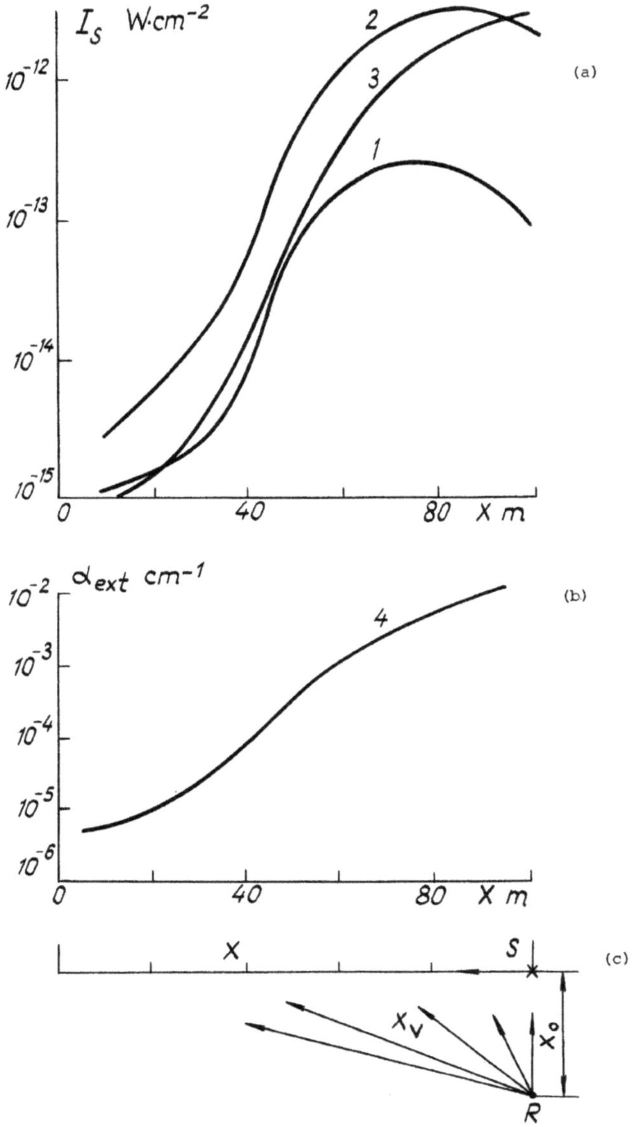


Fig. 8.3.1. Normalized intensities of radiation ($\lambda$ = 10.6 μm) scattered from the high-power beam in fog as functions of the path length (curves 1 to 3) calculated using the Monte-Carlo method. Curve 4 is the profile of the volume aerosol extinction coefficient along the path formed during the course of fog dissipation by

the $CO_2$ laser pulse. $\theta_0$ = 2', $x_0$ = 88.2 m, $\alpha$ = $1.16 \times 10^{-4}$ $cm^{-1}$. $\theta_d$ = 1, 2, and 5 angular minutes for curves 1, 2, and 3, respectively.

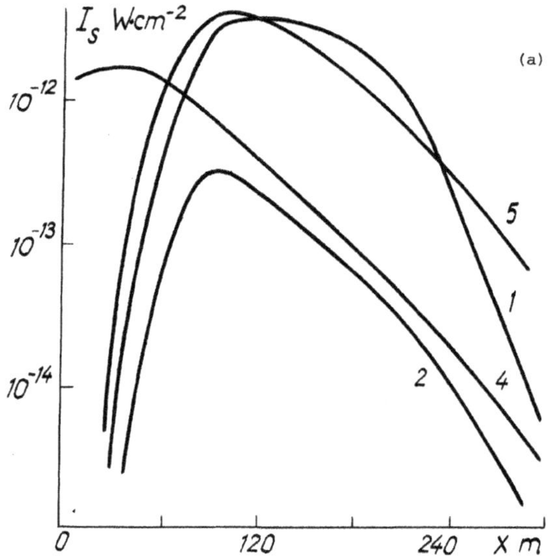


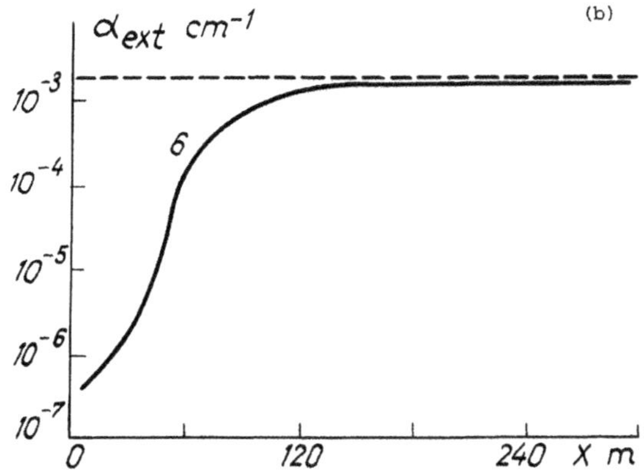


Fig. 8.3.2. Normalized intensities of radiation ($\lambda$ = 10.6 μm) scattered from the high-power beam in fog as functions of the path length, calculated using the Monte-Carlo method. The profile of $\alpha_{ext}$ along the path is presented by curve 6; the dashed line is the initial value of the aerosol extinction coefficient $\alpha_0$. Curve 4 represents the intensity of scattered radiation calculated for

a linear medium with $\alpha_{ext} = \alpha_0$). $x_0$ = 8.82 m; $\theta_0$ = 2'; $\theta_d$ = 1', 2', and 5' for curves 1, 2, and 3, respectively.

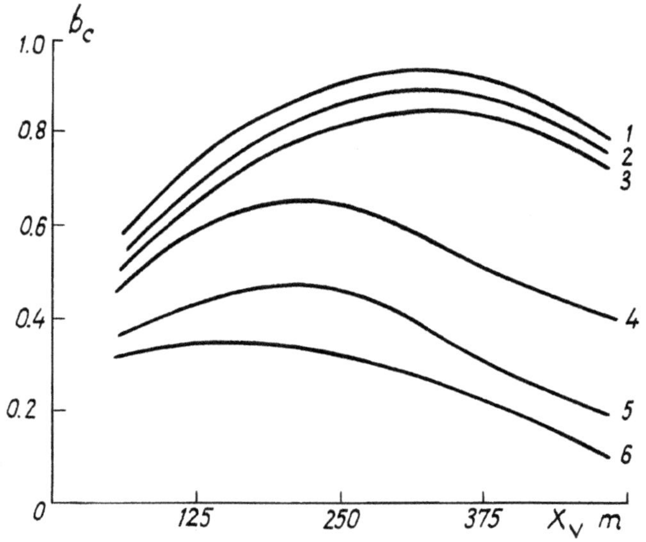


Fig. 8.3.3. Dependence of the contrast coefficient of the sounding beam ($\lambda$ = 0.63 μm) in a water droplet fog on the distance from the receiver $x_v$. x = 510 m; $\theta_0$ = 2'; $x_0$ = 88.2 m. $\theta_d$ = 1', 2', and 5' for curves 1, 2, and 3, respectively ( $\alpha_0 = 2.9 \times 10^{-5}$ $cm^{-1}$); $\theta_d$ = 1', 2', 5' for curves 4, 5, and 6, respectively ($\alpha_0$ = $11.6 \times 10^{-5}$ $cm^{-1}$).

On the whole, the analysis of calculations made for light scattering in the direction of the receiver allows one to arrive at the conclusion [12] that the values and the behavior of scattered light fluxes strongly depend on the path length $s_v$ and on the distance between the beam axis and the receiver $x_0$. Thus, the flux of scattered light decreases with an increase in x ($x_0$ being fixed), this is caused by the increase of the radiation extinction coefficient along the beam's path with a relatively rapid absorption of scattered photons, since in the disturbed zones of the beam channel the probability of photon survival changes from 0.141 to 0.623, while in undisturbed regions of the channel it is constant and equals 0.623. The power of the scattered radiation also decreases with increasing $x_0$ (x being fixed) because of the rapid damping of scattered radiation in the medium (small values of $I_s$).

The comparison of values of $I_s$ calculated both for disturbed and undisturbed media revealed a strong dependence of the behavior of $I_s(x_v)$ on the form of the profile of the extinction coefficient along the beam's path.

The functions $I_s(x_v)$ show that, in the case of disturbed media, the power of the light scattered from the area of the high-power beam channel close to the laser source is too low.

Calculations of $I_s(x_v, x, x_0)$ facilitated the estimation of the background power due to multiple scattering. It was found that, at $\tau_0(x_v) \lesssim 1$, the multiple-scattering background does not exceed 5% of the observed signal. The r.m.s. deviation of the estimates was less than 25%. This enables one to apply some simplified methods for assessing the main flux, neglecting the effect of the multiple scattering of light in the medium.

## 8.4. REMOTE SPECTROCHEMICAL ANALYSIS OF AEROSOL COMPOSITION USING THE EMISSION AND LUMINESCENT SPECTRA INDUCED BY HIGH-POWER LASER BEAMS.

Nowadays, lasers are already used in laboratory spectroscopic analysis for identifying microscopic quantities of matter using the emission spectra induced in vaporized matter, but mainly in a combination of laser-induced melting and vaporization of matter with an electric arc discharge that provides for high intensity and the reproducibility of the spectral lines (see, e.g., [17]). Moreover, advances in high-power pulsed laser technology have paved the way for the solution of the problem of the remote spectrochemical analysis of aerosol substances and gaseous constituents of the atmosphere, including noble gases [4, 5].

The principle of the technique is based on focusing the high-power laser beam into the atmosphere, thus heating the aerosol substance to very high temperatures so that the vaporization of solid particles can take place. The vapor aureoles around aerosol particles facilitate the initiation of optical breakdown processes which transform the vapor aureoles into plasma formations. The presence of free electrons in the plasma results in the excitation of the atoms and molecules of the vapor due to inelastic collisions, thus causing the intense emissions. The intensities of the spectral lines in the emission spectra generated as a result of the optical breakdown process are higher than the intensity of thermal emission of the heated vapors before breakdown has occurred by several orders of magnitude.

The analysis of the emission spectra of vapor provides for a very high degree of selectivity that facilitates identification and provides quantitative information on the elementary composition of aerosol substances and their surrounding gases. The power of the laser source used for spectrochemical analysis must be sufficient for well-developed vaporization of remote aerosol targets, as well as for exciting the emission spectra of vapors. This can be achieved by using the high-power pulsed $CO_2$ lasers available at present.

There are advantages in using, for this purpose, laser radiation of $\lambda = 10.6$ μm as compared with radiation in the visible and near IR ranges; these are the strong absorption of most liquid and solid substances in the region of 10.6 μm and the high efficiency of the avalanche ionization in vapors, which is proportional to $\lambda^2$.

These factors result in a significant (one or two orders of magnitude) decrease in the threshold intensity of lasers with $\lambda$ = 10.6 μm necessary for the optical breakdown of the air near condensed media or of aerosols, as compared to that necessary in the case of lasers emitting radiation with short wavelengths [10, 18].

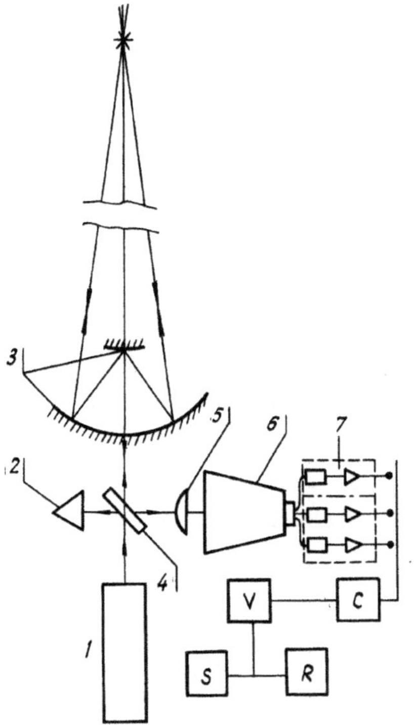


Fig. 8.4.1. Block diagram of the spectrochemical lidar. 1 is the electroionization $CO_2$ laser ($\lambda$ = 10.6 μm); 2 is the power meter; 3 is the Cassegrainian telescope; 4 is the beam splitter; 5 is the focusing lens; 6 is the grating spectrometer; 7 is the receiver block which contains a 20-channel photorecording device; EC is the electromechanical cummutator; DV is the digital voltmeter; SPD is a specialized processor; and P is the printer.

The method of remote spectrochemical sensing is discussed, e.g., in [1, 2, 4-9, 28]. Figure 8.4.1 presents a block-diagram of a mobile spectrochemical lidar [6]. An electroionization $CO_2$ laser [5] is used in this installation. The laser delivers 500 J pulses of 300 μs duration. The Cassegrainian telescope is used for focusing the beam at distances of 50 to 150 m from the laser.

The radiation emitted by laser sparks in the atmosphere is collected using the same telescope and is focused on the entrance slit of the diffraction spectrometer. The spectral separation of output laser radiation in the IR and radiation emitted from laser sparks in the UV and near-IR ranges is performed using a color-selective mirror.

The emission spectra are recorded, using photographic film, by exposing each frame to 3 to 7 laser sparks, or with the use of PMTs. In the latter case, a slit-shaped fiberglass wave guide is used for the transportation of radiation from the spectrometer's exist point to the PMT photocathode. The wave guide block facilitates the selection of the desired spectral region and emission spectrum recording from 20 portions of spectrum simultaneously. Then, output signals from the PMT anodes are amplified, integrated, and memorized. The on-line minicomputer is used for data processing. A chart recorder and a digital voltmeter are used for displaying the output information.

A lidar with the above features facilitated experimental measurements of emission spectra in the atmosphere up to a range of 150 m.

Measurements showed that the efficient excitation of emission spectra takes place due to low-threshold optical breakdown on aerosol particles. Typical photographs of laser sparks in a slightly dusty atmosphere are presented in Figure 7.2.1. In Chapter 7 the reader can find data on the threshold intensities of laser beams required for initiating laser sparks in the atmosphere. As the investigations showed, the plasma decay time is of the order of tens of microseconds, while the emission spectra of aerosols are formed during the laser pulse mainly by inelastic electron-atom collisions since, the plasma involved in optical breakdown, not being in equilibrium, the electrons' temperature is 2 to 3 time the temperature of the gas.

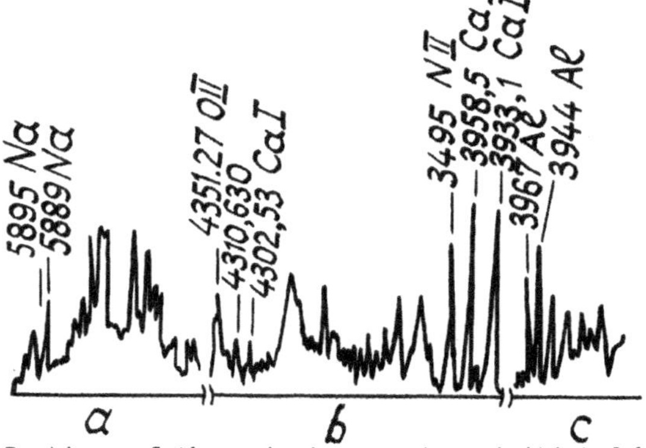


Fig. 8.4.2. Portions of the emission spectrum initiated by a laser spark contaminated with cement particles. The marked lines are spectral lines of CA (I), CA (II), N (I), N (II) and O (II).

Figure 8.4.2. [2] presents a spectrogram of the laser spark emission spectrum of cement dust. The results of spectroscopic and energy measurement in the emission spectra of $Al_2O_3$ and $Na_2CO_3$ are presented in Figure 8.4.3 as functions of the mean laser pulse power in the volume of atmosphere sounded [7].

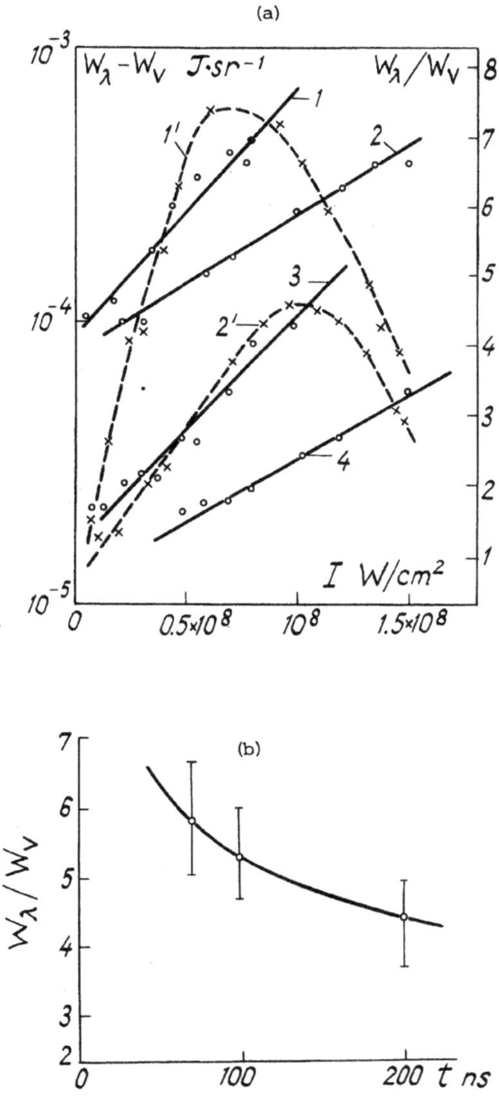


Fig. 8.4.3. The caption for Figure 8.4.3 is to be found on the following page. .

Fig. 8.4.3. Results of measurements of the lines of emission spectra of aerosols. (a) Intensities of spectral lines (solid lines) and the ratio of emission line intensities to the intensity of the background radiation, integrated over the spectral width of the emission lines (dashed curves). Observations were made of the optical breakdown of $N_2CO_3$ and $Al_2O_3$ particles: 1 and 2 are aluminium lines with $\lambda = 394.4$ and $396.1$ nm ($t = 200$ ns for 1, and $t = 70$ ns for 2). 3 and 4 are sodium lines with $\lambda = 588.9$ and $589.5$ nm ($t = 200$ ns for 3 and $t = 70$ ns for 4). (b) The ratio of the sodium doublet's ($\lambda = 588.9$; $\lambda = 589.5$ nm) intensity to the intensity of the continuum in the $Na_2CO_3$ aerosol as a function of laser action duration. $W_0 = 12$ J/cm$^{-2}$.

As follows from the data presented above, the ratio of the intensities of spectral lines of neutral atoms $W_\lambda$ of aerosol substance vapors to the background intensity $W_v$ reaches its maximum at intensities of the $CO_2$ laser radiation of $0.5 \times 10^8$ to $10^8$ W/cm$^2$. At intensities lower than $0.5 \times 10^8$ W/cm$^2$ the ratio $W_\lambda/W_v$ decreases due to thermal emission from the aerosol particles heated by the laser beam but not yet evaporated. The right-hand branches of curves 1 and 2 in Figure 8.4.3 represent the effect of diminishing S/N ratio caused by an increase in the contribution of continuous bremsstrahlung of the plasma formation to the background noise. The shortening of the laser pulses from 200 ns to 50 ns at a fixed pulse energy resulted in an approximate twofold increase in the signal-to-background noise ratio (see Figure 8.4.3(b)).

The emission spectra of laser sparks carry information concerning the elementary composition of atmospheric aerosols, so, using the reference spectra and their combinations, one can determine the initial chemical composition of the aerosol. Such techniques are widely used in metallurgy and mineralogy [17].

Estimates of the relative concentrations $n_i$ of different elements can be made using the empirical relation [2]:

$$n_i = C_{ie} n_e I_{\lambda i}/I_{\lambda e},$$

where $n_i$ and $n_e$ are the number densities of the atoms (to be determined) and of the reference element, respectively; $I_{\lambda i}/I_{\lambda e}$ is the ratio of the intensities of corresponding spectral lines; and $C_{ie}$ is the empirical calibration coefficient.

The main contribution to errors in the measurement of number density is the uncertainty in the determination of $C_{ie}$ caused by the randomness of electron energy $\langle\varepsilon\rangle$ in laser sparks initiated both in the atmosphere and under laboratory conditions. The contribution of this effect to the errors in measurement can be significantly decreased by taking into account the measured ratio of intensities of two fixed spectral lines of the reference gas $I'_{\lambda e}/I''_{\lambda e}$ which can serve as an indirect measure of the mean temperature of the electron gas in the plasma, and hence a measure of the occurrence of excitation due to inelastic electron-atom collisions. Therefore, the parameter $(I'_{\lambda e}/I''_{\lambda e})$ should be measured in field experiments simultaneously with measurements of the spectral line intensities $I_{\lambda i}$, thus forming the input parameter to the nomogram for determining the coefficients $C_{ie}$.

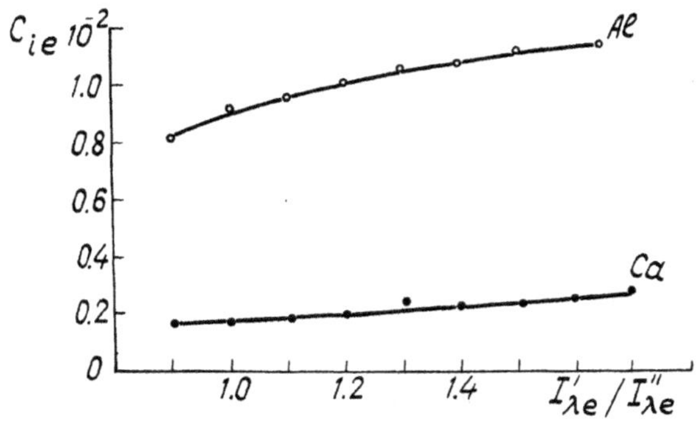


Fig. 8.4.4. Calibration curves for determining the coefficients $C_{ie}$ when carrying out the spectral analysis of an aerosol's chemical composition.

Figure 8.4.4 presents a nomogram of $C_{ie}$ constructed using laboratory measurements of the $I'_{\lambda e}/I''_{\lambda e}$ ratio for calcium emission lines at $\lambda$ = 396.85 nm and $\lambda$ = 393.37 nm, normalized w.r.t. the nitrogen reference line of wavelength $\lambda$ = 399.5 nm. The relative error in the determination of the number densities of aerosol atoms following this technique was 25 to 30%, provided that the error in the intensity measurements was ≈5%.

In field experiments the total number density of atmospheric aerosols can be determined using traditional lidar methods [19] or by photographic methods involving the counting of the number of prime breakdown centers in the beam channel.

Certain problems of the lidar sensing of the atmosphere require one only to distinguish between the main kinds of aerosol substance in the atmosphere (minerals, organic substances, condensed water) and/or to determine their relative abundances. In such cases the method of spectrochemical analysis described above can be replaced by the method of luminescence analysis using UV lasers [20-24]. In spite of low selectivity and sensitivity, luminescence analysis can be of use in the identification of types of condensation nuclei, as well as for the determination of their origin (marine or continental). Luminescence analysis can also be useful in the interpretation of the optical characteristics of seaside hazes and for estimating the percentage of the anhydrous fraction of urban aerosols. This method uses the Raman water line as a reference line for normalizing the intensity of luminescence.

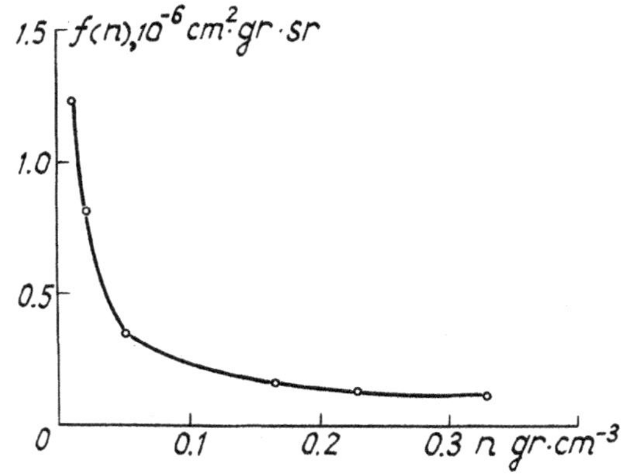


Fig. 8.4.5. Normalized intensity of the signal due to fluorescence f(n) as a function of the relative mass density of sea salt in the sea haze: $f(n) = P_f z^2 \exp[(k_{\lambda 3} + k_{\lambda f})z](ctP_3S)$, where $P_3$ is the laser's output power; $P_f$ is the power of fluorescence recorded with a photodetector; $k_{\lambda f}$ and $k_{\lambda 3}$ are the extinction coefficients of the atmosphere at $\lambda_f$ and $\lambda_3$, respectively; z is the distance to the volume of aerosol sounded; and S is the receiving area.

Figure 8.4.5 shows the power of the luminescence of droplets as a function of the concentration of the sea salt dissolved in the water [20]. These data illustrate the possibility of determining the water content of the sea mists and hazes remotely. The measurements in [20] were made using a molecular nitrogen laser ($\lambda$ = 337.1 μm) delivering pulses of 20 ns duration at a repetition rate of 100 Hz. The mean power of the output beam was about 3 mW. The luminescence signals were recorded by a PMT using the photon counting technique.

Figure 8.4.6 presents the results of a comparison made in [21] between the luminescence spectra of precipitation water and dry aerosols recorded in an intracontinental region far from any source of water. The excitation of the luminescence spectra was accomplished with radiation from a molecular nitrogen laser ($\lambda$ = 337.1 nm) and with the fourth harmonic of a Nd:YAG laser ($\lambda$ = 266 nm). The measurements revealed a strong variability in the luminescence spectra excited using radiation with a wavelength $\lambda$ = 266 nm. In all the observed samples of precipitation water the intensity of luminescence excited with the fourth harmonic of the Nd:YAG laser was only about 5 to 15% of the intensity of the Raman water line. This was about 4 to 7 times lower than the corresponding value for the intensity of luminescence excited using radiation with $\lambda$ = 337.1 nm. When the luminescence was excited with the second harmonic of the nd:YAG laser ($\lambda$ = 532 nm), it reached a maximum in the region from 570 to 590 nm, this did not exceed 2 or 3% of the Raman signal from water in the region of 650 nm.

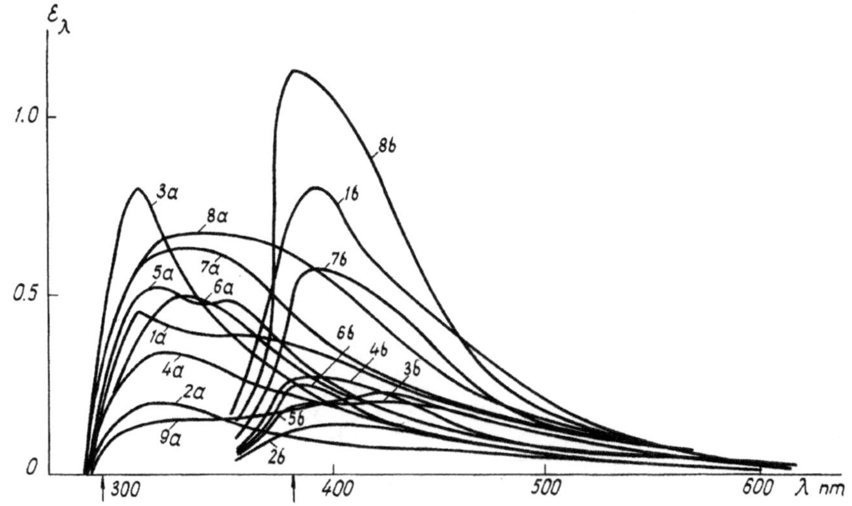


Fig. 8.4.6. Fluorescence spectra of precipitation water (curves 1 to 6), of most types of dry atmospheric aerosols dissolved in water (curves denotes by (a) and (b) represent spectra excited using radiation with $\lambda$ = 266 nm and $\lambda$ = 337.1 nm, respectively. The arrows show Roman lines for water. Curves 1 to 6 correspond to precipitation intensities of 1.3 mm, 1.0 mm, 2.2 mm, 20.4 mm, 6.6 mm, and 5.6 mm, respectively.

The fluorescence cross-sections of aerosol contaminants like diesel fuel, soil dust, and industrial smokes measured in [22-24] using a laser beam operating at $\lambda$ = 266 nm range from $10^{-27}$ to $10^{-24}$ $cm^2 sr^{-1}$, this technique enables measurements of aerosol fluorescence spectra to be carried out up to a range of about 100 m.

## 8.5. AN ANALYSIS OF THE POSSIBILITIES OF SENSING THE HIGH-POWER LASER BEAM CHANNEL USING OPTO-ACOUSTIC TECHNIQUES

In recent years several methods of opto-acoustic sensing of atmospheric parameters, using the effects of remotely initiated optical breakdown, have been suggested [25-27]. The laser spark is the source of the intense, broadband acoustic signal. The arrival time and the Doppler distortions of the spectrum of such signals at the acoustic receiver strongly depend on the prevailing meteorological parameters.

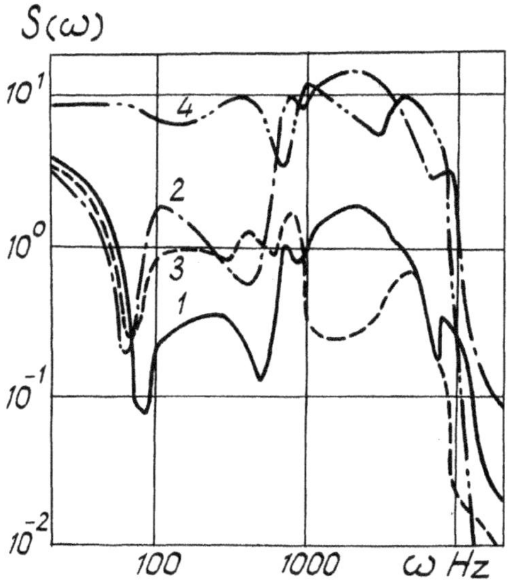


Fig. 8.5.1. Energy spectra of acoustic signals produced in air by optical breakdown initiated by a $CO_2$ laser, as recorded with a microphone at different distances R from the plasma channel. (1) R = 20 cm; $W_0$ = 10 J/cm$^2$; acoustic energy $W_s$ = 1.18 × $10^{-5}$ J; T = 27.5 °C. (2) R = 200 cm; $W_0$ = 9 J/cm$^2$; $W_s$ = 2.9 × $10^{-8}$ J; T = -11 °C. (3) R = 400 cm; $W_0$ = 9 J/cm$^2$; $W_s$ = 2.9 × $10^{-8}$ J; T = -11 °C. (4) A reference spectrum of the gun capsule explosion at R = 50 cm from the microphone.

Figures 8.5.1 and 8.5.2 present, for example, spectra of the acoustic signal from a laser spark and of the efficiency of transforming the laser pulse's energy into the energy of the acoustic wave [25], respectively.

The analysis of the acoustic signal's spectrum and its energy shows that a laser spark produces a broadband acoustic signal with a duration of about 5 to 10 ms, which depends on the spatial length of the spark. The maximum value of the sound pressure, recorded experimentally [25], within the frequency range 20 Hz to 20 kHz varied from 68 to 120 dB, the density of the laser pulse's energy in the beam's caustic ranging from 8 to 16 J/cm$^2$.

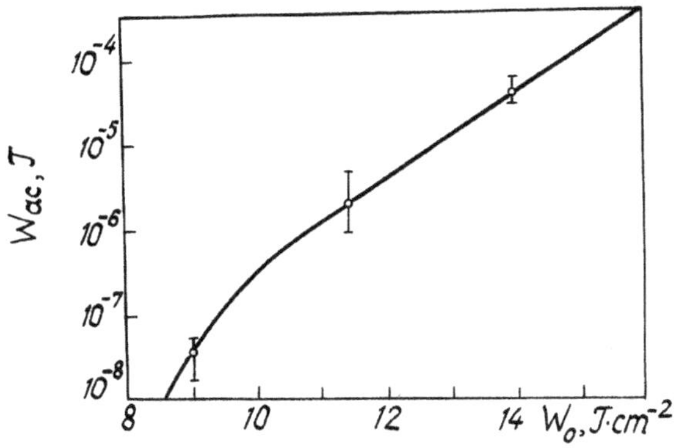


Fig. 8.5.2. The average (over 24 measurements) acoustic energy of the optical breakdown center in air as a function of the energy density of incident laser radiation. $\lambda$ = 10.6 μm; $t_p$ = 0.3 μs.

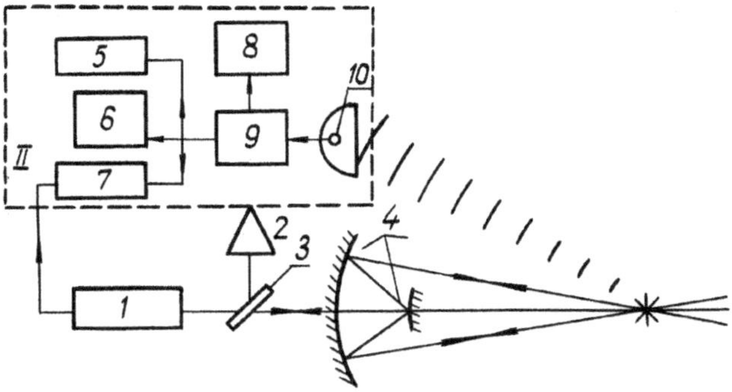


Fig. 8.5.3. Block-diagram of the opto-acoustic radar for atmospheric studies using the effects of laser sparking [27]. 1 is a high-power laser ($\lambda$ = 10.6 μm, t ≈ 1 μs; W = 500 J); 2 is the Cassegrainian telescope (diameter of primary mirror is 500 mm and secondary reflector 120 mm). 3 and 4 are the units used in the experiment for monitoring the laser beam's parameters; 5 is a small mirror, whose movement allows variation of the sounding distance. 12 is the microphone installed at the focal point of the parabolic acoustic antenna; 13 is the wide-band amplifier; 14, 15, and 16 are the devices used in the experiments for memorizing recorded pulses, measuring their durations, and counting the number of acoustic pulses; and 17 is the spectrum analyzer.

The principal set-up for opto-acoustic sensing is depicted in Figure 8.5.3. The high-power laser beam is focused using the telescope 3 at altitude H, where it initiates optical breakdown. The acoustic receiver 10 records the acoustic signal arriving from the spark.

The characteristics of an acoustic signal travelling through a layer of atmosphere contain information on the temperature of the atmosphere, humidity, wind speed, and the spectral transmission of the atmosphere w.r.t. acoustic waves. We will now discuss the development of concepts related to these optoacoustic methods of sensing the atmosphere, as proposed in [26] and [27].

The transmission of the atmosphere w.r.t. acoustic waves $T_{ac}$ can be determined from the relation of the acoustic power emitted by a laser spark to the acoustic power recorded by the receiver:

$$T_{ac}(\omega) = \exp[-\alpha_{ac}(\omega)x] = 4\pi x^2 P_{ac}(\omega)/P_{ac}^{(o)}(\omega), \tag{8.5.1}$$

where $\alpha_{ac}$ is the atmospheric extinction coefficient for sound of frequency $\omega$, x is the length of the path from the laser spark at altitude H to the acoustic receiver;

$$P_{ac}^{(o)}(\omega) = \beta_1(P_o, x)\beta_2(P_o, \omega)P_o \tag{8.5.2}$$

is the spectral power of the acoustic signal produced by the laser spark; $P_{ac}$ is the spectral power of the acoustic signal recorded by the receiver. The absorption coefficient for the laser radiation $\beta_1$ depends on the laser output power $P_o$ and on the height H where the breakdown occurs, as well as on the aerosol's composition; other conditions (laser pulse duration $t_p$ and the wavelength of the radiation) being constant. The coefficient $\beta_2$ is the efficiency of converting the absorbed power of the laser beam into acoustic power at a sound frequency $\omega$, the value of $\beta_2$ also depends on the power output of the laser. As the experiments showed [26], the shape of the acoustic signal's spectrum does not depend, at laser beam intensities $\approx 10^8$ W/cm$^2$, on the chemical composition of the aerosol. This allows one to measure the calibration function $\beta_2(I)$ for the horizontal path and use it when interpreting measurements made along both vertical and slanting paths.

The measurement of the amplitudes of acoustic signals travelling through the atmosphere at three frequencies $\omega_1$, $\omega_2$, and $\omega_3$ facilitates the determination of the temperature and humidity of the atmosphere.

For this purpose the output signals from a spectrum analyzer, at these frequencies, are normalized w.r.t. the corresponding spectral powers of the acoustic signal produced by the laser spark:

$$P_{ac}^1 = P_{ac}(\omega_1)/P_{ac}^{(o)}(\omega_1), \ldots \tag{8.5.3}$$

and the values $A_{12}$, $A_{31}$ are calculated using

$$A_{ik} = \alpha_{ac}(\omega_1) - \alpha_{ac}(\omega_k) = \ln(P_{ac}^k/P_{ac}^i)/x. \tag{8.5.4}$$

These two values are complicated functions of atmospheric temperature and relative humidity h. Then, using the array of values of $A_{21}$ and $A_{31}$ computed for different temperatures and humidities, one can determine the required values of T and h.

Incidently, the measurements of temperature can be performed in a more simple manner. Since the speed of sound is related to the air temperature by $c_s \approx 20.05\ T^{1/2}$, one can asses the temperature of an atmospheric layer by measuring the arrival time of the sound pulse from a laser spark for a fixed atmospheric path.

By measuring the arrival time of the sound pulse simultaneously at several points separated in space, one can determine the vector of the wind velocity.

The experimental verification of the optoacoustic techniques for sensing the atmosphere, described above, was carried out in [27]. The laser installation used in these experiments allowed optical breakdown to be initiated in the atmosphere up to an altitude of 150 m. This optoacoustic facility enables one to obtain information concerning the chemical composition of the aerosol and the meteorological parameters of the atmosphere (humidity, temperature, wind velocity) using the acoustic effects produced by the laser spark.

The possibility of obtaining multiparameter information on the state of the atmosphere makes optoacoustic facilities of paramount usefulness.

Perspectives for the further increase of the operational range of optoacoustic facilities are now connected with the use of lasers delivering pulses with higher energies as, e.g., a $CO_2$ laser delivering pulses with an energy of 3 to 5 kJ and a duration of ≈1 μs.

## REFERENCES: CHAPTER 8


[ 1] V.E. Zuev and Yu.D. Kopytin: 'Lidar and Acoustic Sounding of the Atmosphere', in Conf. Abstracts, Part 2, 5th All-Union Symp. (Inst. Atmosph. Optics, Tomsk, 1978) pp. 88-97, in Russian.

[ 2] V.E. Zuev and Yu.D. Kopytin: 'Application of the Lidar to Atmospheric Radiation and Climate Studies', in Conf. Abstracts, IAMAP Third Scientific Assembly, Hamburg, F.R.G. (1981) p. 87.

[ 3] G.M. Krekov, M.M. Krekova: 'The Estimation of Parameters of the Cleared Zone in a Beam Channel', in Remote Sensing of Physio-Chemical Parameters of the Atmosphere Using High-Power Lasers, ed. by V.E. Zuev (Inst. Atmos. Optics, Tomsk, 1979) pp. 80-97, in Russian.

[ 4] E.B. Belyaev, A.P. Godlevsky, and Yu.D. Kopytin: Kvant. Elektron. 6, 1152-1156 (1978) (Sov. J. Quantum. Electron.).

[ 5] E.B. Belyaev, A.P. Godlevsky, and Yu.D. Kopytin: 'Remote Spectrochemical Analysis of Aerosols', in Remote Sensing of Physio-Chemical Parameters of the Atmosphere Using High-Power Lasers, ed. by V.E. Zuev (Inst. Atmos. Optics, Tomsk, 1979) pp. 3-56, in Russian.

[ 6] Yu.D. Kopytin: 'Nonlinear Optics Methods for the Remote Determination

of Chemical Composition and Microstructure of Aerosols in the Ground Atmospheric Layer', in Investigations of Atmospheric Aerosols Using Lidar Techniques, ed. by M.V. Kabanov (Nauka, Novisibirsk, 1980) pp. 138-166, in Russian.

[ 7] A.P. Godlevsky and Yu.D. Kopytin: Zh. Prikladn. Spektroskopii **31**, 612-617 (1979) (Sov. J. Appl. Spectr.)

[ 8] E.B. Belyaev, A.P. Godlevsky, Yu.D. Kopytin et al.: Zh. Tekh. Fiz., Pis'ma Red. **8**, 333-337 (1982) (Tech. Phys. Lett. Ed.).

[ 9] Yu.V. Akhtyrčhenko, E.B. Belyaev, Yu.P. Vysotsky et al.: Izv. Vyssh. Uchebn. Zaved., Fiz. **2**, 3-13 (1983) (Sov. Phys.).

[10] V.E. Zuev, Yu.D. Kopytin, and A.V. Kuzikovsky: Nonlinear Optical Effects in Aerosols (Nauka, Novosibirsk, 1980), in Russian.

[11] O.A. Volkovitsky, Yu.S. Seduov, and L.P. Semenov: High-Power Laser Beam Propagation through Clouds (Gidrometizdat, Leningrad, 1982) p. 312, in Russian.

[12] V.V. Belov and G.M. Krekov: Izv. Vyssh. Uchebn. Zaved., Fiz., deposited in VINITI, N3992-79 Dep.

[13] G.J. Mullaney, W.H. Christiancen, and D.A. Russel: Phys. Lett. **13**, 145-147 (1968).

[14] Yu.V. Ivanov and Yu.D. Kopytin: Kvant. Elektron. **9**, 591-593 (1982) (Sov. J. Quantum Electron.).

[15] G.A. Mal'tseva: 'The Method of Measuring the Absolute Intensity of a Laser Beam by Studying the Dynamics of the Scattered Radiation', in Remote Sensing of Physio-Chemical Parameters of the Atmosphere Using High-Power Lasers, ed. by V.E. Zuev (Inst. Atmos. Optics, Tomsk, 1979) pp. 98-101, in Russian.

[16] A.P. Abramovsky, V.A. Donchenko, Yu.V. Didenko, et al.: 'On the Question of Measuring the Backscatter of High-Power Optical Radiation', ibid., pp. 202-203, in Russian.

[17] Von. H. Moenke and L. Moenke: Einführung in die Laser-mikroemissions-spektral-analyse (Akademische Verlagsgesellschaft, Leipzig, 1966) p. 250.

[18] D.C. Smith: J. Appl. Phys. **48**, 2217-2225 (1977).

[19] V.E. Zuev: Laser Beams in the Atmosphere (Plenum, New York, 1982).

[20] M.A. Buldakov, Yu.D. Kopytin, S.V. Lazarev, and I.I. Matrosov: Izv. Akad. Nauk SSSR Fiz. Atmos. Okeana **17**, 212-216 (1981) (Izv. Acad. Sci. U.S.S.R. Atmos. Ocean Phys.).

[21] N.P. Romanov and V.S. Shuklin: 'Lidar and Acoustic Sounding of the Atmosphere', in Conf. Abstracts: 5th All-Union Symp. (Inst. Atmos. Optics, Tomsk, 1978) Part 2, pp. 34-38, in Russian.

[22] V.M. Zakharov: ibid., pp. 96-111, in Russian.

[23] V.M. Zakharov and O.K. Kostko: Meteorological Laser Sensing (Gidrometizdat, Leningrad, 1977) p. 215, in Russian.

[24] V.M. Zakharov and V.A. Torgovichev: Trans Am. Geophys. Union **58**, p. 802 (1977).

[25] E.B. Belyaev, A.P. Godlevsky, Yu.D. Kopytin, N.P. Krasnenko, and L.G. Shamonaeva: Zh. Tekn. Fiz. Pis'ma Red. 8, 333-337 (1982) (Tech. Phys. Lett.).

[26] L.G. Shamonaeva, Yu.D. Kopytin, and N.P. Krasnenko: 'Lidar and Acoustic Sounding of the Atmosphere', Conf. Abstracts: 7th All-Union Symp. (Inst. Atm. Opt., Tomsk, 1982) part 2, pp. 126-130, in Russian.

[27] A.P. Godlevsky, Yu.V. Ivanov, Yu.D. Kopytin, V.A. Korol'kov, N.P. Krasnenko, V.P. Muravsky, L.G. Shamonaeva: ibid., pp. 244-247, in Russian.

[28] E.B. Belyaev, N.K. Bortnev, A.P. Godlevsky, Yu.D. Kopytin, and N.P. Soldatkin: 'Spectro-Chemical Lidar for Remote Determination of Elemental Composition of Atmospheric Aerosols', in Problems of Atmospheric Optics, ed. by V.E. Zuev (Nauka, Novosibirsk, 1983) pp. 93-107, in Russian.

## INDEX OF SUBJECTS

GPSR Compliance

*The European Union's (EU) General Product Safety Regulation (GPSR) is a set of rules that requires consumer products to be safe and our obligations to ensure this.*

*If you have any concerns about our products, you can contact us on ProductSafety@springernature.com*

In case Publisher is established outside the EU, the EU authorized representative is:

Springer Nature Customer Service Center GmbH
Europaplatz 3
69115 Heidelberg, Germany

**Batch number: 10761599**

Printed by Printforce, the Netherlands